Handbook of AI-Based Mechatronics Systems and Smart Solutions in Industrial Automation

Artificial intelligence (AI) and mechatronics are booming areas where most of the industrial sectors are becoming smart nowadays. This handbook includes material of multidisciplinary content from the AI, mechanical, and electronics engineering domains, among others. It gives insights into various application sectors discussing current global developments in mechatronics employing AI technology and addressing the complexity of current issues and the effects of diverse mechatronics systems.

Handbook of AI-Based Mechatronics Systems and Smart Solutions in Industrial Automation focuses on system automation, predictive analysis, preventive analysis, and real-time decision-making systems for next-generation automation. It discusses the advancements of mechatronics systems using AI applications along with the global approach toward smart industrial automation and presents the impact of AI on today's work of autonomous and industrial automation. The book discusses future research potential and is beneficial to manufacturing, healthcare, and finance disputes, while it offers AI algorithms to analyze large amounts of data and identifies patterns, trends, and anomalies for accurate predictions and optimization processes. The handbook also addresses use cases and case studies related to AI in mechatronics along with applications.

Scholars in the field of AI in mechatronics and related applications will find this book useful. In particular, attention is drawn to both fundamental ideas and important practical contexts. Readers interested in the most recent findings in the field of problem-oriented processing approaches in mechatronics, including those in academia, data science, industry, research, and graduate and undergraduate students, will find this fascinating handbook extremely interesting.

Intelligent Manufacturing and Industrial Engineering
Series Editors: Ahmed A. Elngar, Beni-Suef University, Egypt
Mohamed Elhoseny, Mansoura University, Egypt

Machine Learning Adoption in Blockchain-Based Intelligent Manufacturing
Edited by Om Prakash Jena, Sabyasachi Pramanik, Ahmed A. Elngar

Integration of AI-Based Manufacturing and Industrial Engineering Systems with the Internet of Things
Edited by Pankaj Bhambri, Sita Rani, Valentina E. Balas, Ahmed A. Elngar

AI-Driven Digital Twin and Industry 4.0: A Conceptual Framework with Applications
Edited by Sita Rani, Pankaj Bhambri, Sachin Kumar, Piyush Kumar Pareek, Ahmed A. Elngar

Technology Innovation Pillars for Industry 4.0: Challenges, Improvements, and Case Studies
Edited by Ahmed A. Elngar, N. Thillaiarasu, T. Saravanan, Valentina Emilia Balas

Internet of Things and Big Data Analytics-Based Manufacturing
Edited by Arun Kumar Rana, Sudeshna Chakraborty, Pallavi Goel, Sumit Kumar Rana, Ahmed A. Elngar

Industrial Internet of Things Security: Protecting AI-Enabled Engineering Systems in Cloud and Edge Environments
Edited by Sunil Kumar Chawla, Neha Sharma, Ahmed A. Elngar, Prasenjit Chatterjee, P. Naga Srinivasu

Intelligent Manufacturing and Industry 4.0: Impact, Trends and Opportunities
Edited by Alka Chaudhary, Vandana Sharma, Ahmed Alkhayyat

Artificial Intelligence Using Federated Learning: Fundamentals, Challenges, and Applications
Edited by Ahmed A. Elngar, Diego Oliva, Valentina E. Balas

Smart Industries in the Gig Economy: Scope, Opportunities, and Challenges
Edited by Elwin Thanakumar Joseph S., Martin Sagayam Kulandairaj, Digvijay Pandey, G. T. Thiru Arooran, G. Bindu, Hatira Gunerhanditor

Intelligent Manufacturing: Exploring AI, Blockchain, and Smart Technologies in Industry 4.0
Edited by Navroop Kaur, Gurpreet S. Dhillon, Sita Rani, Ahmed A. Elngar

Handbook of AI-Based Mechatronics Systems and Smart Solutions in Industrial Automation
Edited by Pushpalatha Naveenkumar, Vandana Sharma, Gunapriya Devarajan, Azeen Abdullah, Ahmed A. Elngar

For more information about this series, please visit: www.routledge.com/Mathematical-Engineering-Manufacturing-and-Management-Sciences/book-series/CRCIMIE

Handbook of AI-Based Mechatronics Systems and Smart Solutions in Industrial Automation

Edited by
Pushpalatha Naveenkumar,
Vandana Sharma,
Gunapriya Devarajan,
Azween Abdullah, and Ahmed A. Elngar

CRC Press
Taylor & Francis Group
Boca Raton London New York

CRC Press is an imprint of the
Taylor & Francis Group, an **informa** business

Designed cover image: Shutterstock

First edition published 2026
by CRC Press
2385 NW Executive Center Drive, Suite 320, Boca Raton, FL 33431

and by CRC Press
4 Park Square, Milton Park, Abingdon, Oxon, OX14 4RN

CRC Press is an imprint of Taylor & Francis Group, LLC

ISBN: 978-1-032-61435-9 (hbk)
ISBN: 978-1-032-61578-3 (pbk)
ISBN: 978-1-032-61579-0 (ebk)

DOI: 10.1201/9781032615790

Typeset in Times
by Newgen Publishing UK

Contents

Preface

In recent years, it has been noticed that mechatronics is a hot research topic, with its rapid development promoted along with artificial intelligence. The optimization of the entire system can be realized when electronics skills, information skills, and mechanical skills are combined through control technology. Human behavior can be imitated by the information known in its environment. The best effect can be obtained with proper analysis, decision-making, and control. The handbook focuses on addressing present-day challenging issues with the aid of artificial intelligence in mechatronic systems.

Due to nonlinear structures and dynamic behavior, the mechatronic systems have control difficulties that are challenging where external disturbances make the situation unpredictable. The existing conventional controllers are found to be ineffective and, therefore, for complex systems design of controllers has great importance in the current book series. To deal with dynamics related to model uncertainties and nonlinearities, this handbook focuses on artificial intelligence techniques such as neural networks and fuzzy systems for the design and optimization of mechatronic systems.

For the performance improvement of mechatronic systems, the design of accurate sensors has to be focused on for a better understanding to measure dynamic variables. The measurement system has to sense the variations in physical parameters such as pressure and displacement and the equivalent changes in voltage or current to be measured. The challenging research is to design a complex system with multiple variables, objectives, and nonlinear relationships where utilizing optimization techniques has to be addressed. State-of-the-art research on new control methods developed for optimal kinematic solutions with the aid of artificial intelligence is covered in the chapters.

Currently, the point of interest is all about monitoring and automating the manufacturing process with the help of networked sensors and equipment machinery with the support of the intranet and the internet, which will be examined here. The big data processing approaches with modeling and simulation of dynamic sub-systems are also covered.

Fault diagnosis and predictive maintenance are expected to become popular in defense and social applications where the process can be made safer and cost-effective. Machine learning algorithms for industrial maintenance and topics related to robotics, healthcare, and automated guided vehicles will also be touched on within this handbook.

Editors

Pushpalatha Naveenkumar is an Assistant Professor in the Department of Electrical and Electronics Engineering at Sri Eshwar College of Engineering, Coimbatore. Her research interests include power electronics and drives, power quality, the Internet of Things (IoT), electric vehicles, and artificial intelligence. She is a life member of the Indian Society for Technical Education (ISTE) and has guided many student projects in the area of IoT and power electronics. She has 13+ years of experience teaching engineering.

Vandana Sharma is recognised in the Stanford list of the top 2% of scientists. She is an accomplished academician and researcher, serving as an Associate Professor at Christ University's Delhi NCR Campus, India. A Senior Member of IEEE, she has an impressive publication record, with over 100 research papers in esteemed SCI and Scopus-indexed international journals and conferences. Dr. Sharma actively contributes to the academic community through various roles, including keynote speaking, session chairing, reviewing, and participating in Technical Program Committees for renowned international journals and IEEE conferences. She holds a postdoctoral degree from Lincoln University College, Malaysia, and adjunct positions at Perdana University, Malaysia, and Beni-Suef University, Egypt, where she is also a Senior Fellow of the Scientific Innovation Research Group (SIRG).

Gunapriya Devarajan is an Assistant Professor with 15+ years of experience at various engineering colleges and is currently associated with Sri Eshwar College of Engineering, Coimbatore. Her areas of interest include electrical machines and drives, artificial intelligence, and renewable energy resources. Her research is with induction motor fault analysis using thermal images. She has published with reputed journals and has book chapters and a patent to her credit.

Azween Abdullah is a Professor at the Faculty of Computing and Digital Technology at HELP University, Malaysia. Dr. Abdullah contributed research, teaching, consulting, and administrative services at the institutions where he has worked. He is a professional development alumnus of Stanford University and the Massachusetts Institute of Technology (MIT), and his work experience includes 30+ years as an academic at institutions of higher learning and as director of research and academic affairs at two institutions of higher learning, vice president for educational consultancy services, 15+ years at commercial companies as a software engineer, systems analyst, and computer software developer and IT/MIS consultancy and training. Dr. Abdullah has several patents under his name and has actively served as an expert reviewer and editorial board member for several high-impact technical journals. He has 150+ publications to his credit published in various technical journals and conference proceedings and has given technical talks at a number of key international conferences, industry summits, and forums.

Ahmed A. Elngar is an Associate Professor and Head of the Computer Science Department at the Faculty of Computers and Artificial Intelligence, Beni-Suef University, Egypt, while simultaneously serving as an Associate Professor at the College of Computer Information Technology, American University in the Emirates, and as an Adjunct Professor at the School of Technology, Woxsen University, India. As the Founder and Head of the Scientific Innovation Research Group (SIRG) and the Director of the Technological and Informatics Studies Center (TISC), Dr. Elngar has established himself as a prolific researcher with more than 150 scientific papers published in prestigious international journals and over 35 books covering diverse topics including data mining, intelligent systems, social networks, smart environments, Internet of Things (IoT), network security, intrusion detection, machine learning, artificial intelligence, big data, authentication, cryptology, and healthcare and automation systems. A collaborative researcher and member of both the Egyptian Mathematical Society (EMS) and the International Rough Set Society (IRSS), Dr. Elngar serves as an editor and reviewer for numerous international journals worldwide. His exceptional contributions to the field have been recognized with several awards, including the Young Researcher in Computer Science Engineering from the Global Outreach Education Summit and Awards 2019 in Delhi, India, and the Best Young Researcher Award (Male) (Below 40 years) at the Global Education and Corporate Leadership Awards (GECL-2018) in Rajasthan, India.

Contributors

Parimala Gandhi A.
Paavai Engineering College
Paavai Nagar, Tamil Nadu, India

Hemananth B.
Department of Electrical and Electronics Engineering
Sri Eshwar College of Engineering
Coimbatore, Tamil Nadu, India

Nandhini B.
Nehru Institute of Engineering and Technology
Coimbatore, Tamil Nadu, India

Malathy Batumalay
Faculty of Data Science and Information Technology
INTI International University
Nilai, Negeri Sembilan, Malaysia

Shadia Yahya Baroud
Department of Computing
University Malaysia of Computer Science and Engineering
Petaling Jaya, Malaysia

Mohan Raj C.
Department of Electrical and Electronics Engineering
Sri Eshwar College of Engineering
Coimbatore, Tamil Nadu, India

Rajasekar C.
Mechatronics Engineering
Akshaya College of Engineering and Technology
Kinathukadavu, Coimbatore, Tamil Nadu, India

Gunapriya Devarajan
Department of Electrical and Electronics Engineering
Sri Eshwar College of Engineering
Coimbatore, Tamil Nadu, India

Sathish Kumar D.
Department of Electrical and Electronics Engineering
Karpagam Academy of Higher Education (Deemed to be University)
Coimbatore, Tamil Nadu, India

Gomathi E.
Coimbatore Institute of Engineering and Technology
Coimbatore, Tamil Nadu, India

Santhiya G.
Department of Electrical and Electronics Engineering
Sri Eshwar College of Engineering
Coimbatore, Tamil Nadu, India

Bhaswanth Reddy Isani
School of Electronics Engineering
Vellore Institute of Technology
Vellore, Tamil Nadu, India

Uma J.
Department of Electrical and Electronics Engineering
M. Kumarasamy College of Engineering
Karur, Tamil Nadu, India

Sumit Kumar Jindal
School of Electronics Engineering
Vellore Institute of Technology
Vellore, Tamil Nadu, India

Kannan K.
Department of Electrical and Electronics Engineering
Annapoorna Engineering College
Salem, Tamil Nadu, India

Muthukumar K.
Department of Electrical and Electronics Engineering
Sri Krishna College of Engineering and Technology
Coimbatore, Tamil Nadu, India

Sabareeshwaran K.
Department of Electrical and Electronics Engineering
Akshaya College of Engineering and Technology
Coimbatore, Tamil Nadu, India

Viswa Brahmana Nanda Kishore
School of Electronics Engineering
Vellore Institute of Technology
Vellore, Tamil Nadu, India

Vinoth Kumar
Department of Electrical and Electronics Engineering
Sri Krishna College of Engineering and Technology
Coimbatore, Tamil Nadu, India

Shubham Mangesh Lad
School of Computer Science Engineering
Vellore Institute of Technology
Vellore, Tamil Nadu, India

Geetha M.
Sri Eshwar Engineering College
Coimbatore, Tamil Nadu, India

Jayaraj M.
Dr. Mahalingam College of Engineering and Technology
Pollachi, Tamil Nadu, India

Meena M.
Department of Information Technology
Vivekanandha College of Technology for Women
Tiruchengode, Namakkal, Tamil Nadu, India

Praveen Kumar M.
Community College (DDU-KK)
The Gandhigram Rural Institute (Deemed to be University)
Gandhigram, Dindigul, India

Vidhaya Datta M.
School of Electronics Engineering
Vellore Institute of Technology
Vellore, Tamil Nadu, India

Praveen Kumar Mani
Department of Computer Science and Engineering
Sri Eshwar College of Engineering
Coimbatore, Tamil Nadu, India

Pushpalatha Naveenkumar
Department of Electrical and Electronics Engineering
Sri Eshwar College of Engineering
Coimbatore, Tamil Nadu, India

Balakumar P.
School of Electrical Engineering
Vellore Institute of Technology
Chennai, Tamil Nadu, India

Pandiyan P.
Department of Electrical and Electronics Engineering
KPR Institute of Engineering and Technology
Coimbatore, Tamil Nadu, India

Hemprasad Yashwant Patil
Faculty of Communication Engineering
Military College of Telecommunication Engineering
Mhow, Indore

Anand R.
Nehru Institute of Engineering and Technology
Coimbatore, Tamil Nadu, India

Senthil Kumar R.
School of Electrical Engineering
Vellore Institute of Technology
Chennai, Tamil Nadu, India

Sivakumar R.
JKKN College of Engineering and Technology
Kumarapalayam, Namakkal, Tamil Nadu, India

Surendar Rangaraju
Faculty of Business Administration
Van Lang University
and
Saigon International School of Business
and
University of Banking, Vietnam
and
Snetel Technologies Company Limited
Ho Chi Minh City, Vietnam

Gowdhamkumar S.
PSG Industrial Institute (PSGCT)
Coimbatore, Tamil Nadu, India

Nithya Shree S.
Department of Electrical and Electronics Engineering
Sri Eshwar College of Engineering
Coimbatore, Tamil Nadu, India

Saradha S.
Department of Computer Science and Engineering
Sri Eshwar College of Engineering
Coimbatore, Tamil Nadu, India

Saravanan S.
Department of Electrical and Electronics Engineering
Amrita School of Engineering
Coimbatore, Amrita Vishwa Vidyapeetham, India

Sheikameer Batcha S.
Department of Electrical and Electronics Engineering
Sri Eshwar College of Engineering
Coimbatore, Tamil Nadu, India

Sivaranjani S.
Department of Electrical and Electronics Engineering
Sri Krishna College of Engineering and Technology
Coimbatore, Tamil Nadu, India

Taruun Prasath S.
Department of Electrical and Electronics Engineering
Sri Eshwar College of Engineering
Coimbatore, Tamil Nadu, India

Tharun S.
School of Electronics Engineering
Vellore Institute of Technology
Vellore, Tamil Nadu, India

Thirunavukkarasu S.
Department of Electrical and Electronics Engineering
Paavai Engineering College
Namakkal, Tamil Nadu, India

Vijayakumar S.
MVI Technologies (S) PTE Ltd
Singapore

Vinoth S.
Department of Mechatronics Engineering
Akshaya College of Engineering and Technology
Kinathukadavu, Coimbatore, Tamil Nadu, India

Suresh Seetharaman
Sri Eshwar Engineering College
Coimbatore, Tamil Nadu, India

Vandana Sharma
Computer Science Department
Christ University
Bengaluru, India

Bharani Prakash T.
Department of Electrical and Electronics Engineering
Sri Krishna College of Technology
Coimbatore, Tamil Nadu, India

Kokilavani T.
Department of Electrical and Electronics Engineering
Sri Krishna College of Engineering College Technology
Coimbatore, Tamil Nadu, India

Santhiya T.
Department of Electrical and Electronics Engineering
Study World College of Engineering
Coimbatore, Tamil Nadu, India

Gomathi V.
Department of Electrical and Electronics Engineering
Dhanalakshmi Srinivasan College of Engineering
Coimbatore, Tamil Nadu, India

Kiruthiga Devi V.
Electrical and Electronics Engineering
KSR College of Engineering
Namakkal, Tamil Nadu, India

Rajanbabu W.
Department of Electrical and Electronics Engineering
Sri Eshwar College of Engineering
Coimbatore, Tamil Nadu, India

1 AI in Mechatronics Engineering

Pushpalatha Naveenkumar, Vandana Sharma, and Gunapriya Devarajan

1.1 INTRODUCTION

Artificial intelligence (AI) is a multidisciplinary field, and the integration of AI with other fields such as mechanical, electrical, computer, robotics, and control systems has led to changes in the progress of the mechatronics industry. AI and its key applications in mechatronics include advanced control, predictive maintenance, strategic design, intelligent robotics, and advanced manufacturing. Electrical engineers can produce more diverse, consistent, and superior designs by means of data analysis, pattern recognition, and AI decision-making capability. This chapter explores the junction of AI and mechatronics combined with how this strong combination might transform the industry. Moreover, it discusses AI concepts and their key applications in mechatronics to improve control systems, predictive maintenance, design optimization, intelligent robots, and sophisticated manufacturing in the modern corporate environment. Using the data analysis, pattern detection, and decision-making powers of AI, electrical engineers may design more rapidly, reliably, and with greater simplicity. This chapter also tackles the moral questions and challenges associated with the integration of AI, thereby providing concepts on likely future directions and trends. Electrical engineers may design more quickly, consistently, and with more ease by leveraging data analysis, pattern identification, and decision-making powers of AI. This chapter also addresses the difficulties and moral issues related to the integration of AI, therefore offering ideas on possible future paths and trends. In the end, this research emphasizes how AI has transformed power to drive mechatronics engineering to unprecedented levels of performance and creativity. By application of the fully active system (FAS) technique, this work addresses event-triggered neural network (NN) adaptive control for uncertain strict-feedback nonlinear systems. Eliminating the presumption that these nonlinearities are always known results in event-based NN for essentially unknown system nonlinearities [1]. This work uses auxiliary systems and fuzzy state observers to address input saturation, enabling an adaptive fuzzy dual-channel event-triggered control for Multiple Input Multiple Output (MIMO) systems and countering deception attacks. It guarantees system stability, reduces computing demand, and achieves Zeno behavior-free [2]. Deep reinforcement learning (DRL) and inverse reinforcement learning (IRL) covering sensing, assembly, uncertain rewards, multitasking, transfer learning, multimodal approaches, and human–robot interaction (HRI) are

DOI: 10.1201/9781032615790-1

applied in robotic manipulation [3]. AI and big data enable building better healthcare systems by means of many patient data evaluations including medical history, genetics, and lifestyle to improve decision-making and patient care [4]. An end-to-end Internet of Things (IoT) system mimicking high-velocity smart meter data uses AI-based anomaly detection for spikes and dips using Microsoft Azure. It encompasses the entire IoT lifecycle using IoT hub, device provisioning service, and lambda architecture for real-time data processing and visualization [5]. A design method to translate centralized NNs into distributed architectures with respect to sensor network restricted bandwidth. By using local classifiers and a compression–reconstruction technique, it allows efficient classification, therefore reducing bandwidth by up to 20× and power by 9× with minimum accuracy loss (2%) [6]. By means of data reduction techniques in wireless sensor networks, problems including energy efficiency, resource utilization, and extended lifespan in environments with limited resources are addressed. It investigates many approaches, their benefits and disadvantages, and their applicability for low-resource settings for efficient data processing and storage [7]. A hybrid control system for robot cars including CNN-based behavior cloning incorporates proportional integral derivative (PID) control for improved driving performance and trajectory monitoring. Under supervised learning, the behavior cloning system mimics human driving with low-cost sensors and actuators [8]. Meta-heuristic optimized deep random neural networks (MH-DRNN) pose cyber security problems for industrial IoT. By employing sunflower-inspired optimization for feature selection, the system achieves a 99.2% prediction accuracy with three Multilayer Perception (MLP) and recurrent layers, hence improving the analysis of heterogeneous data. It increases attack intrusion detection for both antagonistic and catastrophic ones [9]. Emphasizing mechatronics, sensing, control methods, and user performance assessment advances both active and semi-active lower limb prostheses. Analyzing 53 prototypes including ankle, knee, and knee–ankle prostheses helps one to appreciate the challenges in prosthetic research still under progress [10]. Using sensor data to enhance efficiency and reduce costs, a predictive maintenance (PdM) machine learning-based solution for yarn manufacturing plants with one deep learning model outperformed traditional methods with 96% accuracy by means of four machine learning algorithms [11]. Little redundancy for feature selection and reduction, maximum relevancy minimum redundancy (MRMR), and principal component analysis (PCA) enable industrial IoT systems to identify anomalies. Minimum Redundancy Maximum Significance (MRMS) outperformed PCA and reached 99.9% accuracy using 12 features, leveraging decision trees for binary classification. The work highlights how feature engineering can improve intrusion detection by varying model accuracy, complexity, and speed [12]. Reconfigurable machine tools (RMTs) based DRL reconfiguration planning method for digital twin-driven smart manufacturing systems enables adaptive, real-time decision-making and system optimization by leveraging deep reinforcement learning to dynamically adjust machine configurations in response to changing production demands and operational conditions. By leveraging deep Q-network, the approach minimizes reconfiguration, relocation, and processing costs, optimizing flexibility and production sequences [13]. Adaptive Reinforcement learning-based Mobility-aware System (ARMS) based protocol combined with vehicle ad hoc networks (VANETs) for wireless body area networks (WBANs) uses adaptive Q-learning and link lifetime estimate (LLE) to select stable nodes in dynamic networks, therefore guaranteeing consistent and efficient communication [14]. Active neighbor exploitation (ANEX) allows sensor nodes to prioritize nodes with less unscheduled active neighbors and exchange parents, therefore enabling more concurrent transmissions and faster data aggregation in IoT systems [15].

1.2 FUNDAMENTALS OF AI IN MECHATRONICS: AN OVERVIEW

AI can be used for multiple operations like monitoring a system's performance, discovering new possibilities, predicting future requirements, interacting with both machines and humans, and interpreting the data for various decision-making applications in the fields of mechatronics engineering. In any mechatronics system, data are collected from sensors; based on the information collected from the sensors, actuators will perform their unique role for making better system performance. The mechatronics system is a combination of mechanical, electrical, and electronic domains. When AI is integrated with mechatronics systems, it gives a new dimension and various possibilities to improve the system performance, and it also paves the way for the generation of innovative autonomous systems and their flawless operation. AI integration in mechatronics is a fundamental key for the enhancement of system control and automation. Figure 1.1 shows the fundamentals of AI in mechatronics. For any system, data plays a vital role; sensors are used to collect information from various actuators. Here, actuators and sensors are the mechanical parts of the system. The electrical and electronics part of the system consists of input signal conditioning and interfacing, digital control architecture, output signal conditioning and interfacing, and output display. AI is used to monitor, discover, predict, interact with, and interpret the data.

1.2.1 Types of AI Based on Functionality

Artificial Intelligence (AI) can be categorized into four distinct types based on its capabilities and level of advancement

1. Reactive AI
2. Limited memory AI
3. Theory of mind AI
4. Self-aware AI

1.2.1.1 Reactive AI

This type of AI has no memory. It performs only specific tasks based on predefined logic.

1.2.1.2 Limited Memory AI

This type of AI has limited memory, and this memory can be used for decision-making. It is a kind of machine-learning model that learns from facts and events. Self-driving cars use this type of AI.

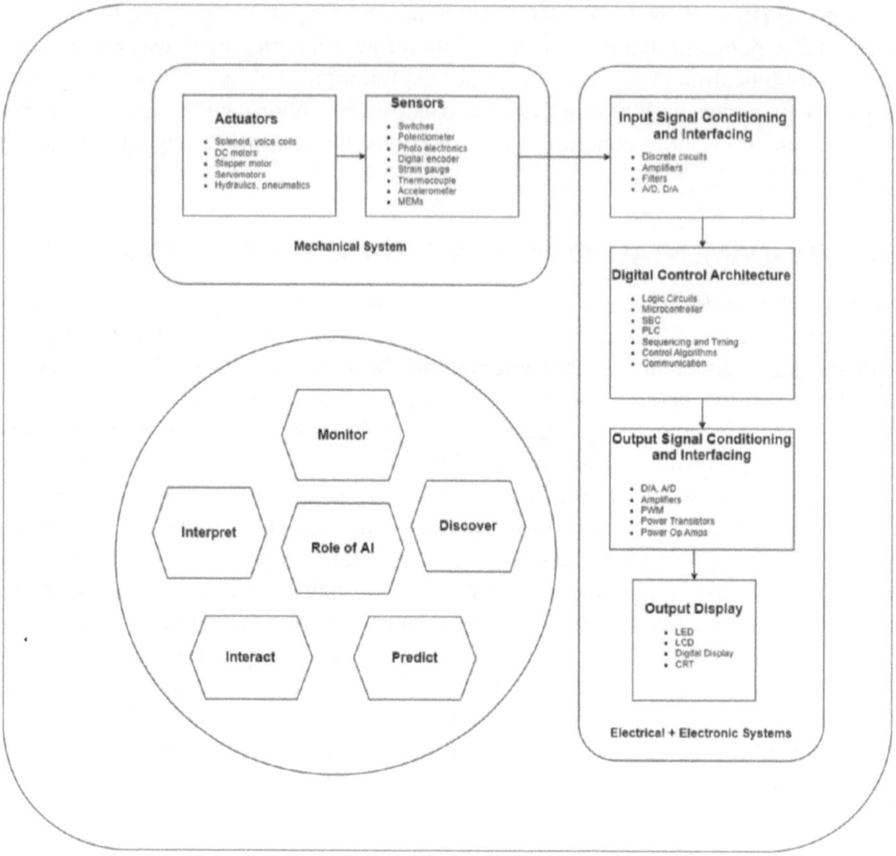

FIGURE 1.1 Fundamentals of AI in mechatronics.

1.2.1.3 Theory of Mind AI

This type of AI understands human emotions, and it is constructed on the basis of social intelligence. It can predict human behavior. The best example for the theory of mind AI is autonomous cars.

1.2.1.4 Self-Aware AI

This type of AI stores data on the basis of human common sense, which can work autonomously like humans. The best example of self-aware AI is autonomous self-aware robots, which can perform any task without human intervention.

1.2.2 Classification of AI Based on Components

Artificial Intelligence (AI) is often classified into three main types based on its level of intelligence and ability to perform tasks:

1. Artificial narrow intelligence (ANI)
2. Artificial general intelligence (AGI)
3. Artificial super intelligence (ASI)

1.2.2.1 Artificial Narrow Intelligence (ANI)

It is widely known as narrow AI. It focuses only on narrow tasks or one problem.

1.2.2.2 Artificial General Intelligence (AGI)

It is a kind of strong AI; its super intelligence suppresses human intelligence. It performs the cognitive abilities of the human brain. This type of AI contains the feature of human-like thinking and reasoning. AGI is a hypothesis, and it is assumed that it will be invented within quantum computing.

1.2.2.3 Artificial Super Intelligence (ASI)

ASI is the logical progression of AGI. It mimics human brain thinking, it presents the task handled by the human brain with greater accuracy and speed.

1.3 MECHATRONICS SYSTEM DESIGN WITH AI INTEGRATION

Mechatronics is a multidisciplinary field that merges mechanical, electronic, and software engineering to create advanced automated systems. Integrating AI into these systems enhances their capabilities, enabling smarter automation, adaptive control, and predictive decision-making. This integration is crucial as industries seek more efficient, intelligent, and responsive solutions to meet the growing demands of modern engineering challenges.

1.3.1 System Objective and User Identification

Defining the system's objective is one of the first and most critical steps in the system design process. This process involves improving the system's understanding of a specific problem and its intended purpose. Depending on the users the objective of the system may vary. The following are the basic mechatronics scenarios where AI can play a major role in the mechatronics industry.

- Automating a manufacturing process to increase efficiency and reduce human error.
- Optimizing the performance of a mechanical device to enhance its functionality and durability.
- Identifying the target users is equally important, as with objective.
- Business owners should prioritize reliable and easy-to-maintain systems.
- Professional engineers will need analytical and management skills to manage complex projects.

1.3.2 Component Selection and Integration

Selection and integration is an important stage of AI integration into any mechatronic system, which determine the overall performance and functionality of the system. Here is a list of systems that need to integrate AI into a mechatronic system. It plays a key role in automation when it is converted into mechanical motion. Electronic devices, including sensors, communication links, and additional devices, allow the body to efficiently collect, process, and manage data. The microcontroller acts as the body's brain, controlling inputs from sensors and actuators accordingly. Ensure a seamless flow of information between different processes, such as providing real-time updates on production activities. This integration helps optimize workflows, increase operational efficiency, and reduce resource consumption.

1.3.3 System Design and Control Integration

Mechanical engineering entails the development of intricate systems that integrate mechanical, electrical, and mechanical components into a unified system. Control systems are essential for the management of the interactions between these devices, the integration of mechanical devices with electronic devices, and the provision of intelligent decision-making capabilities. The AI algorithms integrated into the control system enable adaptive decisions, hence allowing it to change its operation at any given time, depending on sensor input. The flow within the system must be well choreographed: the sensor data must be processed into the required products with control systems and intelligent algorithms. Proper management of data allows the systems to respond to changes in a manner that they function accordingly.

1.3.4 AI Model Development

Apart from integration, another influencing key for AI in mechatronic systems is model development. The first and foremost stage in this model is to collect data; it involves gathering relevant data received from sensors and other sources for training the intelligence of the model. Selection of an AI model is very crucial and depends on the type of tasks the system needs to run. There are many types of models that are used for different tasks, such as predictive maintenance, real-time optimization, and object detection. On-the-fly optimization model tunes workloads for optimal performance. Object detection models can identify and classify objects in the system environment, improving system responsiveness and performance.

1.3.5 Prototyping and Testing

Prototyping and testing are the vital steps in the designing process that ensure the system's intended objectives.

- A prototype is built by incorporating all the mechanical, electronic, and AI components into the final state of the system. This prototype serves as a proof of concept that allows engineers to evaluate the system's design and functionality.
- The prototype is subjected to various testing conditions to validate its functionality, which ensures that the system operates as intended.
- Safety confirmation is the next process that takes place after testing. In this process, the system is tested against some pre-fixed safety norms as per the

requirement. This test ensures the riskless operation of the system to its users as well as its environment.

- Reliability ensuring of the system should also be carried out to understand the system's performance and its consistency over a period of time.
- Testing process and its results often reveal the various areas for improvement of the system design, performance, and operation. It leads to iterative refinements in the design, process improvement, or process fine-tuning of the system to ensure it is robust and performs as expected in real-world applications.

1.3.6 System Deployment and Maintenance

- System deployment involves installing the system in its intended environment and ensuring it is fully operational.
- Installation includes setting up the system, connecting all components, and ensuring they work together seamlessly.
- Calibrating components ensures that they operate within the desired parameters.
- Real-time monitoring tools are often implemented to track the system's performance: AI algorithms continuously optimize operations, making real-time adjustments to improve efficiency and respond to unexpected changes.
- Maintenance is critical to ensure the system's longevity and reliability: AI can predict when maintenance is needed, reducing downtime and preventing costly breakdowns.
- Regular maintenance based on AI predictions can extend the system's operational life and ensure it continues to perform optimally.

1.3.7 Hybrid Mechatronic System with Cloud Integration

- It combines the advanced features of AI and mechatronics' accuracy and adaptability with the scalability of cloud computing.
- These systems leverage AI for real-time decision-making and adaptive control.
- This synergy drives scalable, intelligent solutions across several sectors, remote diagnostics, and improved system efficiency.
- Capabilities of hybrid mechatronics systems are large volumes of sensor data, system behavior prediction, and dynamic performance optimization in various aspects.
- Applications include industrial automation in which robots use AI to modify activities depending on cloud-processed analytics and autonomous autos, which depend on cloud connectivity for updates and route optimization.

1.3.8 Simulation and Modeling

- These systems abound in mechanical, electrical, and software components; AI enhances their performance and value.
- By creating digital copies, duplicating real-world conditions, and evaluating system performance free from the need for actual prototypes, engineers can save a great lot of time and money.

- These advanced simulations support robotics, manufacturing, automotive, and aerospace by streamlining design cycles, extending system lifetime, and allowing accurate system operation.
- AI-based simulations help systems to be more autonomous and efficient by enabling real-time monitoring and control, adaptive learning, and predictive maintenance, that is, by empowering systems.

1.4 CONTROL SYSTEM AND AI-BASED CONTROL

1.4.1 Traditional Control Systems

- Control systems are the backbone of mechatronics engineering, responsible for managing and coordinating the operation of various mechanical and electronic components within a system.
- These systems typically rely on feedback loops to maintain desired outputs: adjusting inputs based on real-time data from sensors to achieve the desired performance.
- PID controllers have been around for a long time. These controllers are simple, reliable, and easy to use; therefore, they can be used in many ways. This makes them suitable for many control applications. As system complexity increases, the limitations of traditional control systems become more apparent. It is also difficult for PID controllers to perform well in controlling complex processes that cause abnormal or anomalous behavior. Getting these under control will help solve these problems.
- Figure 1.2 illustrates the flow of the control systems based on negative feedback.

1.4.2 AI-Based Adaptive Control

AI-based control represents an advancement in control design that allows for more control and tuning strategies. It learns to improve its performance over time. This optimization is tied to the finished product but also increases the power of the system. While the traditional management system has problems, intelligence-based management can always work in unexpected situations.

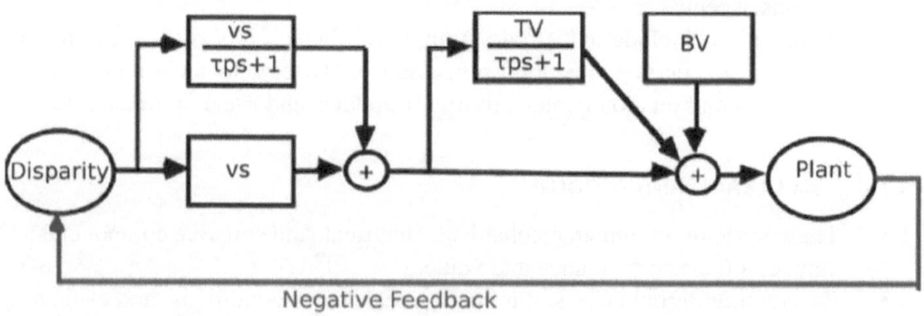

FIGURE 1.2 Negative feedback.

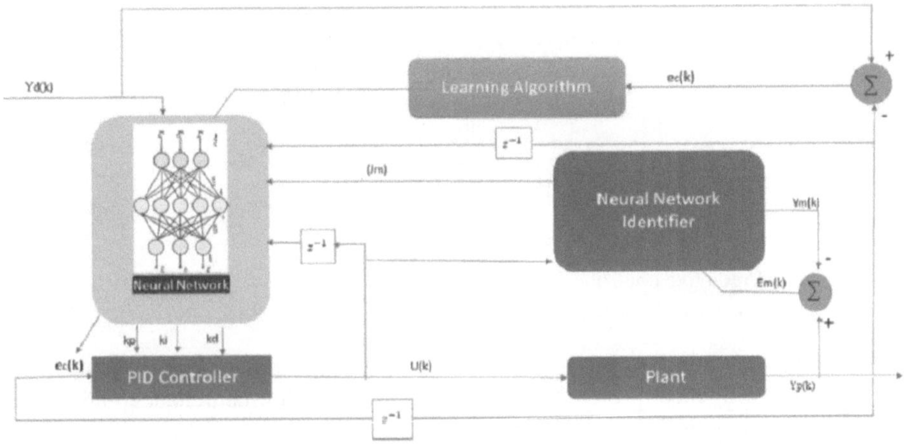

FIGURE 1.3 Adaptive control.

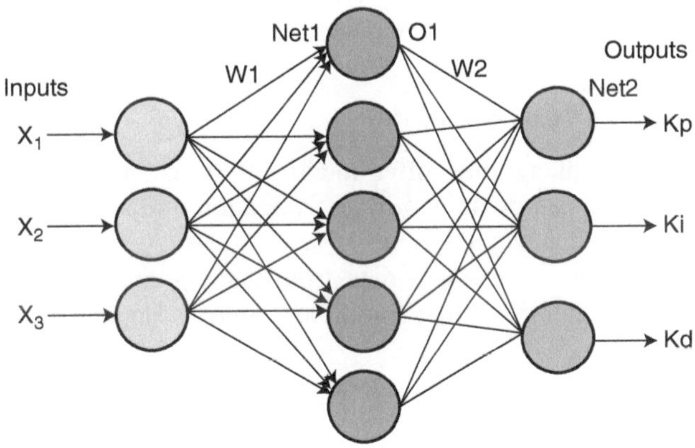

FIGURE 1.4 Flow diagram.

Figure 1.3 shows the idea of NNs in connection with the PID controller. Figure 1.4 presents a diagrammatic picture of NNs.

1.4.3 Predictive Control

- Predictive control is another important application of AI in control systems, enabling the system to anticipate future states based on current and historical data as shown in Figure 1.5.
- This allows the system to make better decisions, reducing errors and increasing efficiency. For example, when it comes to heating, ventilation, and air conditioning (HVAC), AI can predict temperature changes based on past performance and weather conditions.

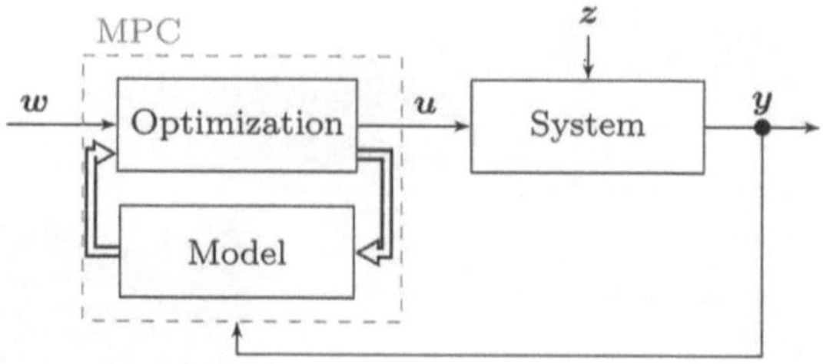

FIGURE 1.5 Predictive control.

- Many other industries, including manufacturing, use real-time management and intelligence to predict equipment and schedule maintenance and repairs before they fail.
- Adjust the system's operation to maintain optimal conditions while minimizing energy consumption.
- In the transportation field, AI has the ability to predict traffic patterns and reroute the traffic to reduce fuel consumption and waiting time.
- Figure 1.5 presents a predictive control model circuit.

1.4.4 Intelligent Decision-Making

- AI-based control systems offer intelligent decision-making, as shown in Figure 1.6, and can do an enormous evaluation of data by seeing trends that traditional systems would ignore.
- AI-based systems can assess data from many sources to fully understand the present situation, unlike traditional controls depending on established policies.
- AI can make real-time decisions to optimize system performance based on the latest data.
- In autonomous cars, AI can, for example, create a complete map of the surroundings of the vehicle by processing input from cameras, Light Detection and Ranging (LiDAR), and Global Positioning System (GPS).
- AI can see prospective hazards like other cars or barriers and act to stop collisions.
- AI can maximize the car's course and speed to ensure efficient and safe functioning of it.

1.4.5 Handling Nonlinear and Complex Systems

- AI-based control systems are particularly effective at managing nonlinear and complex systems, which are difficult to model using traditional methods.
- Machine learning techniques can learn nonlinear relationships directly from data, providing more accurate control than would be possible with conventional approaches.

FIGURE 1.6 Decision-making flow.

- This is especially useful in applications such as robotic control, where the interactions between mechanical components, sensors, and the environment can be highly complex.
- AI can continuously learn and adapt to these interactions, ensuring optimal performance even in dynamic and unpredictable conditions.
- Figure 1.7 distinguishes the kind of nonlinear system solving method.

1.4.6 AI-Powered Feedback Control for Robotics

- This system uses sophisticated AI together with the conventional control system approach to improve robotics performance in a dynamic and challenging environment.
- Often dealing with nonlinearity and uncertain scenarios, the use of PID and state-space controllers is more challenging.
- AI addresses these challenges by means of NNs for system identification, reinforcement learning, and fuzzy logics for uncertainty management.
- In the robotic applications comprising autonomous robots, robotic arms, drones, and humanoid robots, these provide real-time adaptation, precision, and efficiency.
- AI-driven feedback control provides a revolution by allowing systems to learn, adapt, and precisely maximize performance.
- AI-powered feedback control for robotics can be applied for many applications like real-time data analysis, robotic arm movement adjustment, autonomous navigation, and human–machine interactions.

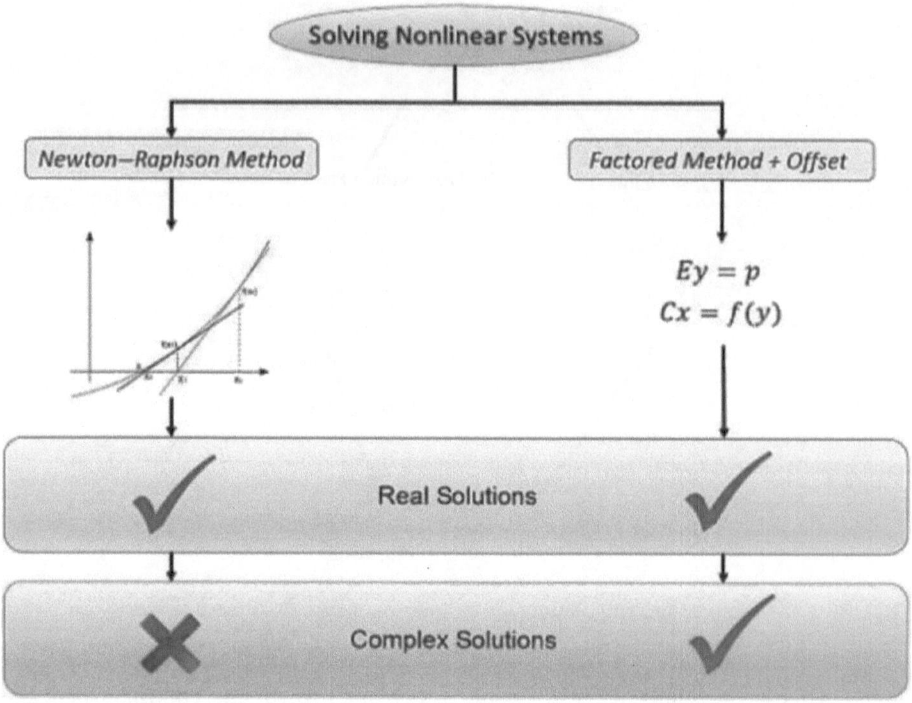

FIGURE 1.7 Solving nonlinear systems.

1.4.7 Quantum Computing in Control System

- Using quantum concepts, quantum computing in control systems offers a breakthrough by solving challenging computational issues in optimization, modeling, and decision-making. Figure1.8 shows the model of quantum computing in the control system.
- High-dimensional, nonlinear, dynamic processes provide difficulties for conventional computer systems where efficiency could suffer.
- Exponential speedups in solving these challenges made feasible by quantum algorithms such as Grover's search and quantum annealing enable faster and more exact control parameter tuning.
- Moreover, by combining quantum computing with AI-based control, the system would be able to give intelligent predictions sufficient to handle big data.
- AI enhances the performance and applicability of mechanical, electrical, and software components in robotics, manufacturing, automotive, and aerospace by speeding design cycles, extending system lifetime, and allowing proper system functioning thanks to these sophisticated simulations.
- Real-time monitoring and control, adaptive learning, and predictive maintenance made possible by AI-based simulations help systems to be more independent and efficient by means of empowerment.

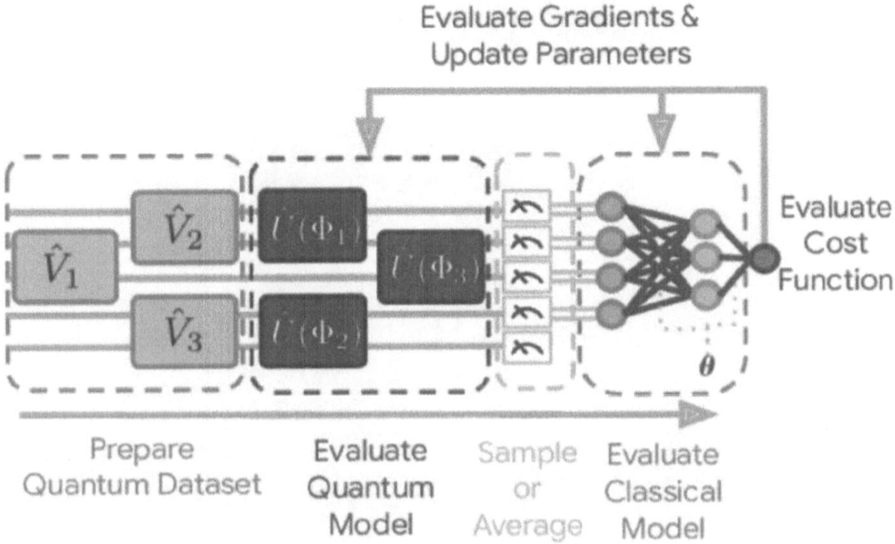

FIGURE 1.8 Quantum computing in control systems.

1.5 SENSOR NETWORK AND AI-BASED DATA ANALYSIS

Sensor networks play a crucial role in collecting data from different areas of the system. With this information, AI can assess the system's current and future conditions. Using the forecasted data, predictive maintenance can enhance the system's dependability.

1.5.1 Role of Sensor Networks

Sensor networks form the backbone of modern engineering systems and play a key role in the collection and transmission of environmental and operational data. These networks consist of interconnected components, each designed to deal with specific parameters such as temperature, pressure, humidity, motion, vibration, and other disturbances in the circuits or systems. It provides information and immediate attention and enables the system to respond immediately to environmental changes. This information is important not only for emergency management but also for long-term decision-making processes that affect maintenance schedules, adjustments, and strategic planning. The data generated can be very large. As knowledge gets more complex, translating becomes more challenging and frequently results in delays or incorrect conclusions. This approach enables fast processing of big data sets and helps to find interesting trends as well as more exact tasks from environmental control to manufacturing.

1.5.2 AI-Based Predictive Maintenance

Predictive maintenance depicted in Figure 1.9 is absolutely fundamental for data intelligence-driven analysis in sensor networks. Basic machinery, motors, and

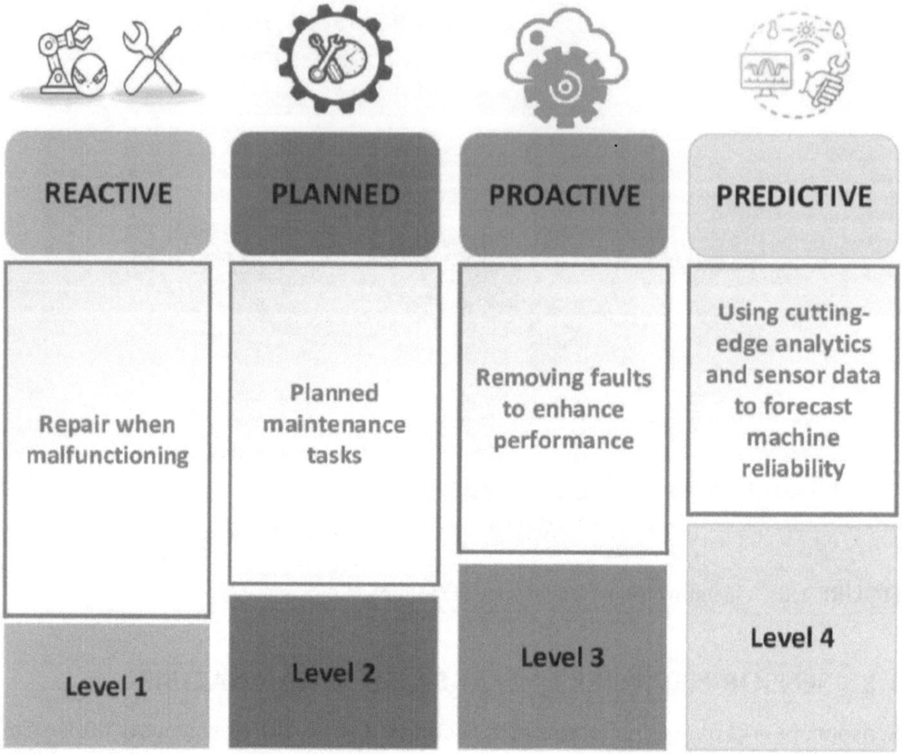

FIGURE 1.9 Predictive maintenance.

electrical equipment used in an industrial environment are many times fitted with sensors measuring a range of elements including vibration, temperature, pressure, and environmental noise. Maintaining careful records of equipment use over time is essential. Predictive abilities enable one to foresee mistakes before they start. This method not only prolongs the lifetime of your tools but also helps to lower the expenses connected to unexpected and emergency repairs.

1.5.3 Anomaly Detection

Fault detection is another vital application of AI in sensor networks, especially in systems generating abundant amounts of data. Sensor networks provide vast amounts of real-time generated data that make it difficult for operators to track and investigate all the data points in their original forms. AI could monitor current trends and find weaknesses implying possible network breakdowns such as overloading, short circuits, or missing equipment malfunction in electrical projects. The more challenging problem is keeping the body in a functioning state. Analysis in sensor networks driven by data intelligence depends on predictive maintenance. Often, basic machinery, motors, and electrical equipment used in an industrial environment fit

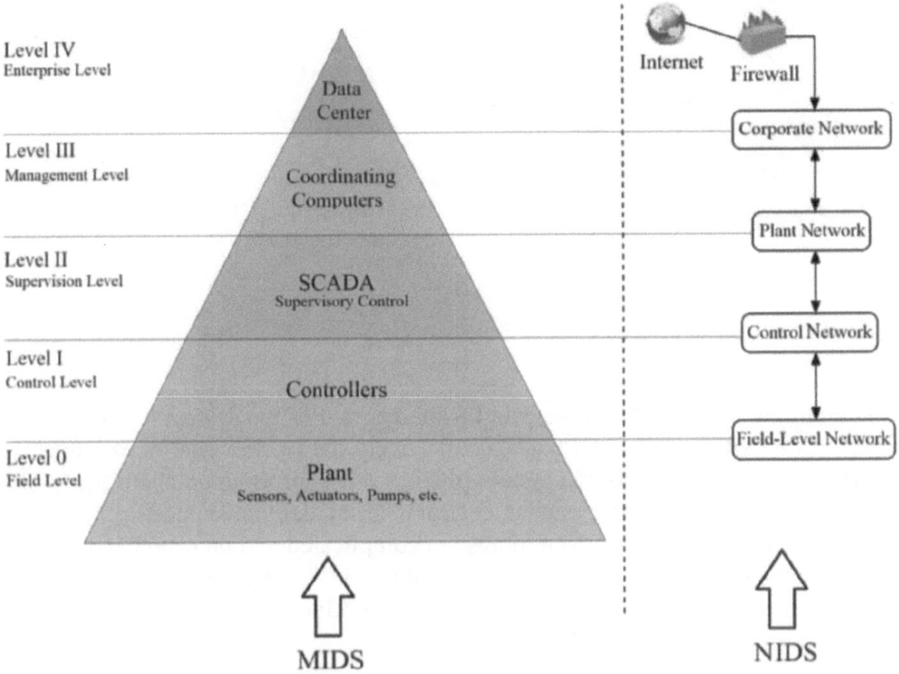

FIGURE 1.10 Anomaly detection.

sensors measuring a range of parameters including vibration, temperature, pressure, and environmental noise. One must keep detailed records of equipment use throughout time. Predictive skills help one to spot faults before they start. This approach not only extends the lifetime of your tools but also helps to cut the costs related to unforeseen and emergency repairs. Figure 1.10 shows the various levels of anomaly detection.

1.5.4 Environmental Monitoring

Environmental monitoring is one field where data-based analytics and sensor networks are clearly proving successful. In smart cities, sensor networks monitor many aspects of the environment, including air quality, noise levels, traffic patterns, water quality, and waste management. Energy and health products improve agriculture and ensure sustainable use of resources. Browse instructions for instant understanding. For example, AI can predict weather based on current data, weather patterns, and historical models, allowing city planners to assess opportunities to reduce pollution. Predict the best irrigation time to increase water efficiency and crop yields.

1.5.5 Data Fusion

Data fusion is a key aspect of AI-based data analytics, where AI systems combine and analyze data from multiple sensors to fully understand the monitoring environment

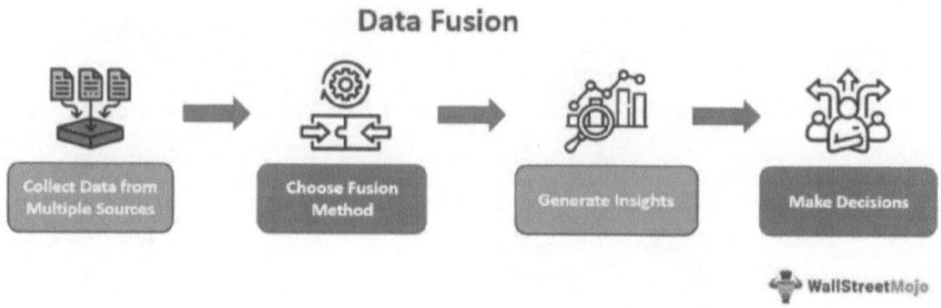

FIGURE 1.11 Data fusion.

or system. Figure 1.11 shows the data fusion diagram of the system. Using sensors, cameras, LiDAR, radar, and GPS taken collectively, a driverless car, for example, can determine where it is. Also, the system predicts other information about the system by using AI systems; it can create an environment model that supports efficient and effective vehicle operation. When things are complicated and precision and dependability are important, data integration is a useful tool. This is not an easy job. The end result is a method that works well in many areas, is reliable, and is more environmentally friendly.

1.5.6 Real-Time Data Analysis in IoT Sensor Network

Edge AI allows quick data processing at or close to the data source, hence reducing latency. Depending on this, the IoT sensor network shown in Figure 1.12 uses edge computing-based AI for various applications like autonomous cars, industrial automation, and healthcare monitoring, where fast decisions can either prevent mistakes or save lives. Edge AI reduces costs by central system transmission of merely actionable insights and local data processing, hence improving IoT sensor networks by minimal bandwidth utilization. Solving problems in sensitive sectors like smart cities and healthcare not only improves local processing but also data security and privacy.

1.5.7 Data Compression for Efficient Sensor Networks

AI-based data compression overcomes bandwidth, energy, and storage limits in sensor networks by means of data size reduction without surrendering significant information. By means of NNs and autoencoders, these techniques preserve only relevant information, therefore matching data patterns for high compression ratios and real-time analysis, hence reducing noise. From environmental monitoring to healthcare to industrial IoT where high-frequency data must be efficiently given and stored, this is basic for applications. AI-based edge devices reduce raw data transportation and energy as they offer local compression.

During crises, adaptive strategies give vital data first priority, therefore enabling speedy judgments. AI-driven compression boosts network efficiency, reduces costs, and extends device lifetime, therefore opening the door for smart, scalable sensor

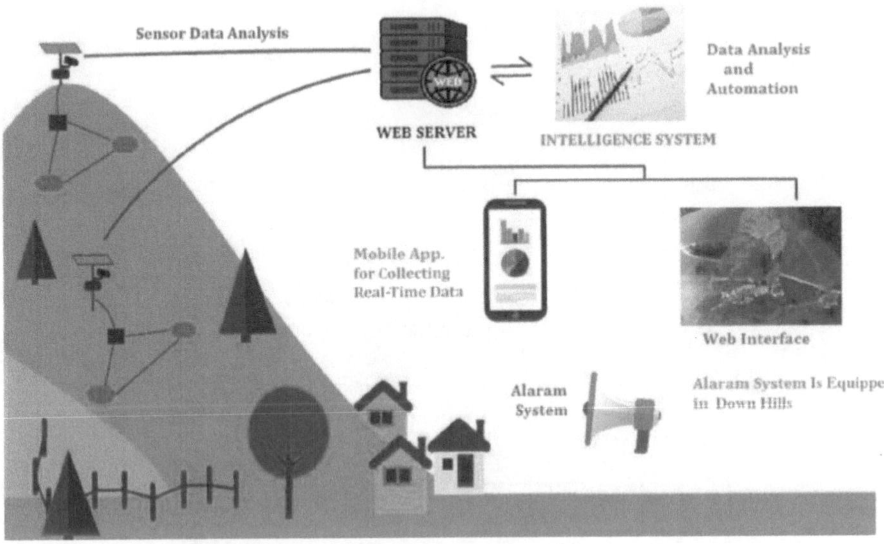

FIGURE 1.12 IoT sensor.

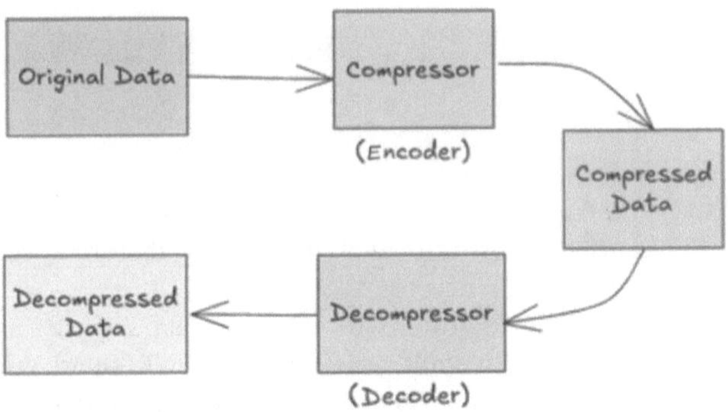

FIGURE 1.13 Data compression flow.

systems, especially in resource-limited places such as underwater sensors or satellites. Figure 1.13 shows data compression diagrammatically.

1.5.8 AI-Enhanced Security in Sensor Networks

Figure 1.14 shows the AI-enhanced security network in sensor networks. In environments like IoT systems, smart cities, and industrial automation, AI safeguards data integrity, confidentiality, and network resilience. By use of real-time sensor data, AI approaches include anomaly detection and machine learning to expose and reduce threats such as illegal access, data manipulation, or cyberattacks. Sensor networks using AI can dynamically detect unusual trends including hostile node behavior

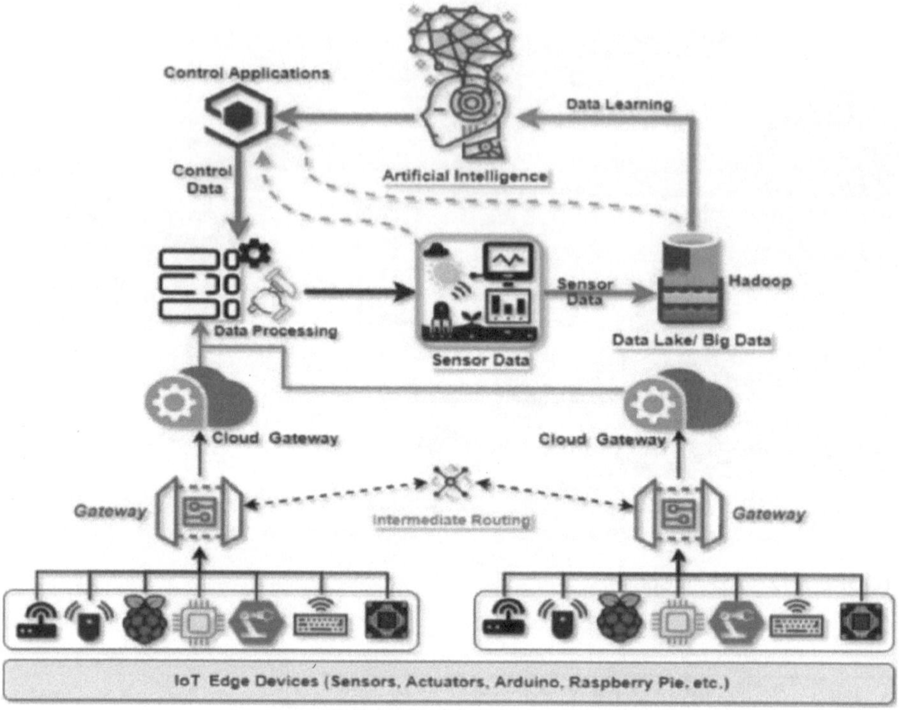

FIGURE 1.14 Security to networks.

or unexpected traffic spikes, therefore enabling preemptive responses. AI not only improves network security but also fits evolving threats; therefore, sensor networks are reliable and robust for crucial purposes.

By combining modern AI algorithms to enhance data security, safe communication, and network resilience, it finds uses including IoT, smart grids, healthcare monitoring, and industrial systems. Using real-time sensor data, deep learning, predictive analytics, and anomaly detection, AI-driven devices examine hazards including cyberattacks, data breaches, or hostile node behavior. Furthermore, edge AI combined with cloud-based analytics offers a scalable and adaptable solution to safeguard sensor networks in various environments, therefore tackling changing cyber security issues exactly and effectively.

1.6 CONCLUSION

Combining AI with mechatronics systems, control systems, sensor networks, and data analysis is difficult but greatly increases capacity, allowing intelligent automation, adaptive control, decision-making, etc. These domains currently bear their effects and the developing needs of the modern corporate environment. The impact of this convergence is across different domains to build smart infrastructure, and it drives innovation in many paths. It supports autonomous systems and predictive technologies. It addresses the need for flexibility, scalability, and intelligent solutions that contribute to the growth and development of the respective sectors.

REFERENCES

1. Y. Wang, G. Duan and P. Li, "Event-Based Neural Networks Adaptive Control of Nonlinear Systems: A Fully Actuated System Approach," *IEEE Transactions on Circuits and Systems I: Regular Papers*, vol. 71, no. 9, pp. 4211–4221, 2024, doi: 10.1109/TCSI.2024.3417013.
2. N. Zhao, Y. Tian, H. Zhang and E. Herrera-Viedma, "Learning-Based Adaptive Fuzzy Output Feedback Control for MIMO Nonlinear Systems with Deception Attacks and Input Saturation," *IEEE Transactions on Fuzzy Systems*, vol. 32, no. 5, pp. 2850–2862, 2024, doi: 10.1109/TFUZZ.2024.3363839.
3. R. Ozalp, A. Ucar and C. Guzelis, "Advancements in Deep Reinforcement Learning and Inverse Reinforcement Learning for Robotic Manipulation: Toward Trustworthy, Interpretable, and Explainable Artificial Intelligence," *IEEE Access*, vol. 12, pp. 51840–51858, 2024, doi: 10.1109/ACCESS.2024.3385426.
4. S. K. Marimekala, J. Lamb, R. Epstein and V. Bhupathi, "Using AI and Big Data in the HealthCare Sector to Help Build a Smarter and More Intelligent HealthCare System," *2024 IEEE World AI IoT Congress (AIIoT)*, Seattle, WA, USA, 2024, pp. 356–362, doi: 10.1109/AIIoT61789.2024.10578989.
5. S. D. BhavaniPeri, A. Ravi and M. Supriya, "Design and Implementation of a Real-Time IoT Solution for Smart Meter Data Analysis in Microsoft Azure," *2024 15th International Conference on Computing Communication and Networking Technologies (ICCCNT)*, Kamand, India, 2024, pp. 1–5, doi: 10.1109/ICCCNT61001.2024.10725804.
6. T. Strypsteen and A. Bertrand, "Bandwidth-Efficient Distributed Neural Network Architectures with Application to Neuro-Sensor Networks," *IEEE Journal of Biomedical and Health Informatics*, vol. 27, no. 2, pp. 933–943, 2023, doi: 10.1109/JBHI.2022.3225019.
7. G. D. Arora, N. K. Chopra, N. Gopinath, S. K. Ravichandran, M. K. Chakravarthi and D. Gangodkar, "Data Reduction Techniques in Wireless Sensor Networks with AI," *2022 5th International Conference on Contemporary Computing and Informatics (IC3I)*, Uttar Pradesh, India, 2022, pp. 48–51, doi: 10.1109/IC3I56241.2022.10073380.
8. H.-G. Kim, J. Myoung, S. Lee, K.-M. Park and D. Shin, "Design of AI-Powered Hybrid Control Algorithm of Robot Vehicle for Enhanced Driving Performance," *IEEE Embedded Systems Letters*, vol. 16, no. 2, pp. 94–97, 2024, doi: 10.1109/LES.2023.3257774.
9. X. Li, C. Xie, Z. Zhao, C. Wang and H. Yu, "Anomaly Detection Algorithm of Industrial Internet of Things Data Platform Based on Deep Learning," *IEEE Transactions on Green Communications and Networking*, vol. 8, no. 3, pp. 1037–1048, 2024, doi: 10.1109/TGCN.2024.3403102.
10. L. Fagioli, A. Mazzarini, C. Livolsi, E. Gruppioni, N. Vitiello, S. Crea and E. Trigili, "Advancements and Challenges in the Development of Robotic Lower Limb Prostheses: A Systematic Review," *IEEE Transactions on Medical Robotics and Bionics*, vol. 6, no. 4, pp. 1409–1422, 2024, doi: 10.1109/TMRB.2024.3464126.
11. T. Akyaz and D. Engın, "Machine Learning-Based Predictive Maintenance System for Artificial Yarn Machines," *IEEE Access*, vol. 12, pp. 125446–125461, 2024, doi: 10.1109/ACCESS.2024.3454548.

12. A. Houkan, A. K. Sahoo, S. P. Gochhayat, P. K. Sahoo, H. Liu, S. G. Khalid and P. Jain, "Enhancing Security in Industrial IoT Networks: Machine Learning Solutions for Feature Selection and Reduction," *IEEE Access*, vol. 12, pp. 160864–160883, 2024, doi: 10.1109/ACCESS.2024.3481459.
13. J. Huang, S. Huang, S. K. Moghaddam, Y. Lu, G. Wang, Y. Yan and X. Shi, "Deep Reinforcement Learning-Based Dynamic Reconfiguration Planning for Digital Twin-Driven Smart Manufacturing Systems with Reconfigurable Machine Tools," *IEEE Transactions on Industrial Informatics*, vol. 20, no. 11, pp. 13135–13146, 2024, doi: 10.1109/TII.2024.3431095.
14. M. Y. Arafat, S. Pan and E. Bak, "An Adaptive Reinforcement Learning-Based Mobility-Aware Routing for Heterogeneous Wireless Body Area Networks," *IEEE Sensors Journal*, vol. 24, no. 19, pp. 31201–31214, 2024, doi: 10.1109/JSEN.2024.3440412.
15. V.-V. Vo, D.-T. Le, S. M. Raza, M. Kim and H. Choo, "Active Neighbor Exploitation for Fast Data Aggregation in IoT Sensor Networks," *IEEE Internet of Things Journal*, vol. 11, no. 8, pp. 13199–13216, 2024, doi: 10.1109/JIOT.2024.3354730.

2 AI Frameworks in Mechatronics Engineering

Gunapriya Devarajan, Pushpalatha Naveenkumar, Malathy Batumalay, and Vinoth Kumar

2.1 INTRODUCTION

Mechatronics is defined as the multifaceted field of engineering that involves three major fields, namely mechanical, electrical, and computer science. Artificial intelligence (AI) plays a significant role in the transformation of traditional mechatronic systems into intelligent systems [1, 2]. AI acts as the essential tool in creating the brain of intelligent machines. The field of computer science and engineering known as "artificial intelligence" focuses on developing computational models with capabilities comparable to those of human beings. AI facilitates the ability of machines to see, understand, behave, and acquire human-like skills [3]. The major advantages of implementing AI in mechatronics are as follows:

- Human resources can focus on more sophisticated activities by automating many repetitive and monotonous robotics operations with AI, machine learning (ML), and deep learning (DL).
- These technologies can improve system accuracy and precision.
- AI-powered robots can adapt to changing environments, making them helpful in many sectors and applications.
- AI/ML algorithms enable robots to make informed decisions and take suitable actions by analyzing massive volumes of data.
- Automation of harmful tasks via AI, ML, and DL reduces workplace accidents and injuries.
- Prevents wastage of energy and minimizes the time of production by providing optimized control of electrical and mechanical components.
- Provides human–machine collaboration.
- Significantly reduces the labor and maintenance cost.

The software platform that provides the necessary tools and libraries required for AI deployment in various applications called an AI framework. Application of AI framework in mechatronics has created a significant transformation in the mechatronic field. The different types of AI frameworks available in real time are TensorFlow (Google), PyTorch (Facebook), Robot Operating System (ROS), Keras (Open Source), Scikit-learn (Open Source), Caffe (Berkeley AI Research

DOI: 10.1201/9781032615790-2

Lab), Apache MXNet (Apache Software Foundation), Microsoft Cognitive Toolkit (CNTK), OpenAI Gym (OpenAI), Fast.ai (Open Source), etc. The major advantage of AI frameworks in real-time applications is to promote automation and quality control in industrial processes, treatment planning and patient monitoring in healthcare processes, and finance and marketing in business sector.

2.2 ML ALGORITHMS IN MECHATRONICS ENGINEERING

ML is the branch of AI that utilizes statistical techniques to make the computer learn from data without explicit programming [4, 5]. The major role of ML algorithms in mechatronics engineering is to learn the pattern between input and output data and utilize the knowledge for advanced decision-making. Some of the applications of ML in the mechatronics sector are as follows:

- Development of intelligent robots that learn from the environment, adapt to changing conditions, and have the capability to take advanced decisions.
- Performance optimization of mechatronics control systems such as in controller design and in system identification.
- Development of an autonomous vehicle that has the capability to perceive the environment, plan the routes, and make abrupt decisions.

ML is basically classified into three types, namely supervised learning, unsupervised learning, and reinforcement learning.

- *Supervised Learning:* It is defined as training the algorithm using a labeled dataset. In mechatronics, supervised learning is used for various applications such as optimization of control systems, image processing, and object recognition.
- *Unsupervised Learning:* It is defined as training the algorithm using an unlabeled dataset. In mechatronics, unsupervised learning is used for various applications such as detection of anomaly, system identification, and clustering.
- *Reinforcement Learning:* It is defined as training the algorithm based on trial-and-error-based approaches. In mechatronics, it plays a major role in robot navigation, optimization of control systems, and advanced decision-making.

Various types of ML algorithms have been proposed for mechatronics applications such as linear regression, support vector machine, and Gaussian process regression.

- *Linear Regression:* It is a type of supervised ML algorithm that maps the linear relation between the dependent variable and one or more independent variables using a mathematical equation. It is used in the mechatronics sector to predict the generated motor torque, which is crucial for precise control of the motor. It is also used to optimize the manufacturing process and improve its efficiency.
- *Decision Tree:* It is also a type of supervised ML algorithm that predicts the outcomes based on the input data. It is a tree-like structure consisting of nodes

and branches. It is suitable for both regression and classification. It is used in the mechatronics sector to aid preventive maintenance by identifying the faulty machines and healthy machines. Decision trees play a crucial role in robotic applications. It is used for object recognition.

- *Support Vector Machine:* It is a powerful supervised ML algorithm used for both classification and regression. They have the capability to deal with high-dimensional and highly nonlinear data. Hence, they are preferred in various mechatronic sectors, such as image recognition in robotic application and aiding predictive maintenance.

2.3 DL IN MECHATRONICS ENGINEERING

Although ML algorithms have significant advantages, DL algorithms play a significant role in mechatronic systems due to their capability to deal with complex systems involving highly nonlinear and high-dimensional data [6]. It is defined as the multi-layered neural network that stimulates the capability of the complex decision-making power of the human brain. The DL algorithms are used in various mechatronics applications such as advanced robotic control, smart manufacturing, human–machine collaboration, sensor data processing in real time, intelligent condition monitoring, and prosthetic and assertive devices [7].

- *Advanced Robotic Control:* The robots perform the intrinsic task with high accuracy and precision with the help of a DL algorithm that analyzes a huge dataset of robotic movements and trains the robots according to it.
- *Smart Manufacturing:* Optimize the production process and assures the product quality.
- *Processing of Real-Time Data:* Process the real-time multitude sensor data and help the system in accurate decision-making. Due to this capability, they are highly preferred in autonomous vehicles and ship and robot navigation.
- *Human–Robot Collaboration:* This helps the robot to understand human gestures and respond to them.
- *Intelligent Condition Monitoring:* By analyzing sensor data, the DL algorithm identifies changes and alerts the mechatronic systems to impeding failure, thereby preventing the system breakdown. Hence, the performance of the system can also be enhanced by increasing its reliability and minimizing downtime.
- *Enhanced User Experience:* Improves the user experience by providing prosthetic limbs with personalized control systems.

2.4 REINFORCEMENT LEARNING IN MECHATRONICS ENGINEERING

ML- and DL-based AI algorithms play a crucial role in mechatronic systems only when the learning dataset is available. During the absence of a training dataset, reinforcement learning is used. It is generally defined as the subset of AI algorithms where the model is trained using the trial-and-error-based approach by interacting with the environment [8, 9]. The various applications of reinforcement learning in mechatronics engineering are the multiobjective learning capability of robots, human-robot interaction, and advanced control systems.

- *Multiobjective Learning Capability of Robots:* Robots mostly operate in the intrinsic environment with multiple objectives. Reinforcement learning helps the robots to prioritize the work and schedule them simultaneously.
- *Human–Robot Interaction:* RL algorithm promotes the easier interaction of robots with humans. By the application of RL algorithm, the robot can learn to adapt to the environmental changes and cooperate with human operation seamlessly in the shared workplaces.
- *Advanced Control System:* RL algorithms provide the solution for the unpredictable and dynamic real-world environment. It interacts with the environment and receives feedback to adapt the control system in real time and enhance the system's response to changing conditions. Due to this nature, RL-applied mechatronic systems find application in industrial automation systems and walking robots.

Although RL provides numerous advantages in mechatronic systems, they also suffer some major drawbacks, such as the requirement of clear objectives and accurate defining of reward structure. They also require extensive computer resources.

2.5 EVOLUTIONARY COMPUTING IN MECHATRONICS ENGINEERING

Along with ML, DL, and reinforcement learning, evolutionary computation is also a powerful AI technique that plays a major role in mechatronics engineering [10]. These algorithms are inspired by natural selection and mimic the process of evolution in mechatronic system design and control strategy. The flowchart of the evolutionary algorithm used to analyze the behavior of a mechatronic component is shown in Figure 2.1.

The four major steps involved in evolutionary algorithms are evaluation, selection, variation, and iteration.

- During the evaluation phase, based on a predefined fitness function that considers factors like cost, time, and efficiency, each and every solution is evaluated.
- In the selection phase, the solutions with high fitness values are selected and passed to the next generation.
- To create diversity in the population, the operations such as crossover and mutations are carried out within the selected solutions, which are known as variations.
- Until the highly improved solutions are obtained, the phase of evaluation, selection, and variation is carried out multiple times. This phase is called iteration.

The major advantage of evolutionary algorithms is that they do not require in-depth knowledge of system dynamics and provide a broad range of solutions for a

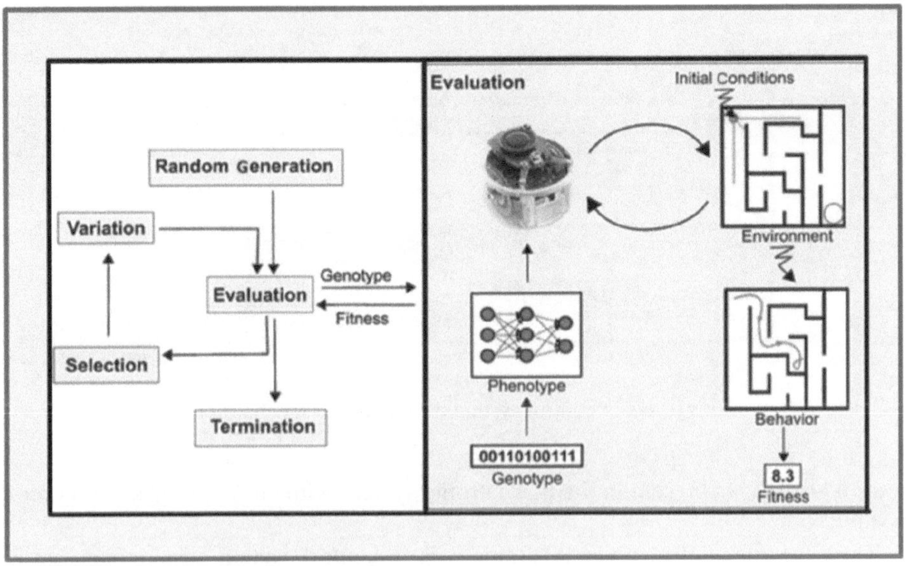

FIGURE 2.1 Application of evolutionary algorithm in mechatronic sector.

single problem. In the mechatronics field, evolutionary algorithms are applied for various applications such as tuning of control parameters, optimization of the design of various mechatronic components, and motion planning.

- Evolutionary algorithms have the capability to identify optimal tuning parameters of the mechatronic system, which enhances the system performance, such as accuracy, reliability, adaptability, and stability, by analyzing thc various control parameter combinations.
- The optimal paths of the robots can be selected by the evolutionary algorithm with minimum energy usage by simulating various movement trajectories of the robot and estimating their performance accuracy.
- By continuous iteration of various geometric configurations and properties of materials, the suitable design of mechatronic components such as gears and linkages that meet all the performance requirements, such as size, weight, and cost, is selected by the evolutionary algorithms.

The main challenge with evolutionary algorithms is that identifying an effective fitness function can be difficult and computationally expensive.

2.6 AI FRAMEWORKS FOR MECHATRONICS CONTROL SYSTEMS

In the mechatronics sectors, AI plays a crucial role in designing the brain of intelligent machines [11, 4]. AI offers huge resources, especially for the design of control systems. The various levels of controls offered by AI techniques for a mechatronic robot are shown in Figure 2.2. These techniques help the engineers to focus on unique

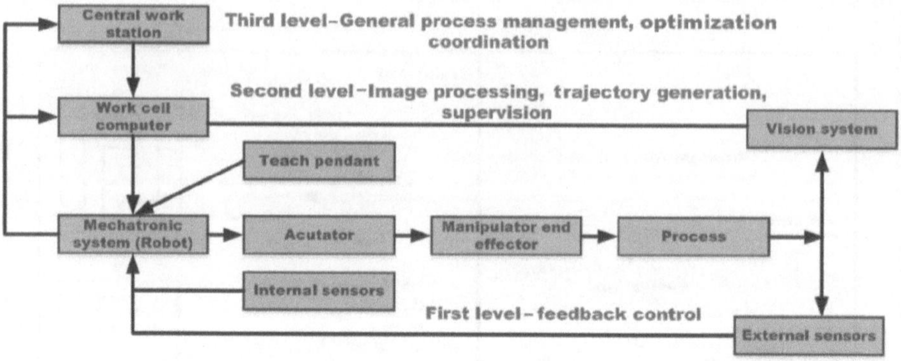

FIGURE 2.2 Various levels of control system developed for mechatronic system.

control challenges present in the mechatronic system with an AI-based smart control system.

The advantages of using an AI framework in control system development are as follows:

- Most of the AI frameworks include various libraries and functions that are used for common control tasks such as feedback control, system identification, and signal processing, which promote the rapid development of prototypes. This reduces the model development time and promotes quicker time-to-market of the mechatronic systems.
- AI framework supports different AI algorithms like ML, DL, and reinforcement learning, which help the mechatronic engineers to select the appropriate algorithm for specific applications. For example, for model predictive control, ML is utilized to analyze the future system behavior, and reinforcement learning is used for control system design.
- AI framework also provides tools and libraries to interface hardware components such as sensors, actuators, and microcontrollers. This promotes real-time decision-making and system control.

The various types of AI frameworks that are used for mechatronics control systems are as follows:

- *TensorFlow:* It is an open-source framework introduced by Google that has high flexibility and scalability. It provides numerous tools and libraries for the design of control systems using different AI algorithms.
- *PyTorch:* It is a research friendly framework that promotes rapid prototyping. It also promotes different AI techniques such as ML, DL, and RL for the design of control systems.
- *MATLAB:* It is a commercial platform that provides tools for data analysis, control systems design, and engineering computations. It offers specialized toolboxes—such as the Control System Toolbox, Reinforcement Learning

Toolbox, and Deep Learning Toolbox—that enable the integration of AI techniques into control system modeling, simulation, and optimization.

2.7 AI FRAMEWORKS FOR MECHATRONICS SENSING AND PERCEPTION

Intelligent machines not only involve the design of control systems but also have the capability to perceive and understand the environment [12]. The application of various AI algorithms to mechatronic system sensing and perception is shown in Figure 2.3.

AI frameworks have the tools to transform the sensor raw data into mechatronic system with real-world interaction capability. The roles of AI framework in mechatronics sensing and perception are as follows:

- AI framework provides libraries for advanced image processing tasks like recognition, image detection, and computer vision. These capabilities make them useful for applications for robot navigation and industrial quality control.
- AI framework employs DL models for various tasks, including data processing for autonomous vehicles and environmental sensing for robotics.
- AI systems enable sensor fusion by integrating and jointly analyzing multimodal sensor data, thereby enhancing environmental perception.
- Promotes advanced decision-making by allowing real-time perception.

The various AI frameworks that used mechatronics sensing and perception are as follows:

- *OpenCV (Open Source Computer Vision Library):* It offers a vast library of functions for image processing, image detection, and real-time video analysis.
- *TensorFlow and PyTorch:* This utilizes DL models for complex perception tasks.
- *ROS:* ROS combines with AI frameworks like open CV and builds robust perception systems for robots.

Using these frameworks, intelligent mechatronic systems can be created that interacts with real world more effectively.

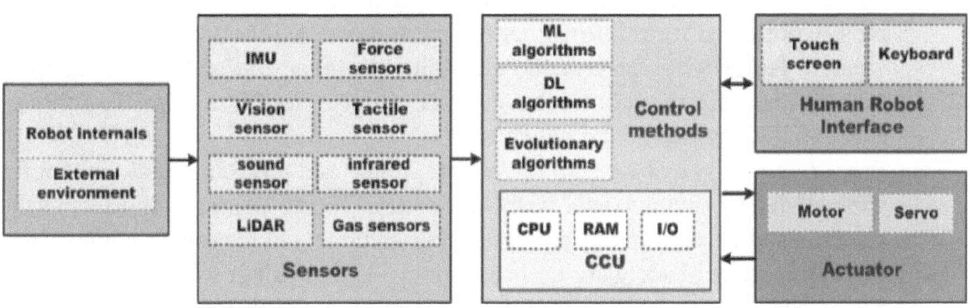

FIGURE 2.3 AI application in mechatronic sensing and perception.

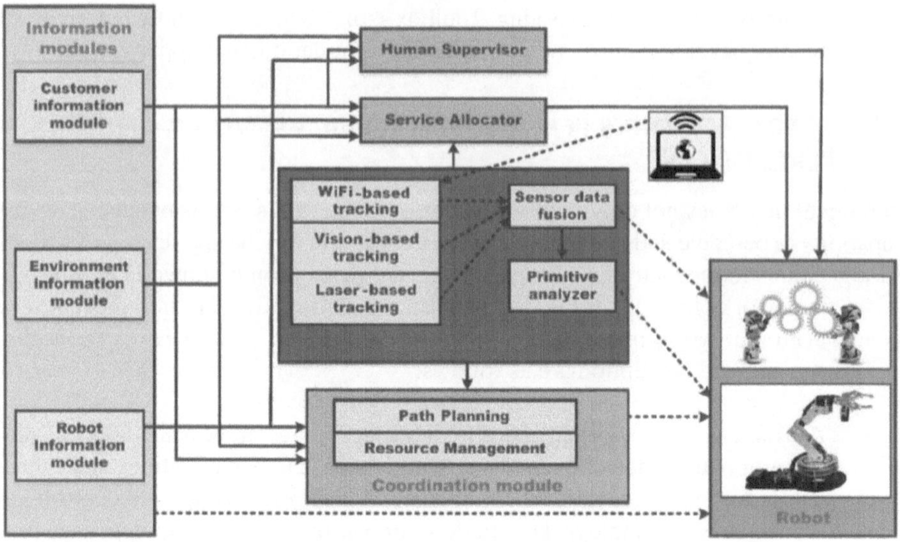

FIGURE 2.4 AI application in mechatronic robot application.

2.8 AI FRAMEWORKS FOR MECHATRONICS ROBOTICS

AI plays a significant role in the development of smart robots that have the capability to continuously learn and update, interact with humans, and solve complex tasks [13]. The various applications of AI in robots are shown in Figure 2.4.

To provide communication between the physical robot and AI model, the AI framework should verify the compatibility between the model and robot components such as sensors and controllers. The various AI frameworks introduced for robotic application are classified into three types, namely DL, ML, robotic framework, and high-level framework. The various AI frameworks used for robotic application are as follows:

1. *TensorFlow (Google):* The image recognition capability in TensorFlow enables the robot to interact with the environment based on identified objects. Similarly, it is used to identify the component failure, thereby ensuring smooth operation and reduced downtime.
2. *PyTorch (Facebook):* It is known for its dynamic computational graph. It is mostly preferred for reinforcement learning-based AI models. The AI models created using PyTorch have the capability to sense the real-time data from sensors and vary the robot behavior accordingly. Hence, it helps the robot to adapt to the environmental changes.
3. *Scikit-Learn (Open Source):* It provides a simple and predictive analysis tool for various ML algorithms to analyze the sensor data. Thereby it promotes detecting the anomalies in the data and identifies the malfunctioning of the robot's component.
4. *ROS:* It is a set of software tools and libraries that bridges the AI framework and hardware component. It promotes communication between the various robotic components and manages the sensor data.

5. *Keras (Open Source):* High-level interface that promotes fast development of DL models. The quickly developed model can be used for various robotic tasks such as controlling and data analysis.

2.9 AI FRAMEWORKS FOR MECHATRONICS PREDICTIVE MAINTENANCE

Preventive maintenance plays a crucial role in the smooth maintenance of the mechatronic system [14, 15]. The traditional methods require scheduled inspection, increased downtime, and rudimentary fault detection. An AI-incorporated mechatronic system creates a revolution in predictive maintenance by proactively identifying faults and reducing downtime. The application of AI in predictive maintenance is shown in Figure 2.5.

The functions of various AI frameworks used for predictive maintenance are as follows:

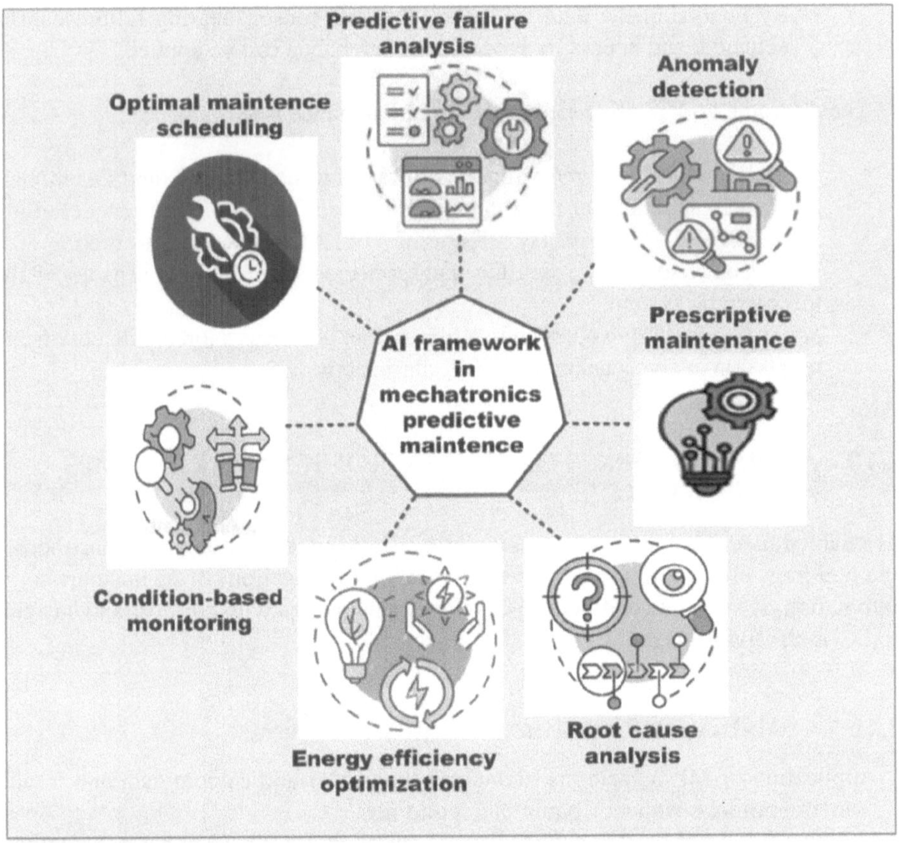

FIGURE 2.5 AI-based predictive maintenance in mechatronic systems.

Numerous raw data are generated from mechatronic system sensors. Using the AI tool present in AI infrastructure, anomalies in the sensor data are analyzed and impending faults can be indicated. This proactive method minimizes the downtime and helps in applying rectification measures before the occurrence of downtime.

A ML-based predictive model identifies the faults and enables the system to incorporate predictive measures by analyzing the history of sensor data. The processes involved are monitoring the components near to failure, implementing an optimized maintenance schedule, replacing of failure components before catastrophic breakdown, and promoting smooth operation of the system. This measure helps to increase the remaining useful life (RUL) of the mechatronic component.

DL algorithms have the capability to handle complex, nonlinear, high-dimensional data. The AI framework incorporates DL algorithms for the application, such as image-based fault detection and early detection of bearing faults using vibration analysis.

- In image-based fault detection, images are analyzed to identify the fault, such as cracks and failure of components, enabling the implementation of proactive maintenance.
- By analyzing the data from the vibration sensor, bearing failure can be detected, and necessary proactive maintenance can be applied.

The AI framework applied for predictive maintenance is as follows:

- *TensorFlow and PyTorch:* Contains libraries to develop ML and DL models that can be applied to analyze the sensor data and detect the equipment failure.
- *Industrial IoT Platform:* By integrating with AI framework, promoting analyses, and visualizing real-time data collected from various sensors of the mechatronic system.
- *Scikit-Learn:* It is an open-source library that performs anomaly detection and regression analysis using many ML algorithms.

2.10 AI FRAMEWORKS FOR MECHATRONICS SMART HOMES AND BUILDINGS

AI-enabled mechatronic systems play a crucial role in revolutionizing smart homes and buildings [15, 8]. Figure 2.6 shows the various applications of AI in smart home automation. It helps the engineer to develop smart homes with comfort and sustainability in the living space.

2.10.1 ML-Enabled Personalized Comfort Control

Application of ML algorithms to analyze the sensor and camera data and implement optimal occupancy patterns in buildings.

Utilizing AI algorithms to analyze the weather condition and home or building's heating and cooling requirement. This helps in optimal use of home appliances with minimized energy consumption.

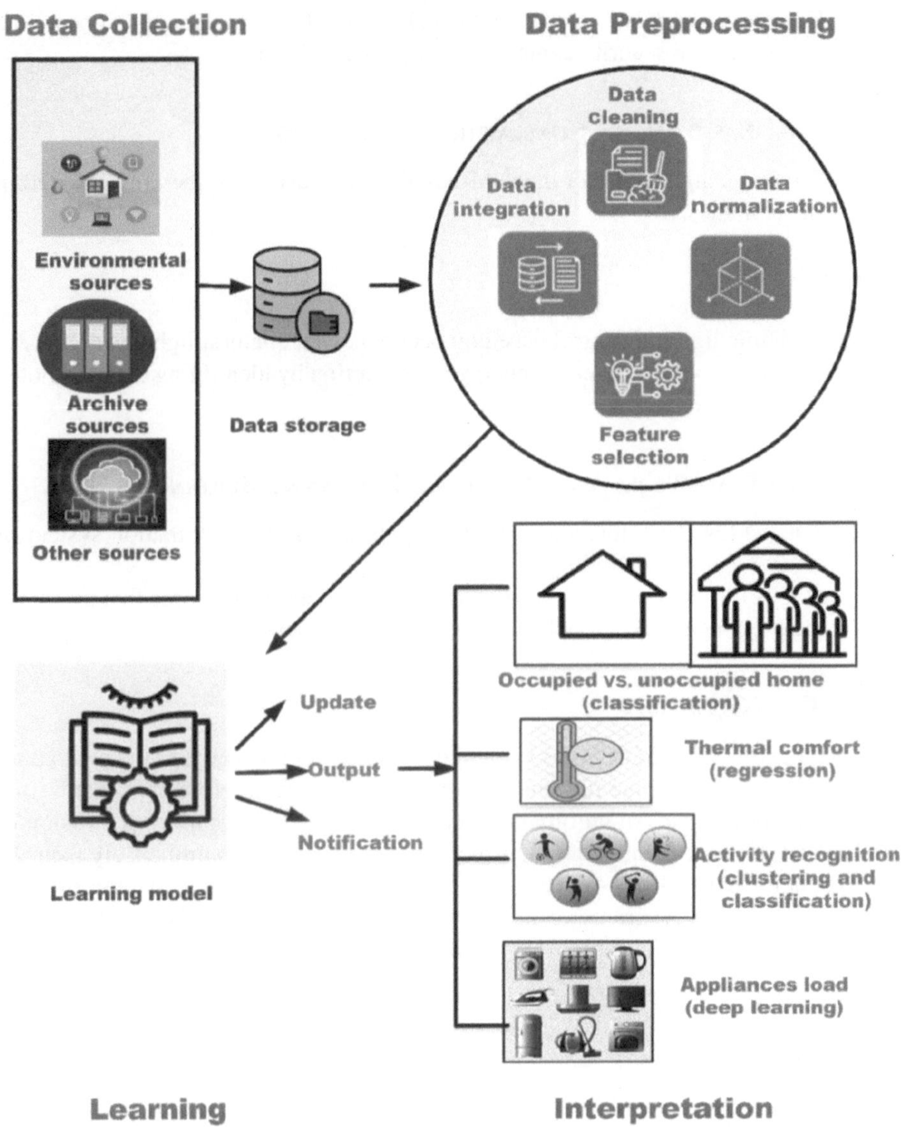

FIGURE 2.6 Application of AI-based mechatronic system in smart home and buildings.

2.10.2 DL Algorithm-Based Home Security Enhancement

- Utilizing pattern-identifying capability, DL algorithms can be used to enhance the security and automation in smart homes and buildings.

Example:

- Improving building security by implementing DL algorithm-based face recognition system.

- Detection of potential intrusion by a DL-trained system using the sensor data obtained from security cameras and triggering alarms.

2.10.3 AI-Based Energy Consumption Optimization

- By analyzing the sensor data, AI-based energy usage can be implemented in smart homes.

Example:

- Dynamic lighting control based on occupancy and natural light availability
- Promotes the adjustment of energy consumption by identifying the peak hours in smart grids.

2.10.4 AI Frameworks Applied in Smart Homes and Buildings

- *TensorFlow Lite:* Integrates the AI capabilities in the automation system of smart homes and buildings.
- *ONNX (Open Neural Network Exchange):* It is an open source that promotes the exchange of AI models between various frameworks.

2.11 CONCLUSION

The integration of AI frameworks into mechatronics engineering has brought about transformative advancements, enabling the creation of intelligent systems that redefine the boundaries of automation and efficiency. Mechatronics, as a fusion of mechanical, electrical, and computational disciplines, benefits immensely from AI by introducing capabilities such as learning, decision-making, and adaptability into traditional systems. AI frameworks serve as the backbone for this transformation by providing the tools and libraries needed to implement complex algorithms and models.

REFERENCES

1. Kuru, K., and Yetgin, H. (2019). Transformation to advanced mechatronics systems within new industrial revolution: A novel framework in automation of everything (AoE). *IEEE Access*, *7*, 41395–41415.
2. Hashemi, A., and Dowlatshahi, M. B. (2023). A review on the feasibility of artificial intelligence in mechatronics. In: *Artificial Intelligence in Mechatronics and Civil Engineering: Bridging the Gap* (pp. 79–92). Springer.
3. Bohušík, M., Stenchlák, V., Císar, M., Bulej, V., Kuric, I., Dodok, T., and Bencel, A. (2023). Mechatronic device control by artificial intelligence. *Sensors*, 23(13), 5872.
4. Zaitceva, I., and Andrievsky, B. (2022). Methods of intelligent control in mechatronics and robotic engineering: A survey. *Electronics*, *11*(15), 2443.
5. Pandey, R. (2021). Integration of machine learning algorithms in mechatronic systems for predictive maintenance. *Mathematical Statistician and Engineering Applications*, *70*(2), 1822–1829.

6. Salahzadeh, Z., Rezaei-Hachesu, P., Gheibi, Y., Aghamali, A., Pakzad, H., Foladlou, S., and Samad-Soltani, T. (2021). A mechatronics data collection, image processing, and deep learning platform for clinical posture analysis: A technical note. *Physical and Engineering Sciences in Medicine*, *44*(3), 901–910.
7. Ayankoso, S., and Olejnik, P. (2023). Time-series machine learning techniques for modeling and identification of mechatronic systems with friction: A review and real application. *Electronics*, *12*(17), 3669.
8. Staessens, T., Lefebvre, T., and Crevecoeur, G. (2022). Adaptive control of a mechatronic system using constrained residual reinforcement learning. *IEEE Transactions on Industrial Electronics*, *69*(10), 10447–10456.
9. Chen, W., Hu, J., Xu, C., Zhou, H., Yao, J., and Nie, W. (2023). Optimal tracking control of mechatronic servo system using integral reinforcement learning. *International Journal of Control*, *96*(12), 3072–3082.
10. Bech, M. M., Noergaard, C., Roemer, D. B., and Kukkonen, S. (2016). A global multi-objective optimization tool for design of mechatronic components using generalized differential evolution. In: *IECON 2016—42nd Annual Conference of the IEEE Industrial Electronics Society* (pp. 475–481). IEEE.
11. Lila, M. K. (2021). Integration of artificial intelligence techniques in mechatronic systems for smart manufacturing. *Mathematical Statistician and Engineering Applications*, *70*(1), 432–439.
12. Shi, Q., Sun, Z., Le, X., Xie, J., and Lee, C. (2023). Soft robotic perception system with ultrasonic auto-positioning and multimodal sensory intelligence. *ACS Nano*, *17*(5), 4985–4998.
13. Borkar, K. K., Aljrees, T., Pandey, S. K., Kumar, A., Singh, M. K., Sinha, A., Singh, K. U., and Sharma, V. (2023). Stability analysis and navigational techniques of wheeled mobile robot: A review. *Processes*, *11*(12), 3302. https://doi.org/10.3390/pr11123302
14. Atashzar, S. F., Carriere, J., and Tavakoli, M. (2021). How can intelligent robots and smart mechatronic modules facilitate remote assessment, assistance, and rehabilitation for isolated adults with neuro-musculoskeletal conditions? *Frontiers in Robotics and AI*, *8*, 610529.
15. Ramoly, N., Bouzeghoub, A., and Finance, B. (2018). A framework for service robots in smart home: An efficient solution for domestic healthcare. *IRBM*, *39*(6), 413–420.

3 Extendable AI in Mechatronics Engineering

Pushpalatha Naveenkumar, Surendar Rangaraju, Kokilavani T., Kannan K., and Praveen Kumar M

3.1 INTRODUCTION

Artificial intelligence (AI) is significantly improving mechatronics, which combines mechanical engineering, electronics, computer science, and control systems. This allows machines to learn, adapt, and function at their best in real time. Figure 3.1 shows the integration of AI in mechatronics. This powerful combination is revolutionizing traditional manufacturing techniques and driving progress in robots, autonomous vehicles, healthcare, and advanced infrastructure. AI-driven mechatronic systems are facilitating industries in achieving exceptional efficiency, precision, and creativity, fundamentally transforming human involvement with technology and shaping the future of engineering, enhancing it toward a more intelligent and automated era.

3.2 TRANSFORMATIVE POWER OF INDUSTRY

Mechatronic engineering, in combination with AI, is a revolutionary concept in the industry, which opens up a new epoch of fabricating production systems that are flexible, robust, and innovative. This confluence has provided opportunities for the prospecting of diverse industries toward realizing hitherto unimagined quotas of profitability optimization, efficiency, and environmental friendliness.

Within manufacturing industry, AI integrated mechatronic systems are the modern concept of smart factories. These intelligent systems are capable of controlling specific manufacturing processes and self-optimizing and this considerably reduces waste and time. The enhancement of the use of AI in the predictive maintenance field also strengthens the effectiveness of the business because it is designed to help the organization predict equipment failures and reduce their frequency while also increasing the longevity of equipment. The increasing production rate that constitutes this decisive strategy is also the rational use of resources to protect the environment.

In addition, the probability of realizing high benefits from the symbiosis of AI and mechatronics is also high for healthcare businesses. Intelligent diagnostic equipment broadens the possibilities of how doctors can diagnose diseases and treat their patients with significantly higher effectiveness. In this case, surgical robots are more accurate and a lot less risky than their partners. This revolutionary improvement has prospects in giving a large populace equality to quality healthcare services.

 DOI: 10.1201/9781032615790-3

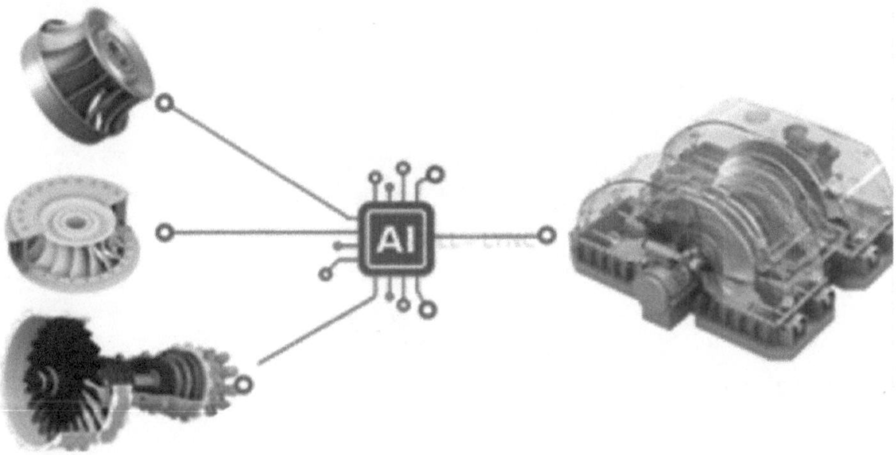

FIGURE 3.1 Integration of AI in mechatronics.

As more businesses adopt this revolutionary asset, they are not only improving businesses' productivity but also creating a brand-new future for work. This kind of transition calls for a shift and transformation, which means the workforce needs to be trained in the new skills as they relate to the new industrial landscape. AI integration in the workforce is enhancing even learning and flexibility, through the formation of climate of innovations and change.

The integration of mechatronic engineering and AI marks the dawn of a new era in the industrial revolution. This bold collaboration paves the way for businesses to achieve unprecedented levels of productivity and efficiency, bringing to life a future where intelligent, adaptive practices fulfill the evolving sustainability needs of organizations. As various operational sectors continue to embrace this powerful synergy, it is increasingly clear that what was once seen as a paradigm shift is now actively transforming—and has already reshaped—the industrial landscape. Ultimately, the key to unlocking these benefits lies in developing human capital equipped to navigate this emerging domain shaped by the fusion of mechatronics and artificial intelligence.

3.3 AI IN MECHATRONICS

AI in mechatronics has created highly advanced and endurable holo-architecting robotic systems that are able to closely imitate human and function autonomously. Such features are complemented by machine learning which adds the capability of the machines processing large amounts of data, identifying patterns, and behaving in constantly changing environments. Self-optimization of extensible AI widens not just the application of minimizing down time but also enhances the overall performance of mechatronic systems. Figure 3.2 shows the integration of AI in mechatronics domain.

AI innovation has impacted the industrial and mechanical segments considerably; newness has been enacted in robots, automation, structural changes in mechanical maintenance, quality control, and process enhancement strategies. In the first

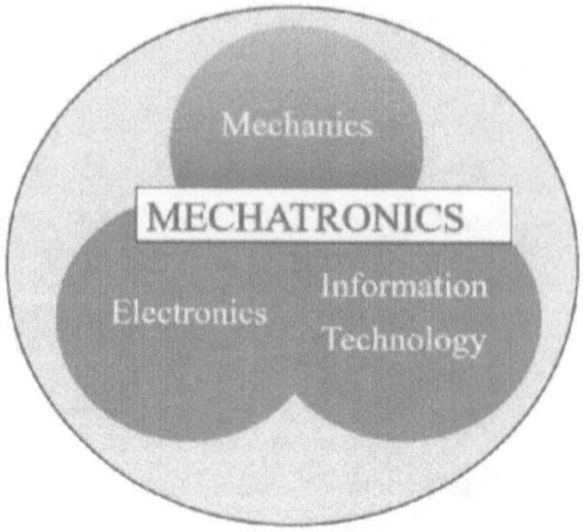

FIGURE 3.2 Mechatronics types.

category of AI systems, the application was mechanical and navy, able to function only as intelligent and rule-based systems and expert systems. Nevertheless, machine learning, or more specifically supervised and unsupervised learning methods, has become an essential step in the domain, as they enable a system to make the decision based on the data received in the past.

Integration of AI and robotics has advanced the formation of intelligent robots that are advanced to solve tasks independently with efficiency, flexibility, and accuracy. Such breakthroughs have been most helpful in some fields such as automotive, electronics, and logistics as safety and productivity have recorded an upward trend. In addition, collaborative robots have been evolving for the purpose of working with humans to enhance worker performance and condition.

Looking into the future, AI systems shall advance their capability to analyze and decide, interpenetrate human social circle, and be more communicative in ways they convey information. With respect to the developments of AI over the years in the field of mechatronics, there is a theme of great potential in later advancements of future AI in the mechatronic system in terms of the scope of improving communication and teamwork between human and robotics, problem-solving strategies, and the capacity to exist and functioning in diverse and dynamic environments. Therefore, the incorporation of AI in mechatronics has opened new frontiers for intelligent and advanced robotic systems with capacity for self-operations that are incredibly efficient, and efficient in enhancing continuous improvement in many fields. These applications will cover healthcare, logistics, agriculture, and further sectors. To facilitate widespread adoption, it is necessary to tackle issues such data protection, ethical constraints, and the necessity for specialist labor, notwithstanding the considerable promise. The amalgamation of AI with mechatronics is poised to catalyze substantial changes across diverse sectors, yielding intelligent systems competent of adaptation, learning, and efficient collaboration.

3.4 AI-ENHANCED HUMAN–MACHINE COLLABORATION

Figure 3.3 shows the possible ways of human–machine interaction. In mechatronics in the future, it will be simple and standard for people and robots to work together.

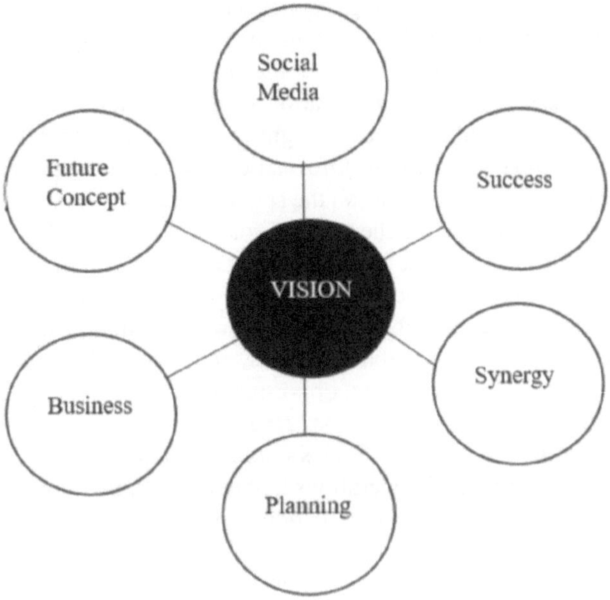

FIGURE 3.3 Human–machine collaboration.

This is because AI makes it feasible. AI-powered machines will not get rid of jobs, but they will help people do their jobs better. These robots, which are also known as "cobots," will work with people. They will have progressed AI that can learn how people act. Workers in a wide range of business fields will be more productive, safe, and healthy because of this relationship.

Adding AI to mechatronics will accelerate the process of making systems that can work on their own. Unmanned aerial vehicles (UAVs) and self-driving cars will get smarter and be able to see and react to their surroundings in real time. This will make shipping, transportation, workplace automation, and production safer and more consistent.

Putting mechatronics and AI together will help make robots and technologies that are more equipped to address people's needs. Robots will help with daily chores, track patient health, and provide company in nursing homes and for the elderly. This will enhance people's quality of life and help more inclusive communities for all to grow.

AI in mechatronics depends on fusion of complex systems data from several devices and using AI models for interpretation. These devices can simultaneously feel, hear, see, and look at data about their surrounds. This helps people make smart choices and take real steps based on tools. In many areas, being creative is helped by combining awareness, decision, and action. AI could be used in various methods in mechatronics. Machines can move around, pick up things, and interact with other machines with the help of AI. The complexity embedding of mechatronics systems, AI, and other sophisticated technologies in different industries is changing the course of healthcare and business. Integrating mechatronics involving the use of safe sensors such as lasers and cameras, and AI is making the realization of more self-driving cars and enhanced industrial robotization possible. In the automotive industry, mechatronics systems are finding great application in the production of self-driving cars through safe operations and navigation. Light Detection and Ranging (LiDAR) and camera act as detectors that capture information about the environment, AI which then uses it to make correct decisions on the spot and manage the vehicle. This has the ability of considerably enhancing the faces of road safety, traffic flow, and minimization of carbon footprint. Likewise, the implementation of AI in industrial automation is changing the manufacturing industry. AI-based systems can definitely improve the process flow, increase operational speed, and improve the quality of production. This can mean increased productivity, reduced costs, and product homogenization to suit markets for companies. In this context, opportunities for solving healthcare problems by improving safety and accuracy in operating on a patient's body with the help of mechatronics systems are revealed. For example, in robotic surgery where the AI is integrated, AI can ensure that, in operations that demand high accuracy, human errors would not occur. They also can result in better patients' lives and an overall lessening of expenses in the healthcare system. Applying AI to mechatronics in this separable way can dramatically change industries or our surroundings for good. Their integration can also come with added benefits in the areas of advancements in self-sufficient technologies, integrated facilities, and tailored medical prescriptions. With reference to the topics mentioned above, we are to come to the conclusion that AI, mechatronics products, and technologies will become increasingly entwined with the lives of people while provoking sustainable development.

3.5 MECHATRONICS IN AI AND ITS IMPACT ON HUMAN LIFE

When applying AI in mechatronics, such as a digital embroidery technique, one is guaranteed of reliable and safe mechatronic systems, provided some considerations are made. Some potential solutions to these issues include the following:

1. *Comprehensive Testing and Validation:* Before designing and developing AI-driven mechatronics systems for commercial use and free market operation, it is recommended that they should undergo certain rigorous tests and validation to provide greater safety and reliability. This should include both software and hardware testing as well as simulation of the equipment for the likely failure modes.
2. *Collaboration between Stakeholders:* Consult an engineer, ethicist, lawyer, scientist, and other like-minded professionals so that a list of risks could be established and, more importantly, necessary measures could be developed to prevent harm and guarantee the best outcomes toward the enhancement of mechatronics through AI.
3. *Transparency and Explainability:* Regulatory compliance and auditable decision-making processes should form part of AI systems, which are easy to interpret and explain by the system users and other stakeholders. This helps build trust, thus increasing the path of AI technologies.

4. *Continuous Monitoring and Updating:* Like any other system, AI systems should be checked frequently to guarantee their safety, security, and efficiency. There may be a need to develop a feedback mechanism by which it is possible to obtain data that will help to determine any risks that are likely to occur and how they can be handled.
5. *Educating the Public:* The use of AI-driven mechatronics in the treatment of diseases should be accompanied by an increase in public consciousness concerning the possibility and impossibility of utilizing intelligent technologies. This can be effected through opinions from leaders and policy makers through awareness creation, education, and effective information dissemination on maps.
6. *Developing Industry Standards:* Establishing guidelines on the practical application of intelligent mechatronics systems that are acceptable across the industry and integrating ways of checking compliance to these guidelines together can go a long way toward the promotion of the right use of intelligent mechatronics systems in diverse applications. It should be seen, however, that many challenges may have to be faced in integrating AI into mechatronics, especially in the integration process, but the advantages of the integration process are numerous. Protection of the future mechatronics systems based on AI is possible through testing, integration, reporting, and mostly through constant assessments and development of appropriate guidelines and standards that will ensure the mechatronics system based on AI cannot become insecure, unpredictable, and antisocial. Utilizing robots, however, can see better work output, protection, and healthy individuals everywhere in the different sectors and usages of robotics technology in return.

3.6 TOPOLOGICAL INSULATORS IN NANOPHOTONICS

Topological insulators (TIs) are a novel category of electronic materials distinguished by a bulk band gap and robust conducting states on their edges or surfaces that stay intact despite disorder. These materials have significant possibilities for quantum computing and other sophisticated applications. Surface plasmon polaritons (SPs), photonic systems, light confinement, and other fields have all made use of the principles of topological physics in recent times. SPs, arising at metallic–dielectric interfaces, have attracted interest in nano-optics due to their capability to confine electromagnetic radiation in sub-wavelength regions. Figure 3.4 shows the architecture of neural networks.

Meta-surfaces including diverse topologies represent a possible method for facilitating plasmonic edge states, applicable in localized light manipulation. Recent tests and simulations demonstrate that line defects and point singularities inside these structures may sustain both dark and brilliant modes, with point singularities demonstrating significant light localization. This platform of topologically protected plasmonic states provides a straightforward and efficient method for improving waveguiding and light localization, which is advantageous for the advancement of future nanophotonic devices.

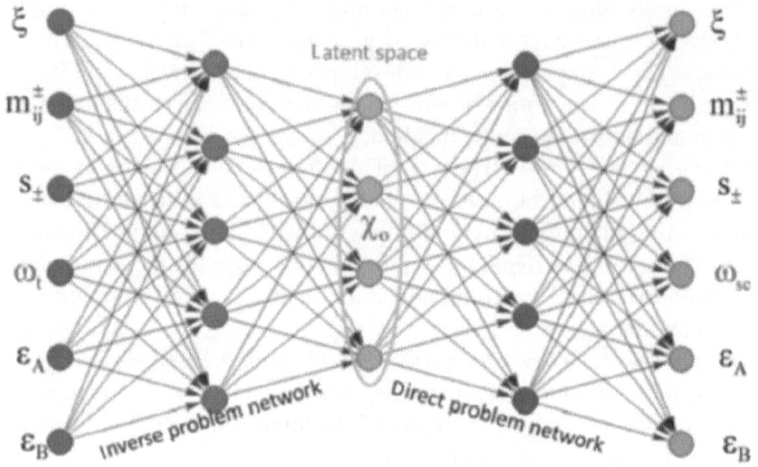

FIGURE 3.4 Neural network architecture.

The Internet of Things (IoT) is evolving the communication and functionality of objects within a network. The proliferation of IoT introduces heightened security risks, as devices are often found in hostile situations and are susceptible to assaults. Intrusion detection systems (IDS) are essential for safeguarding IoT networks through the surveillance of activities and the identification of intrusions. There are two major categories of IDS: signature-based, which contrasts fresh data with established attack signatures, and anomaly-based, which detects abnormalities from typical behavior in intelligent settings.

3.7 APPLICATIONS OF AI IN MECHATRONICS

The integration of AI into mechatronics engineering, as discussed at the 2021 International Conference on Multi-modal Information Analytics (MMIA, 2021), is shown in Figure 3.5. It highlights a transformative shift in building intelligent systems. Key areas include the following.

3.7.1 Machine Vision and Natural Language Processing

The engineering discipline involves robotics and automation and signal processing discipline. Mechatronics engineering has been advancing to work with robots and automation to design and manufacture complex and efficient systems. When it comes to robotic and automation technology, the industries can enhance production and overall efficiency; they can also cut costs and enhance safety. Signal processing is the essential process for analyzing the data from the sensors and other control or monitoring devices for improving the decision-making and control processes in the automated systems. This may include audio–video data signal processing, which are important in enhancing interaction between machines, humans, and the environment. Furthermore,

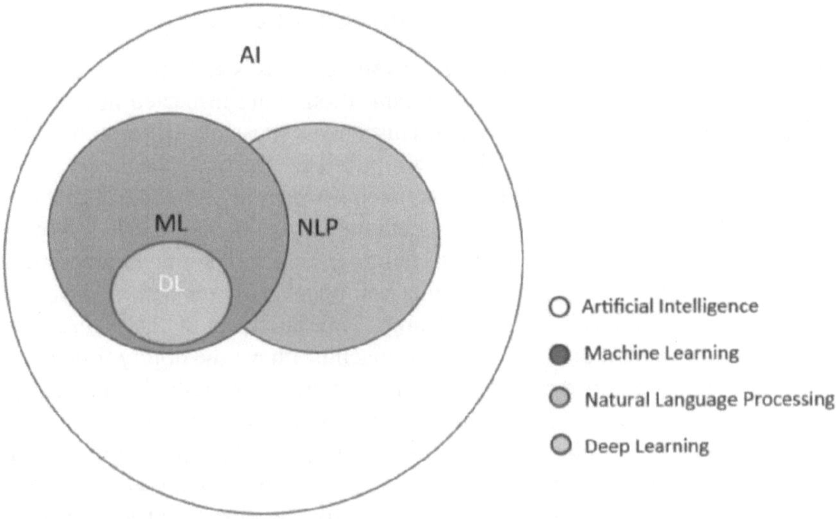

FIGURE 3.5 Applications of AI.

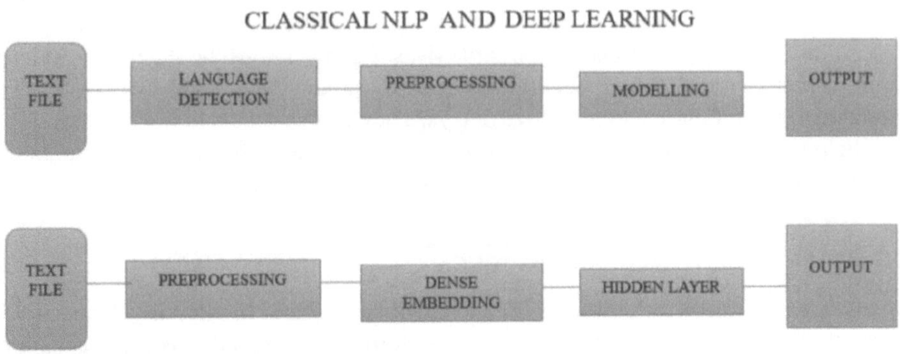

FIGURE 3.6 Classical NLP and deep learning in text processing.

the developments in signal processing technology have shown promise in enhancing systems which are applied to extract and analyze information from multiple sources, including IoT gadgets, wearable devices, smart homes gadgets, among others.

Figure 3.6 shows the classical natural language processing (NLP) and deep learning in text processing. By blending AI, robotics, automation, and signal processing mechatronics engineering can and will create new frontiers into fields such as healthcare, transportation, and energy. Therefore, MMIA 2021 [1] focused on showcasing the employment of applications such as AI, Machine Vision (MV), NLP, robotics, automation, and SP to fashion the future of mechatronics engineering. However, when these technologies are blended and integrated into a new system, novel intelligent systems are created to improve quality assurance, robotic mobility, human interfaces, and productivity in various fields and aspects of work [2].

3.7.2 Predictive Maintenance and System Optimization

Predictive maintenance as a reputable business strategy has been implemented in many industries, with mechanical assets of an organization being inspected in real time to identify possible failures before they occur. This makes it possible for corrective actions to be initiated before a breakdown, thereby minimizing the time machinery is out of use and costs of operation, as well as lengthening the productive life of machinery. Predictive maintenance is another facet of data mining being facilitated by AI to analyze monitoring data for future equipment failure. Machine learning engines are able to perform high sophistication analyses that can uncover trends and inconsistencies which may point to a failure. Machine learning is one such section that falls under the category of reinforcement learning in which machines have the ability to learn from reinforcement learning and rectify errors. In the context of predictive maintenance, robots with reinforcement learning abilities can decide on their further actions in real time. This enables the machines to increase accuracy in their predictions resulting in in enhanced maintenance practices because of new information acquired in new conditions. For instance, in the transportation sector, reinforcement learning can be used to further refine real-time choices of automated automobiles. AI systems help in understanding the environment, adapting to it, and ensuring hassle-free operations on any road. From the case studies, it is observed that these AI systems can also learn in a real-time environment by reinforcement learning which helps them in decision-making with efficient avoidance of possible risks. Finally, it must be stated that AI and reinforcement learning are especially important for predictive maintenance to identify future equipment failure and minimize delays. They enable robots to learn and evolve, thus helping in closer real-time decision-making, for instance, self-driven automobiles like driverless cars. Figure 3.7 shows the AI-driven error analysis method.

3.7.3 Decision-Making and Autonomy

Involvement of these technologies. Sensor fusion, which is the key in advanced AI mechatronics, is the use of two or more sensors to collect data that is more accurate

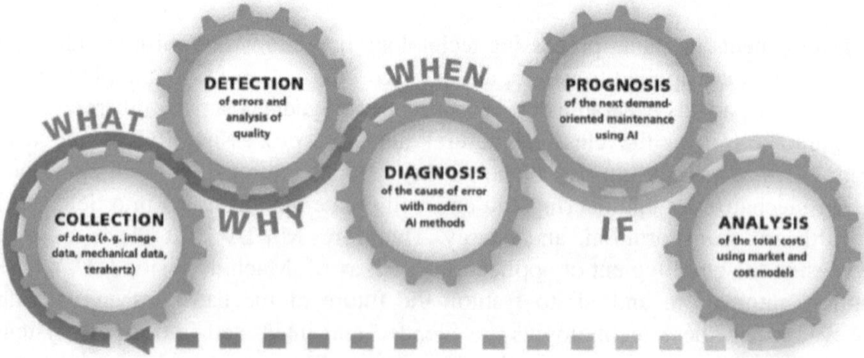

FIGURE 3.7 AI-driven error analysis.

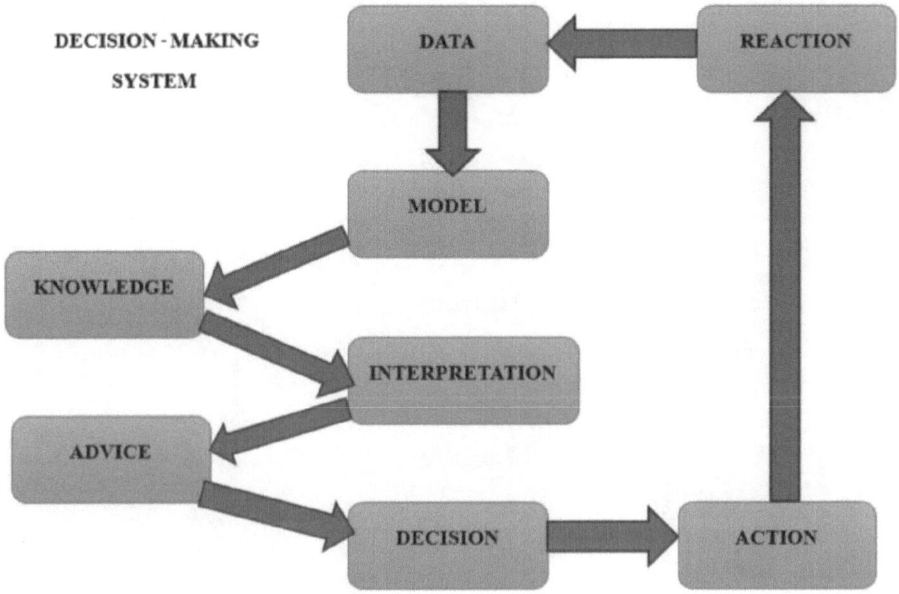

FIGURE 3.8 Decision-making workflow.

and reliable than the individual data collected by each sensor. In this way, because of the information collected from various sources like cameras, accelerometers, gyroscopes, and ultrasonic sensors, the system can recognize the milieu, make a conclusion, thereafter decide and perform the appropriate action [3]. Figure 3.8 shows the decision-making system and its workflow.

Computer vision, the other important technology, helps mechatronics systems to make sense of visual data from cameras and other visioning equipment. This enables the system to see, improve pattern recognition, and make decisions from information supplied by vision. The processes of designing and implementing the mechatronics systems, which combine mechanical, electrical, control, and software aspects, has been accelerated in the breath with the help of the developments in the AI. The involvement of AI technologies like machine learning, neural networks, and deep learning algorithms enhances the ability to make intelligent decisions and enables flexible operation. These systems benefit from the application of AI-driven computer vision, which helps mechatronic systems interpret and respond more effectively in complex environments. The flexibility incorporated by mechatronics system through AI is valuable in industry automation and auto-navigation. In industrial automation, AI offers flexibility and diversification of such systems, to perform numerous functions. Such robots can acquire knowledge from working, can modify according to their environment, and do a better job for better output.

Another area where AI-driven mechatronics systems can provide is autonomous navigation. For example, self-driving cars rely on sensor fusion and computer vision for control and decision-making in traffic conditions based on real- time data. This technology can be applied to different types of vehicles including aerial, ground, and underwater vehicles, increasing the spectrum of application of autonomous

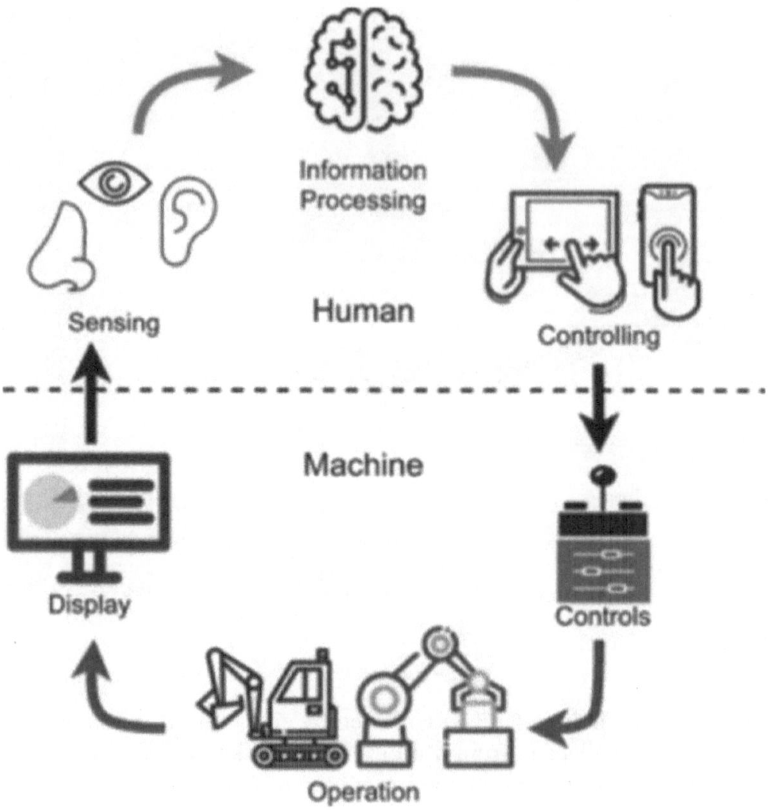

FIGURE 3.9 Human–machine interaction cycle.

navigation. Finally, it is concluded that mechatronics combined with the implementation of AI technologies, such as sensor fusion and computer vision, improves mechatronics operational flexibility as well as decision-making. This has significant impacts on industrial automation and autonomous navigation, resulting in smarter systems with better performance, improved reliability and capability to learn and improve upon their operations as they encounter challenging scenarios.

3.7.4 Human–Machine Interaction and Sustainability

AI enhances human–machine cooperation and promotes sustainability by means of energy economy. Data-driven insights help mechatronic systems to be optimally aligned, thereby improving general performance and reducing environmental impact. Figure 3.9 shows the human–machine interaction model for sustainability.

3.7.5 Intelligent Control Techniques

Figure 3.10 shows various ways through which intelligent control techniques can be accomplished. Using advanced control strategies such as intelligent

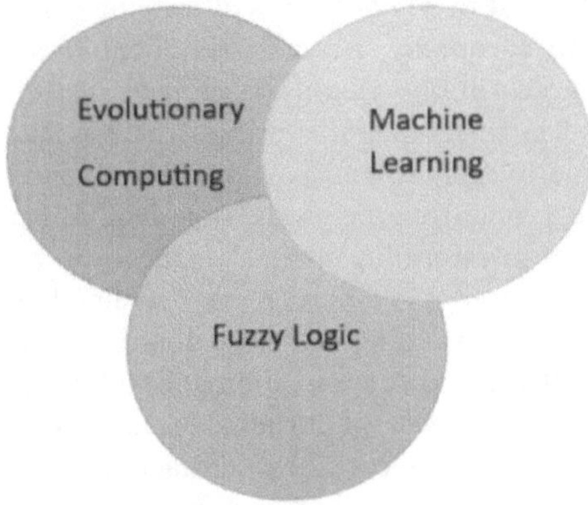

FIGURE 3.10 Intelligent control techniques.

Proportional-Integral-Derivative (PID) tuning with fuzzy logic and genetic algorithms improves the capability of mechatronics systems. These approaches let systems change with the times. Technologies such as swarm intelligence and model predictive control (MPC) help make multi-robot systems to be more efficient and organized.

3.8 ENGINEERING NUMERICAL SIMULATIONS

Numerical computing is widely used in engineering mathematics because solutions which are achieved with its help are approximate solutions to problems in which the methods based upon theoretical mathematics are inapplicable [4]. This is because most of the engineering problems include systems of equations, nonlinear and partial differential equations, which cannot be solved by analytical methods. Numerical computation is an essential tool in design optimization and outcome simulation and prediction in modeling. That way engineers can make better decisions and optimize the design to achieve a better performing and safer product/system. Finite element analysis (FEA) and computational fluid dynamics (CFD) are two critical methods of numerical analysis used in numerical computing which concern with the study of fluid dynamics, heat transfer, and mechanical strength. FEA and CFD are used to predict behavior of solid objects under set conditions and for fluids or gas, respectively (these are important in aerodynamics and heating ventilation and air conditioning systems). Besides, the numerical optimization technique finds its application in enhancing the efficiency in different engineering disciplines such as signal processing, medical imaging, and electrical engineering to minimize the cost factors. Using superior algorithms and reducing errors, engineers can improve the function of several gadgets, including machines, systems, and procedures and achieve more efficient outcomes and lower further expenditure.

3.9 AI-POWERED DECISION-MAKING

AI can also be used in smart manufacturing for other factors such as supply chain as it makes it easier to manage stocks, provide means of moving items around, and even determine what is most in demand. This allows the manufacturers to minimize wastage, cut costs, and eventually increase the clients' satisfaction due to short delivery and product accessibility. Quality assurance is also one of the important areas where AI is used in smart manufacturing. It is possible to correct a mistake right away, reduce time to market, and prevent defects or abnormality that an AI system can detect in real time from the production data. This makes product quality better, and as we are not performing manual inspection, it is more efficient and cheaper. Moreover, AI- based systems can be beneficial in the management of resources in the network, for example, in the distribution of the equipment, material, and power. In fact, the effect of that application goes a long way toward eradicating waste and contributing to the cause of sustainable manufacturing. Last but not the least, IT solutions such as computer vision and machine hearing can be utilized on the manufacturing operations as well. For example, machine vision can be applied to a defect detection problem, while machine hearing can be applied to kilohertz range for monitoring health of the equipment or studying abnormal vibration sounds that may indicate a problem. This, therefore, combines AI with mechatronic systems to improve smart manufacturing by optimizing on the utilization of robotics and automation in production. In manufacturing, AI-based systems allow for efficient prognosis and prevention of equipment failures, supply chain and resource utilization, quality and safety, thereby effectively helping to chart a more sustainable and productive future, and enhancing operational efficiency. This partnership between humans and machines not only boosts operational efficiency but also cultivates a secure work environment. AI-powered data analysis of extensive datasets discloses insights for ongoing enhancement and quality assurance. The incorporation of AI results in enhanced energy efficiency, thus reducing operating expenditures and mitigating environmental effects. Furthermore, AI improves cybersecurity, safeguarding manufacturing facilities against cyber attacks and maintaining data integrity. The onset of Industry 4.0 represents a transformative change in production, propelled by the integration of mechatronics, AI, and cybernetics. This change provides potential for enhanced efficiency, production, and creativity, while also posing obstacles such as data security, integration complexity, and talent deficiencies. Organizations must reconcile these difficulties and possibilities by implementing a strategic strategy, allocating resources in workforce development, and cultivating a culture of innovation to effectively traverse the Industry 4.0 landscape. Numerical computing is a crucial aspect of engineering mathematics that tackles complex issues across various disciplines. It achieves this by estimating solutions to mathematical equations in scenarios where analytical approaches are impractical. It plays a crucial role in simulations and modeling, facilitating the formation of predictive insights that can be utilized to enhance design. In structural analysis, fluid dynamics, and thermal transport, methodologies such as FEA and CFD are essential. Technologies including sensors, drones, and autonomous vehicles are employed in precision agriculture for the meticulous monitoring and management of resources. Automated technology enhances precision and reduces labor in crop planting, harvesting, and irrigation. Mechatronics enhances intelligent control and asset management, facilitating more efficient utilization of water and fertilizer while concurrently minimizing waste. Agriculturalists can enhance proactive decision-making by employing real-time meteorological data, while training efforts guarantee the effective implementation of technology.

3.10 HARMONIZING AI INTEGRATION

A modern Canadian engineering design program has integrated AI while augmenting human cognitive abilities. Proficiency in AI is crucial for prognostic modelling, optimization, and data analysis, enabling students to apply knowledge, enhance design processes, and increase productivity.

The simultaneous cultivation of cognitive abilities, such as innovation, adaptability, problem-solving, and ethical reasoning, is crucial [5]. Future engineers must precisely review AI recommendations, identify biases, and make ethical choices. Through the automation of routine tasks, AI allows engineers to focus on more complicated and socially relevant design aspects. The program must incorporate interdisciplinary collaboration and practical applications, allowing students to interact with AI systems, understand their limitations, and unleash their capabilities. Initiatives emphasizing human–AI interaction, sustainability, and societal relevance can bridge technological advancement with human-cantered design is shown in Figure 3.11. The program must emphasize ongoing learning and adaptation, preparing graduates to employ current AI technology and adapt to upcoming advancements.

3.11 MECHATRONICS

Utilization of AI, leveraged by deep learning techniques in the field of mechatronics, has brought in the possibilities which have never been witnessed before in

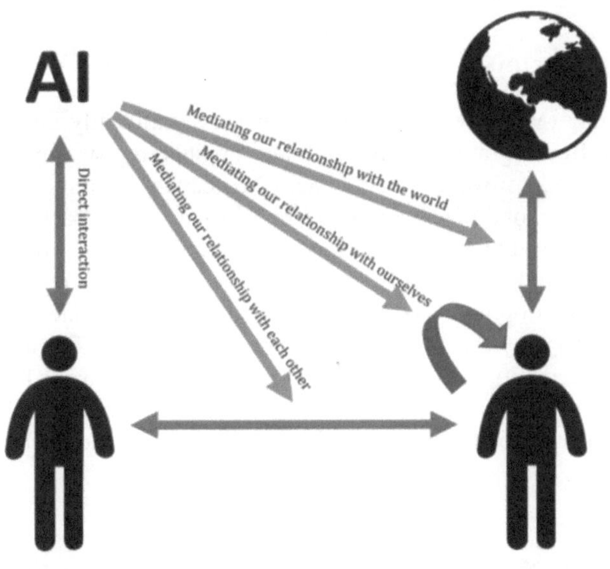

FIGURE 3.11 Harmonizing AI integration.

the engineering domain. These machines can be programmed to perceive and reason through their environment at a growingly higher level with the help of neural networks.

The benefits of deep learning in mechatronics are very profound and have the potential to revolutionize the industry [6]. For instance, self-driving cars, working autonomous robots and drones can decide on the spot and perform in an environment packed with obstacles due to deep learning algorithms. This capability becomes useful in making these systems safer and more efficient, especially where the operating environment is complex and volatile. Moreover, deep learning enables the machine and human interactions to be easier between the two through better human–machine interaction [7]. This, in turn, enhances reliable mechatronic systems' reliability and efficiency by enabling precise anticipations of the most possible malfunctions through deep learning. As we progress forward, the applicability and possibilities of deep learning is in the best position to redefine intelligent automation, robotics, and pronounced innovation modifications in the experiences of mechatronics. Appropriate partitioning and processing methods will be required to improve the next generation of machines with higher forms of AI in different contexts of usage [8–14].

Therefore, the interactions between AI, especially deep learning and mechatronics domain, will keep pushing further the enhancement of intelligent machines. The combination of these technologies has the potential of developing machines that work independently, enhanced interaction between man and the machine, and optimal scheduling for maintenance [15–17]. Therefore, the interactions between AI, particularly deep learning, and the mechatronics domain will continue to drive the advancement of intelligent machines. The combination of these technologies has the potential to develop machines that operate independently, enhance interaction between humans and machines, and optimize maintenance scheduling. With further advancements in technology, deep learning may become even more significant for the future of mechatronics and intelligent automation.

3.12 CONCLUSION

Machine learning (ML) is so revolutionary that it significantly impacts the mechatronics domain by integrating interdisciplinary engineering streams across mechanical, electronic, and software systems. With ML capabilities, mechatronic systems can become smarter, optimized, and more responsive. The impact of ML in mechatronics can be assessed in various ways to achieve optimization. ML algorithms have the ability to analyze large datasets, learn from them, and draw conclusions or make recommendations to optimize system performance. In this context, ML algorithms can be used to design systems with flexibility and adaptability. The integration of ML in mechatronic technology allows systems to adjust their behavior based on varying conditions or inputs. Innovation in ML within the mechatronics field can enhance performance criteria, flexibility, self-sufficiency, reliability, and interaction quality.

REFERENCES

1. Wang J., Based on the application of artificial intelligence technology in mechatronics engineering. In *Application of Intelligent Systems in Multi-Modal Information*

Analytics: 2021 International Conference on Multi-Modal Information Analytics (MMIA 2021), Springer International Publishing, 1, 254–260, 2021.

2. Xiong Q., Overview of the Relationship between Mechatronic Engineering and Artificial Intelligence. In *2021 International Conference on Wireless Communications and Smart Grid*, IEEE, 532–535, 2021.
3. Zhang J. and Wang L., A Novel Approach to Self-Tuning Control in Mechatronics Using Artificial Intelligence, *IEEE Transactions on Industrial Electronics*, 55(2), 586–593, 2008. doi: 10.1109/TIE.2007.911943.
4. Kamalov F. and Leung H.H., Numerical Computing in Engineering Mathematics. In *2022 Advances in Science and Engineering Technology International Conferences*, IEEE, 1–5, 2022.
5. Vadakkepat P.K., Bio-Inspired Robotics: Artificial Intelligence in Mechatronics. In *Proceedings of the IEEE/RSJ International Conference on Intelligent Robots and Systems, Vilamoura, Portugal, Oct. 7–12, 2012*, pp. 4269–4274, 2012. doi: 10.1109/IROS.2012.6386143.
6. Lila M.K., Integration of Artificial Intelligence Techniques in Mechatronic Systems for Smart Manufacturing, *Mathematical Statistician and Engineering Applications*, 70(1), 432–439, 2021.
7. Frochte J., Lemmen M. and Schmidt. M., Concerning the Integration of Machine Learning Content in Mechatronics Curricula. In *Research Anthology on Machine Learning Techniques Methods and Applications*, IGI Global, 379–395, 2022.
8. Liagkou V., Stylios C., Pappa L. and Petunin A., Challenges and Opportunities in Industry 4.0 for Mechatronics, Artificial Intelligence and Cybernetics, *Electronics*, 10(16), 2001, 2021.
9. Holland J. H., *Adaptation in Natural and Artificial Systems: An Introductory Analysis with Applications to Biology, Control, and Artificial Intelligence*, Cambridge, MA: MIT Press, 1992.
10. Trivedi M., Autonomous Vehicles: Seminal Events, Present Challenges, and the Future, *IEEE Intelligent Transportation Systems Magazine*, 9(1), 6–14, 2017.
11. Smith M. L. and Milgram P. J. R., Intelligent Manufacturing Systems, *International Journal of Production Research*, 38(3), 463–476, 2000.
12. Yan. S., Zhang J. and Li. S., A Review of Artificial Intelligence in Maintenance: Challenges, Opportunities, and Path Forward, *Journal of Intelligent Manufacturing*, 31(3), 767–792, 2020.
13. Tian G. N. Q., Yang S. Z., and Newman S. T., Real-Time Quality Control in Manufacturing Systems Using Artificial Intelligence Techniques, *Journal of Intelligent Manufacturing*, 15(3), 321–329, 2004.
14. Md Ali. A. M. H., Lu. C and Wu. X. W., A Review of Artificial Intelligence-Based Robots in Manufacturing, *International Journal of Precision Engineering and Manufacturing-Green Technology*, 8(1), 223–241, 2021.
15. Sharma A., Kumari A., Sharma V., Al-Farouni M. and Mishra S., Consolidation of Cloud Computing in Smart and Sustainable Environment. In *2023 International Conference for Technological Engineering and Its Applications in Sustainable Development (ICTEASD), Al-Najaf, Iraq*, 86–91, 2023, doi: 10.1109/ICTEASD57136.2023.10585245
16. Sharma S., Pandey A., Sharma V., Mishra S. and Alkhayyat A., Federated Learning and Blockchain: A Cross-Domain Convergence. In *2023 3rd International Conference on Technological Advancements in Computational Sciences (ICTACS), Tashkent, Uzbekistan*, 1121–1127, 2023, doi: 10.1109/ICTACS59847.2023.10390227
17. Nielsen M., *Neural Networks and Deep Learning*, Determination Press, 25, 2015.

4 The Resilience of Artificial Intelligence in the Mechatronics Sector

Challenges and Future Directions

Praveen Kumar M., Saradha S., Meena M., Vijayakumar S., and Pushpalatha N.

4.1 INTRODUCTION

The integration of artificial intelligence (AI) in the area of mechatronics is a current approach that is slowly gaining considerable importance. Flexibility in AI technology is the key characteristic in mechatronics. This chapter describes the highlights of the rapid development in all the fields of engineering and their simulations in the real-world applications through AI. The discussion goes into AI, especially resilience, which plays a vital role in the mechatronics sector.

By mid-20th century, the application of AI has become relatively extensive. The main goal in this chapter is to familiarize mechatronics, which is the multidisciplinary branch of engineering, and supports various domains such as digital electronics, telecommunications, control systems, and computer science to fully analyze the implementation of AI in the mechatronics.

Industrial, socioeconomic, and environmental reliance on automation is increasing. The continuous improvement in the area of machine learning (ML), especially deep learning (DL), has made it part of our life – a vital tool for designing, modeling, and controlling the systems. In recent days, AI has become the core part of industrial development in the field of mechatronics, and it is generally termed mechatronics system. This leads to new scientific and application encounters such as system and communication security, decentralization, cost, prediction, energy efficiency, predictive maintenance (PdM), etc.

The continuous increase of flexibility in the automation process in the other sectors such as industrial, socioeconomic, and environmental ranges is unavoidable. The rapid advancements show the realm of ML, especially DL, have proved its position as an unavoidable tool for the processes of systems such as designing, modeling, and controlling.

DOI: 10.1201/9781032615790-4

AI becomes a vital part in industrial development, particularly in the vast area of mechatronics. Moreover, this kind of progress naturally introduces unique scientific and application challenges, including harmonic filtering, predictive capabilities, and smart energy efficiency and maintenance protocols that require complete exploration to truly understand and harness the complete potential of AI in the mechatronics realm.

This chapter also casts a critical look at the dangers of integrating AI technologies into mechatronics, which require careful consideration, and suggests potential future directions in this area. Robotic systems now incorporate AI in a variety of ways, but we argue that efficient and reliable AI systems need to be resilient. Our definition of AI resilience includes desirable and measurable qualities such as robustness, fault tolerance, adaptability, survivability, and safety. This work delves into potential routes to creating AI systems that are both efficient and robust. To achieve this, an examination of existing AI technologies or techniques can help in determining the aspects of existing algorithms that require attention or improvement in a modern dynamic framework. Moreover, it reveals how developers and researchers might address AI from a perspective that enhances its efficient and robust operation as part of mechatronic systems. The remainder of this chapter is organized as follows. Section 4.2 provides an overview of AI, general history in the mechatronics field, and recent developments. Section 4.3 describes AI resilience attributes and metrics and also juxtaposes the potential pitfalls and the benefits emergent from this field. Sections 4.4 and 4.5 present potential research directions moving forward.

In discussing the dangers of integrating AI technologies into mechatronics, it becomes paramount to thoroughly examine the potential implications that arise. Consequently, it is imperative to analyze existing AI technologies and techniques to identify areas that require attention or improvement in the context of a modern dynamic framework. By doing so, developers and researchers can adopt a perspective that elevates the efficiency and robustness of AI as an integral component of mechatronic systems.

Followed by the introduction, this chapter consists of four sections. The final section provides the future research direction along with the summarization of the discussed areas.

4.2 BACKGROUND AND CONTEXT

The background of this chapter started from the combination of two fields, namely mechatronics and AI. This chapter starts its discussion from understanding the AI in mechatronics. Then it moves to current applications of AI in the mechatronic sectors followed by challenges faced by AI in mechatronics and finally concludes with future directions and innovations.

4.3 UNDERSTANDING AI IN MECHATRONICS

AI is the branch of computer science, a game-changing technology that has emerged in recent years, focusing on developing machines and computers to learn, reason, and act in the ways that normally require human intelligence. It has an unparalleled capability to raise machines to thinking levels that overtake human capabilities in various domains.

As found in the field of mechatronics, with the combination of mechanical systems, AI has high potential to provide considerable improvements in terms of performance, flexibility, and comfort, so that a revolution in the machines happened in operations and interactions with the world.

It is also notable one that AI incorporates various types other than its traditional model and each provides unique and innovative methods to solve problems. It comprises artificial neural networks (ANN), fuzzy systems, expert systems, genetic algorithms, and intelligent systems. All these AI approaches focus on the optimization and enhancement of the mechanical systems by allowing them to overcome their integral limitations to reach a new stage of efficiency.

Because of the inclusion of AI elements, machines are not just mechanical systems, but also became intelligent systems. This combination of AI and mechanical systems demonstrates the dynamic area of research, commonly known as intelligent machines.

Actually, AI supports the mechanical systems to develop and hold remarkable behaviors such as autonomy, adaptability, and ability to emulate human capabilities. These intelligent systems are considered by their horizons, and compared to traditional counterparts, they are less confined and limited. The integration of AI into mechanical systems enable them to make informed decisions, learn from the environments, predict the future outcomes, self-organize, self-optimize, diagnose faults, and also enhance the overall reliability and robustness.

The core of intelligent mechanical systems lies in the vast amount of data they rely on. These systems can analyze extensive measurements and other data sources, which are then processed to integrate AI algorithms at their core. The data plays a vital role in driving the mechanical system toward optimized performance and serves as the foundation for training and refining the AI algorithms.

Even though it is important to mention the potential challenges that arise during the implementation of AI algorithms and resources, their effectiveness is highly augmented when accompanied by an abundant amount of representative data. This kind of relationship between data and AI is essential to unlock the full potential of intelligent machines.

Finally, the implementation of AI has resulted in a huge transformation in the area of mechanical systems, converting their evolution into intelligent systems that achieve high capabilities and showcase behaviors that were only possible for humans once. The combination of AI and mechanical systems, especially in the field of mechatronics, has created a way to remarkable advancements in high performance, reliability, and comfort. With the ongoing research and incessant combination of huge data, AI-powered mechanical systems is all set to restructure industries, empower innovation, and enlarge boundaries of what machines can achieve.

4.4 RELATED WORKS

In paper [1], the relationship between sustainability and resilience of AI systems is discussed, covering various sectors such as agriculture, smart health and medicine, smart cities and police, justice, and military-based projects.

A precise approach is proposed for analyzing the versatility of AI frameworks, accompanied by a review of relevant scholarly publications [2]. The strategy involves establishing a set of flexibility components, structuring and categorizing both hierarchical and ontological relationships among versatility variables in AI systems, and examining key flexibility strategies and associated challenges.

The paper [3] discusses the key aspects of Industry 4.0 and explores how its insights can be integrated into manufacturing to develop smart factories. It provides a brief overview of the core components of Industry 4.0 and emphasizes the security challenges posed by its fully interconnected environment, including the risks associated with each component. The importance of ensuring security is underscored as a critical factor for the success of Industry 4.0. Additionally, the paper outlines prominent research directions aimed at protecting the confidentiality and integrity of data exchanged among Industry 4.0 components.

In paper [4], the challenges of measuring the versatility of AI and ML frameworks with the current set of strength evaluation devices are described. Whereas the discussion in this paper applied to all AI frameworks, whether driven by ML or not, it primarily focused on ML-based frameworks, as they present the most significant set of challenges.

Paper [5] offered a global overview of the progress of such technologies in segments with high potential for maintainability such as agriculture, forestry, and the extraction of marine resources.

In paper [6], the advances in AI in mechatronics is discussed. The future intelligent mechatronics is also included in this. Intelligent approaches such as human behavior imitation and optimizing mechatronics are mentioned.

In paper [7], an overview of current advancements in AI, ML, and DL in advanced mechanical technology frameworks is presented. along with an examination of different applications of the frameworks in robot adjustment.

In paper [8], a conceptual system is proposed where organizational values and structural components work together to enable speedy and satisfactory trade choices in complex and questionable disruptions. The system presented the connections among AI, data frameworks, and supply chain disruptions. Finally, the suggestions for hypotheses and the limitations and scope for future research are discussed.

4.5 CURRENT APPLICATIONS OF AI IN THE MECHATRONICS SECTOR

AI has been playing a key role in enhancing the level of automation in handling and controlling different manufacturing processes. These advanced AI techniques have huge potential to improve performance, accuracy, and fault tolerance for several manufacturing technologies and can also be applied in robotics and intelligent mechatronic systems. Ranging from minimizing human intervention to increasing system reliability, there are several applications of AI that make the mechatronic system more efficient, productive, and easy to operate. Regardless of an ultimate goal-oriented application, this section reviews studies where practical testing of AI implemented through mechatronic applications in challenging real-world contexts has shown benefits in improving operational and energy efficiency; reduced maintenance, repair, and overhaul costs; and increased organizational innovation [9].

AI is gaining significance particularly due to its support in PdM. The paradigm shift from preventive maintenance to PdM is also possible with the help of AI tools. It not only monitors and manages individual system functions, but also diagnoses the system as a whole and helps in predicting maintenance time, which is also essential from an efficiency point of view. The application of AI tools and techniques in smart transportation involves elements of forecasting, optimization, and extraction of vehicle traffic patterns that have been widely studied in recent times. Mechatronic circuits have advanced in the healthcare industry using the design of biomedical devices and also contribute to the area of bionic–human interfacing, such as advanced prosthetics and tactile sensors. AI tools are crucial for developing autonomous systems to diagnose, control, or monitor the parts of a broader system that involves both physical and virtual components [10–14].

Some of the applications of AI in the mechatronics can be analyzed as discussed below.

4.5.1 AI in Robotic Technologies

In robotic technologies, AI plays a vital role in enabling sensing, positioning, grasping, and other robotic tasks. Also, the AI process has been implemented into various industrial robot functions. The feature of AI to enhance human-like reasoning becomes the backbone behind the development of automation and intelligent robotics.

4.5.2 AI in Smart Factories

By providing online sensor data and real-time decision-making, AI is applied in smart factories to support intelligent machines. Additionally, the quality control and inspection procedures of products have been enhanced by strengthening AI. AI is applied in smart factories to support intelligent machines by leveraging online sensor data and enabling real-time decision-making. Furthermore, AI has enhanced product quality control and inspection processes by detecting subtle variations that are difficult to identify using traditional methods.

4.5.3 AI and Robotics Integration

The self-learning robots are integrated with AI and robotics. The real-time application of AI in autonomous vehicles received high acceptance that makes vehicles safer and more convenient. Developing robust algorithms and changing tactical behaviors are the current AI-related actions in autonomous vehicles.

4.5.4 Smart Systems in Mechatronics

AI works with the mechatronic principles to create smart systems. These intelligent systems can be categorized based on their ability to be intelligent, green, smart, and robotic. The collaboration between the above four systems guarantees the advantageous outputs. In their applications, both the research groups and industries correspond to the need to initiate various AI concepts.

4.5.5 Impacts and Advancements of AI in Mechatronics

In recent times, AI has become a unique part in the mechatronics sector, in finding applications across a wide range of areas. Its main contribution can be found in the realm of automation, reforming processes in production and manufacturing. The human–machine collaboration enabled by the conversion of robotic welding into intelligent systems. AI plays a vital role in advancements of robotic technologies and enables them to process tasks with remarkable precision in areas such as sensing, positioning, and grasping.

4.5.6 AI in Smart Factories and Quality Control

The ability of AI is the backbone in the progress in automation and intelligent robotics. In the reign of smart factories, AI is implemented to help intelligent machinery by giving real-time sensor data and making fast decision-making. Also, the quality control and inspection procedures are highly benefitted from AI. It can avoid human inspectors and guarantees high efficiency and accuracy in the production process [15].

4.5.7 AI and Robotics Integration for Self-Learning Robots

The mixture of robotics and AI has led to growth to self-learning robots that can familiarize and increase their performance over time. The development of self-driving autonomous vehicles is a real-time example of the considerable acceptance of AI [16].

4.5.7.1 Future Possibilities in Mechatronics with AI

By connecting vigorous algorithms and changing behaviors, AI has not only resulted in vehicles' safety but also made it more convenient for users. Research groups and industries are keenly implementing the high potential of AI, including a different range of concepts to open up new possibilities and resize the future of mechatronics.

4.6 MECHATRONICS INTO THE FIELD OF AI – A EVOLUTIONARY JOURNEY

Mechatronics, which blends mechanics, electronics, and control systems, has significantly evolved with the integration of artificial intelligence. Initially centered around automation through pre-defined logic, modern mechatronic systems now incorporate AI to enable learning, adaptation, and real-time decision-making.

Advancements in machine learning, sensor technologies, and data processing have transformed mechatronic devices into intelligent systems. These improvements are evident in areas like smart manufacturing, autonomous vehicles, and robotics, where systems must respond dynamically to changing environments.

However, this integration brings challenges such as increased system complexity, the need for real-time performance, and ensuring the reliability of AI decisions. Addressing these concerns is essential for further progress.

TABLE 4.1
Mechatronics into the Field of AI

Sl. No	ERA	Key Milestones
1	1960s	Emergence of mechatronics as a field integrating mechanics, electronics, and computing.
2	1970s	Pioneering work on computer numerical control (CNC) machines, early automation efforts.
3	1980s	Advancements in sensor technology and robotics. Initial AI applications in manufacturing.
4	1990s	Integration of AI in robotics for increased autonomy and decision-making. Evolution of smart manufacturing.
5	2000s	Growth of IoT, connecting mechatronic systems. Expansion of ML in mechatronics.
6	2010s	Rise of Industry 4.0 with AI-driven automation. Ethical and security concerns gain prominence.
7	2020s	AI resilience in mechatronics highlighted. Sustainability and AI's role in green practices. Ongoing exploration of AI's impact on competitiveness.

The evolution of mechatronics with AI marks a major step forward—shifting from static automation to systems capable of intelligent behavior and autonomous operation (see Table 4.1).

4.7 AI IN MECHATRONICS – CHALLENGES

The implementation of AI comes with a number of challenges in the field of mechatronics. Truly autonomous decision-making and action implementation using AI processes are possible only if technical issues of AI are resolved. The biggest challenge in AI has been identified as having data of good quality as a basis for process development. AI processes deal with statistical patterns, which is why outliers or the statutory triplet 'Garbage in, garbage out' are elements that have to be considered. AI and its impact on the application of ML and DL only convolute the complexity involved in the implementation of AI because AI is a highly pragmatic field and has a clear connection to practice and the production of goods. The enhancement of computing power is also crucial for making DL truly viable for productive usage.

Another correlation in technological development, apart from computing power, is the change in function of Radio Frequency (RF) technology from a pure value distributor to an enabling technology for mechatronic systems. One of the urgent problems is skilled personnel. AI, being a cross-disciplinary field, is a clear indicator of the pressing need for qualified experts who are both experienced in AI and capable of building bridges to practical production, unlike many leading frontrunners in AI research who are more interested in the technological and algorithmic side of AI. Economic challenges are costs, which include acquiring the technology and processes, maintaining such fields of innovation, and retraining personnel to operate the new systems. Security challenges manifest in the integration of AI and mechatronics. On one hand, mechatronic systems use fine-grained electronic architecture penetrated by cyber attackers after establishing a breach in the web. On the other hand, manual operation of maintaining data security is expensive and tiresome, producing errors and, furthermore, not really feasible, given the growing attacks on cyberspace and the lack of physical manpower to coordinate saving the manifold IT systems from cyber threats. Developed societies have launched regulatory initiatives aimed at stymying the spread of AI in mechatronic systems. It is too early for legislation to provide guidelines, but it is clear that AI and a broad range of concepts employing AI may soon form the centerpiece of extended court cases involving product and operation faults, limited liability, and the onus of operators of mechatronic products and systems. Nevertheless, it is our argument that coherence between these many views and careful and progressive legislation should be treated with caution. It is crucial for any kind of sustained development of AI in industry and society that the dangers and pitfalls listed are properly researched, understood, and, if possible, controlled.

DL as a ML method and type of AI has an overwhelming potential for altering knowledge regarding data presented. It does not work efficiently for lower consumer hardware, but it has been expected to considerably help in many fields. It is said to streamline procedures for the implementation of ML to some point. While AI cumulatively features mechatronic principles that have already been established in the early stages of mechatronic development, AI differs from solutions that underline mechatronics, is more independent of initial variables after manufacturing, and is learned from data. The exploitation of AI shall yield more adaptable and playful mechatronic products to suit the final user. To take full advantage of this feature, extensive research should be performed to clear the way for a host of potential challenges that impede the piloting and implementation of AI. All of these will need to be examined in a multifaceted, multi-tier process.

4.8 ETHICAL CONSIDERATIONS IN AI MECHATRONICS

Due to the integration of AI with mechatronics, the field is constantly developing but a lot of challenges are also there. Most of the AI systems require a high quantity of qualified data to derive reasonable performance, so that only a few organizations are capable of providing those kinds of facilities. Also, wrong data can lead to poor predictions in real-life applications, endangering human and environmental security. In the technical aspect, the existence of a greater number of tools, systems, technologies, and standards results in information isolation. The above are considered as barriers to AI adoption.

Additionally, AI may become the reason for the difficulties in ethical, legal, and societal issues. For the above issue, AI systems need to connect with their operational principles, a way of working with transparency, responsibility, ethics, and the law, thus moving toward responsible AI. Any further improvement in this field should aim to considerable development in terms of safeguarding privacy and security when allowing for the normal use of AI in everyday lives of people [17–21].

4.8.1 Data Quality and Quantity

The trained data are deeply linked as a result of AI models. One of the basic limitations in the acceptance of AI applications is the quality of data used. The accurate prediction becomes a common problem due to the low amount of available data. Due to this, the disposal of large amount of data and to have data as a representative with respect to the applications is critical. Despite the availability of vast amounts of data, differences may arise in other ways due to the heterogeneity of the collected databases and the application's domain. AI models rely heavily on well-connected and structured training data. One of the fundamental limitations in the adoption of AI applications is the quality of the data used. Accurate prediction becomes challenging when only a limited amount of data is available. Therefore, having access to large volumes of data that are representative of the target application is crucial. Even with abundant data, inconsistencies may arise due to the heterogeneity of the datasets and differences in application domains. Consequently, ensuring both high-quality and sufficient quantities of data is essential for AI models to achieve strong generalization and accurate predictions.

4.8.2 Interoperability and Integration Issues

The system complexity itself became a major challenge to adopt AI in a mechatronics system. In the case of Internet of Things (IoT), organizing embedded AI in a mechatronic environment led to compatibility issues with legacy hardware and software. The hardware used in manufacturing systems can be too costly, time-consuming, and not safe to replace. Also, the continuous 24/7 operation required in production and industrial systems increases the risk of disruption when installing new AI solutions.

The most effective solution to this issue is to use reliable, well-established devices that are easily integrable into mechatronic systems. This approach offers significant advantages for manufacturers and system integrators, as these devices can be seamlessly incorporated into products, thereby enhancing the overall capabilities of the mechatronic systems they produce.

4.8.3 Ethical and Legal Concerns

The increasing usage of AI in the mechatronics sectors also increases the number of ethical issues. For example, what happens when an intelligent system designer cannot understand why the system arrived at a particular decision? Should someone be still accountable for the decision? Nowadays, intelligent systems such as AI and big data are widely used in smart healthcare and smart cities. While these systems facilitate decision-making, they should also provide warning data to ensure adherence to 'best practices'.

There are problems as well with AI: the algorithms are only as good as the data they operate on. There has been considerable media attention to facial recognition algorithms, which are not able to recognize black people as well as white people. Systems such as the Dutch Traffic Jam System show shorter delays in the vicinity of a high-speed road, such as a highway. At the same time, there are no traffic jams indicated on the highway itself. The system assumes that the average driver has a mobile phone with Global Positioning System (GPS). However, this bias in the data forms a serious societal problem: the information on the shorter traffic jam might increase traffic in that specific area, which could cause the normal traffic flow to become congested at that specific spot as well. This causes the whole highway to clog without a clear reason, resulting in unnecessary traffic jams. Data and algorithms can also be used for manipulative purposes, leading to misuse of data by actors who have access to individuals' private data. Furthermore, often, the data used is of a sensitive and personal nature. Therefore, data owners may also suffer moral and/or legal harm. Data privacy can be seen as the right to be let alone. Both physical and informational privacy are considered to be fundamental rights in many legal systems. Next to the question of data privacy and worldwide, many AI organizations are struggling with the correct interpretation of newly available laws. Within these regulations, explanations are split into two subdivisions: lawfulness, fairness, and transparency, and data protection principles. There are two areas of AI that can potentially breach these regulations: automated individual decision-making and profiling. The principle of data protection by design and by default is based on the concept of privacy settings and the minimization of personal data. This legal right contradicts strategies applied in the field of AI, as more data increases the system's accuracy, which provides a strong contradiction. It has been proposed to instill transparency by making sure that the purpose of the algorithms is based on the beliefs of the public, followed by discussion and legislation. It is the responsibility of policymakers to avoid violations and instill tools that avoid unintended use of data. The development of proactive countermeasures at three levels – audit, measuring countermeasures, and compliance checks – has been suggested. Regulative countermeasures at each level, such as worldwide legal cooperation, are proposed to increase the impact on individuals or organizations. The follow-up action should be the establishment of a consortium where different stakeholders play a significant role in developing a 'responsible AI' mindset. The importance of AI ecosystems has been stressed and invites feedback and suggestions for avoiding AI-induced moral and legal harm.

4.9 FUTURE DIRECTIONS AND INNOVATIONS

The previous sections highlight the application of AI approaches to the mechatronics sector and how these have been exploited for effective mechatronic system design and control. In this final section, a discussion on future directions and the emerging trends in AI for mechatronics is first provided, indicating the potential areas where we anticipate significant advancements and transformative changes in the mechatronics landscape in the near future. Various emerging technologies, such as quantum computing, blockchain, and next-generation robotics, have the potential to offer a paradigm shift in the development of mechatronic systems, allowing the engineering community to tackle highly sophisticated problems in an improved and effective manner. Integrating AI with the IoT is also likely to result in the improvement of system connectivity, control, diagnostics, and predictions. This is then followed by a discussion on the ideal research and innovation factors that we should focus on addressing to develop effective AI-driven mechatronics systems. We aim to share ideas and thoughts with the broader AI and engineering communities to encourage and stimulate open discussions on the resilience and the development of AI systems within dynamic mechatronics applications. Even though we have highlighted the challenges facing key requirements such as system reconfiguration and predictability in some emerging technologies, numerous technological developments and breakthroughs are predicted, offering promising potential in the future of AI mechatronic systems.

Emerging technologies and trends: The areas that are set in the mechatronics sector offer emerging technologies with considerable potential are AI, quantum computing, blockchain, and next-generation robotics. Other solutions currently emerging include the knowledge extraction of AI with action execution of machines in the fast-growing field of the IoT industry. Emphasizing or utilizing emerging technologies will offer a competitive edge to the AI investments in mechatronics and robotics and facilitate speedier growth in the areas of technology, policy, and/or funds that support AI advances in innovation. The good news is that AI is already offered by over 40% of EU industrial companies yearly in different stages in their development projects according to technological advancement and innovation consumers' missions. AI will continue to change the physical understanding of technical systems in the future for the better and will focus on the virtual design and virtual experimentation or what-if prediction for previously installed communication networks that might not be detected by those in charge of system security loops for research and development needs.

ML is undergoing a transition from intuitive algorithms to DL. It is estimated that the current trend will become the default method for the future. In DL, neurons are added according to the needs of the input. They keep learning data or findings and provide good predictive accuracy in system performance. Predictive accuracy is reduced compared to other systems with normal learning data. Therefore, we can say that human life is pledged to AI to this extent today in the next generation. This trend can be predicted by the following points:

- The role of predictive accuracy in AI systems is necessary in the field of Industry 4.0. For example, AI will evolve inside different mechanical systems, device systems, remote and cloud systems within a computing process.
- People are looking to work in a collaborative way with AI, namely, human and robot cooperation.

It is given a different direction in a new way. Indigenous research and the use of new databases are required in professional terms. Robotics is highly reflected in AI. The trend of AI is moving from predictive accuracy toward more accurate values, to work with human practical aspects to deliver non-dangerous and real-time messages to people with profound human knowledge for future industrial requirements inside Industry 4.0. It should be useful to people in all aspects. It is geared toward the marriage with the IoT. The innovation mode cannot be implemented individually, but in collaboration with different industry innovator sectors. Ongoing training will help the industry and the public use the apprenticeship. It may be a challenge for business and industry. Due to the aging of the popular internet cycle through 2011, using AI technology to satisfy the potential in the mechatronics sector, the revolution of technology is necessary.

4.9.1 Advancements in ML Algorithms

Provoked by the ideas of mimic and heuristicists, in the past few years, ML has made the most achievements in AI. For instance, convolutional neural networks (CNN), which are based on DL, have surpassed human capabilities to recognize images and have become the most representative research for image classification and reading road images. Besides, stacked autoencoders, the most representative research in DL, reconstruct the input after distortion for data compression. Moreover, long short-term memory (LSTM), the most representative research in DL, can restore an understanding of the contribution of memory for tasks involving robotic object handling and mobile robot navigation. Furthermore, collaborative reinforcement learning has shown good learning performance when acquiring real-time information. Notably, the observation and state estimation usually have lags and failures in complex mechatronic systems, which need small-scale inherent fault tolerance.

ML has developed better learning algorithms for complex and unstructured data in terms of violation, chaos, perturbation, smell, timeliness, uncertainty, volubility, and large quantities of data under big data architecture. For instance, the Monte Carlo tree search algorithm supports reinforcement learning; stochastic gradient descent enhances the learning model with a large quantity of data; and genetic algorithms employ the top parent algorithms to improve learning. A single-layer or multilayer neural network is more than five times more accurate than a phone-based gyroscope and reaches a 96% accuracy rate in noise reduction approaches. However, it is common for the algorithm not to constantly process big data in real application environments, and another technique that processes entire feature values is the most popular. Moreover, there is a strong need to process the data many times to promote learning methods depending on real-time data and online training strategies. Furthermore, ML is predicted to make more significant advances, such as a continuous ML system that will learn new experiences continuously. This implies that mechatronic systems can adapt well over time to environmental changes and operational conditions. Moreover, each value-based algorithm is optimal for separate problems. Notably, a specific method can manage large quantities of data and accurately depict states. Thus, we should devise an algorithm for each mechatronic to analyze many aspects of how to relate to mechatronics tasks to empower the machine to perceive how to interact more efficiently with mechatronics.

4.9.2 Enhancing Human–Machine Collaboration

Adopting AI in mechatronics manufacturing can facilitate replicability, modularity, robustness, scalability, and integrality, offering fast reconfigurable mechatronic systems adaptable to changing production models. To promote these capabilities, we need to encourage AI-based Human-Machine Collaboration (HMC) in such a way that collaborative interaction becomes a harmonious unity able to improve productivity and safety, without undermining the full potential of experience and creative competencies that distinguish human operators. In this context, different useful methodologies and tools were addressed to promote an effective human–intelligent system interaction. In particular, intuitive interfaces based on gestures and speech have become capable of wrapping technological aspects to provide operators with natural tools allowing them to share their thoughts and intentions with intelligent systems and possibly avoiding inappropriate commands. Furthermore, it is essential to employ real-time feedback mechanisms to constantly assess if the HMC is proceeding in a correct way. Companies and scientific organizations should also invest in appropriate educational and training programs to prepare operators today to be able to collaborate with such intelligent technologies. Besides, small and medium enterprises also need training programs on AI because their workers usually have experience in the automotive field and little knowledge of AI. Various circumstances underline the need to make decisions in an ineffective and incomplete informative manner, and the recommendations issued by AI systems need to be fully justified and comprehensible to decision-makers. In general, AI should support decisions and not replace them. The aim is to assist decision-makers in enabling them to influence the decision process through consulting AI recommendations and collective intelligence, as well as steering the design of the system and its functionality within the detailed specifications. In addition, some cases may require an analysis of AI when previous scientific theories differ. AI may also verify new hypotheses. Decision support systems can, in fact, be a support tool in industrial applications involving HMC. This solution is based on information systems and mathematical decision models that are used to combine analysis and decision aids to establish human intervention in automatic procedures. To understand the benefits gained by automating mechatronic systems, a set of successful applications were developed in which AI systems were established to provide decision and control support to decision-makers, with high satisfaction and positive feedback from the persons involved.

4.9.3 Integration of AI with IoT and Robotics

Championing the advancements in the field of mechatronics, the integration of IoT, AI, and robotics holds the promise to evolve the sector. The amalgamation of AI with IoT can be used to churn enormous data gathered by sensor-packed machinery and physical configurations. The sensory data uncover the efficiency, performance, and health of machinery and systems. Analyzing information in real time results in a greater degree of speed, flexibility, and autonomous operational efficiency. The integration of IoT and AI forms smart devices and systems. Empowered by AI, robots can automate such mundane and complex tasks, becoming robust and intelligent. The marriage of robotics and AI, thus, impacts mechatronics, where sophisticated movements are essential to operations in aerospace, manufacturing, logistics, automotive, and more. One of the bottleneck challenges in AI, IoT, and robotics integrations is the standardization and interoperability problems. The system connectivity and data management are critical issues in establishing system collaboration. The AI, IoT, and robotics integration demand a connected and networked environment in terms of dense data transmission and data management. These standardization and system collaboration challenges demand an interdisciplinary approach. Collaboration between technologists and manufacturers, along with the active involvement of regulatory bodies, will boost system acceptability and adoption by end users across the globe. The demonstrated results from IoT devices, AI, and robotic technologies would attract many takers worldwide. The AI market is less likely to slow down due to the potential it wields and its integration with IoT and robotics, and the mechatronics sector will benefit from it. The AI–IoT–robotics convergence would lead to innovation, which shall be beyond the traditional scope of Industry 4.0. The suggested integrated approach would help frame the future of AI, IoT, and robotics together.

4.9.4 Explainable AI in Mechatronics

Contemporary AI systems are capable of making sophisticated predictions and decisions about environments that are difficult to understand. This level of predictive capability, at times even outperforming human capabilities, motivates interest in them being embedded in decision support systems. One of the most important difficulties in this regard, specifically in safety-critical systems where the quality of decision-making is fundamental, lies in the fact that the inner workings of these models are not transparent, hence their characterization as 'black box'. Without access to system parameters, mechanisms, models, and data, it is impossible for a user to trust any input, output, or decisions that arise from it. This lack of transparency may discourage the use of data-driven decision support systems and motivate reliance on manual or rule-based AI. For this reason, there is an increasing demand for AI systems to be capable of providing clear rationales for their decisions, explaining predictions and the reasons behind their recommendations. In other terms, contemporary AI systems should aspire to be explainable, or similarly, interpretable, transparent, or understandable. This demand is particularly significant in the context of mechatronics forums, as the intersection of mechatronic systems and AI should enhance, rather than disrupt, users interacting with them. For this reason, it has been referred to as explainable AI. Typically, operationalizing human understanding requirements into AI systems is difficult because tuning transparency levels could be challenging in general, given the trade-off between the different algorithms utilized. In safety-critical applications, understanding the limits of AI could also be important. Mechatronics particularly would benefit from such forms of operationally explainable AI systems in which intentions and insights are explainable. Research on such models is also important when working in applications impacted by formal regulations and internationally acceptable standards relevant to transparency. The need for Explainable Artificial Intelligence (XAI) research assumes an important dimension when designing systems based on the concepts of human-centered computation. This is a broader term that also encompasses human emotions, understands the limitations of humans, and ensures that automation works for humans primarily by understanding the local context and explanatory models.

4.9.5 AI-Enabled PdM

PdM is the key to Industry 4.0 and industrial AI systems, providing an overlap between the mechanical and computational parts of mechatronic systems. Such an approach is followed to predict critical failures in the mechatronic system, thereby avoiding unscheduled downtimes and reducing the maintenance costs of an industrial plant. In general, all types of data captured from the normal functioning of the mechatronic system can be used to build predictive models for PdM. Through data preprocessing and AI algorithms, the PdM system can identify the indicators of catastrophic failures and make estimations on when they will occur. Continuous data collection is of utmost importance, as it allows for evidence-based model-based maintenance. The onset of faults can be predicted more accurately if additional sensor data is considered, hence dictating the importance of simultaneously supporting sensor data preprocessing. PdM provides an additional optimization layer for Industry 4.0 paradigms, adopting ML models to predict machine breakdowns. Successful large-scale technologies for PdM deployment can save up to 12% for advanced manufacturing, 30% for the energy and utilities industry, and 20% for the transportation industry. Thus, the adoption of such AI-driven PdM technology offers a significant competitive advantage. We suggest future research on how the outputs of the implementation could be communicated to ensure operator buy-in and create adaptable neural network techniques that can assess values by defining the reliability of a chosen alert. *Complete Data Acquisition*: A mandatory requirement of such a model is the ability to collect data from a wide range of sensors to guarantee that the entirety of the mechanical components is represented. Unlike many maintenance strategies, PdM makes broad inferences from across the whole machinery system, reducing the need for extensive measurement device planning. Routine checks like lubrication system quality, the gear's temperature, and lubricant quality are all collected and modeled in many different studies. These already widely available variables are therefore less intrusive, but with the ability to use them repeatedly could become important in trending not only the gearbox's health with a series of well-acquired bearing sensors but also other unprecedented areas. The complexity of a progressively trained AI, which can link these types of systems and summarize outputs of the mechanical components more coherently, means that increasingly detailed data acquisition expertise is required, as opposed to discrete system assessments. Through experience, in this AI scenario, all system elements are modeled in full, and traditional condition monitoring performance is seen replicated accurately by learning machines similar to this description. Variables that represent the outcomes of AI model outputs from the condition of different tasks can be used to link slowly varying conditions of low-frequency vibrations and suddenly disrupting faults more effectively. Therefore, this represents a need to look at additional data generated both corresponding to those used in a well-monitored subsystem and those that can reflect the indicators typically used to demonstrate whole systems, such as the multitude of available chain sprocket or a cooling fan's data. In addition, it may also allow for the identification of effects on the overall system of subcomponents, which could observe differences in the subsystem that do not have widely interpreted signals. Both types of data sources will be authentic signals that need to be collected to ensure model operation.

4.10 CONCLUSION

The potential and importance of AI have been emphatically shown in the preceding sections. AI's significance in the field of mechatronics is just as well understood and widely studied. Despite the recent increase in popularity, AI is also beset by challenges that are likely to create hurdles for its wide-scale adoption. Low data quality, issues about data interoperability, and ethical considerations are problems that have been addressed, and they underscore the need for extensive research. This chapter has discussed some reasons why AI for mechatronics is experiencing resilience. However, many successful applications can already be found. The ongoing digital transformation of manufacturing and the application of digital technologies in the industrial setting clearly show that AI has the potential to revolutionize industry by providing more flexible and effective approaches to mechatronics. The disruptions from AI can be due to significantly higher productivity and quality in knowledge augmentation because of its ability to understand and react to broad input signals. This holds particularly for the planned convergence with the large market of the IoT, potentially affecting day-to-day life. In its area of application, this can lead to more personally intelligently integrated and more efficient mechatronic systems. Developments in ML will make advances in handling highly dynamic, uncertain, and complex real-world scenes possible. Moreover, progress in sensor technology combined with more sophisticated ML advancements will also have a huge impact on medical diagnostics. Research institutions, universities, knowledge centers, and industrial corporations will be at the forefront of these trends, and companies must, therefore, closely follow the field of AI research, as some of the approaches will become increasingly feasible over the next few years. The current research and applications reflect just the tip of the AI pencil. Ongoing research and technology advances will lead to the development of more integrated and autonomous human–machine collaborative manufacturing environments in the future. The number of human interactions with autonomous systems and co-bots will thus keep rising. Furthermore, AI is going to be a significant component in the integration of the concepts of IoT, Industry 4.0, and cyber–physical systems. Critical infrastructure, reliability, and dependability are important characteristics for the successful crosslinking of information systems. Finally, on the road to human-centered cyber–physical production systems in 2030, it is hypothesized that AI should anticipate the changing context of production according to the cooperation between machines and workers. In conclusion, significant improvements in efficiency and adaptability of mechatronics can be achieved in the future. Consequently, there is every need to pursue basic as well as application research in the field of AI. However, for AI to reach its full potential to revolutionize systems, the challenges of low data quality, lack of interoperable data schemes and standards, and ethical considerations should be addressed. It is hoped that these issues, once tackled, will lead us to a new level of resilience and sustainability in future mechatronic systems.

REFERENCES

1. Priebe, T., Kieseberg, P., Adrowitzer, A., Eigner, O., and Kovac, F. (2023). Sustainability Effects of Robust and Resilient Artificial Intelligence. In: Holzinger, A., Kieseberg, P., Cabitza, F., Campagner, A., Tjoa, A.M., and Weippl, E. (eds.), *Machine Learning and Knowledge Extraction. CD-MAKE 2023*, Lecture Notes in Computer Science, vol. 14065. Springer, Cham. https://doi.org/10.1007/978-3-031-40837-3_12
2. Moskalenko, V., Kharchenko, V., Moskalenko, A., and Kuzikov, B. (2023). Resilience and Resilient Systems of Artificial Intelligence: Taxonomy, Models and Methods. *Algorithms*, 16, 165. https://doi.org/10.3390/a16030165

3. Liagkou, V., Stylios, C., Pappa, L., and Petunin, A. (2021). Challenges and Opportunities in Industry 4.0 for Mechatronics, Artificial Intelligence and Cybernetics. *Electronics*, 10, 2001. https://doi.org/10.3390/electronics10162001
4. Petrilli, A., and Lau, S. (2019). *Measuring Resilience in Artificial Intelligence and Machine Learning Systems*. Carnegie Mellon University, Software Engineering Institute's Insights (blog), accessed October 3, 2023, https://insights.sei.cmu.edu/blog/measuring-resilience-in-artificial-intelligence-and-machine-learning-systems/
5. Galaz, V., et al. (2021). Artificial Intelligence, Systemic Risks, and Sustainability. *Technology in Society*, 67, 101741, ISSN 0160-91X. https://doi.org/10.1016/j.techsoc.2021.101741
6. Guo, K. (2023). Special Issue on Application of Artificial Intelligence in Mechatronics. *Applied Sciences*, 13, 158. https://doi.org/10.3390/app13010158
7. Soori, M., Arezoo, B., and Dastres, R. (2023). Artificial Intelligence, Machine Learning and Deep Learning in Advanced Robotics, A Review. *Cognitive Robotics*, 3, 54–70, ISSN 2667-2413. https://doi.org/10.1016/j.cogr.2023.04.001
8. Gupta, S., Modgil, S., Meissonier, R., and Dwivedi, Y. K. (2024). Artificial Intelligence and Information System Resilience to Cope With Supply Chain Disruption. *IEEE Transactions on Engineering Management*, 71, 10496–10506. https://doi.org/10.1109/TEM.2021.3116770
9. Aljafari, B., Devarajan, G., Subramani, S., and Vairavasundaram, S. (2023). Intelligent RBF-Fuzzy Controller Based Non-Isolated DC–DC Multi-Port Converter for Renewable Energy Applications. *Sustainability*, 15(12), 9425. https://doi.org/10.3390/su15129425
10. Sharma, V., Balusamy, B., Thomas, J. J., and Atlas, L. G. (eds.). (2023). *Data Fabric Architectures: Web-Driven Applications*. Walter de Gruyter GmbH & Co KG.
11. Pushpalatha, N., Murugesan, M, Ravi, A., Pandian, P. P., Kannan, K., and Gunapriya, D. (2024). A Review on Leveraging Reinforcement Learning for Enhanced Decision-Making in Power and Energy Systems. In: *2024 International Conference on Science Technology Engineering and Management (ICSTEM), Coimbatore, India*, pp. 1–8. IEEE. https://doi.org/10.1109/ICSTEM61137.2024.10561146
12. Singh, A., et al. (2022). Blockchain Technology in Biometric Database System. In: *2022 3rd International Conference on Computation, Automation and Knowledge Management (ICCAKM)*. IEEE.
13. Aruchamy, P., Gnanaselvi, S., Sowndarya, D., and Naveenkumar, P. (2023). An Artificial Intelligence Approach for Energy-Aware Intrusion Detection and Secure Routing in Internet of Things-Enabled Wireless Sensor Networks. *Concurrency and Computation: Practice and Experience*, 35(23), e7818. https://doi.org/10.1002/cpe.7818
14. Ryalat, M., Franco, E., Elmoaqet, H., and Almtireen, N. (2024). The Integration of Advanced Mechatronic Systems into Industry 4.0 for Smart Manufacturing. *Sustainability*, 16(19), 8504. https://doi.org/10.3390/su16198504.
15. Sathya, D., and Saravanan, G. (2024). Reinforcement Learning for Adaptive Mechatronics Systems. In: *Computational Intelligent Techniques in Mechatronics*. Scrivener Publishing, pp. 135–184.
16. Cai, W. (2024). Research on the Integration of Artificial Intelligence Technology and Mechatronics Engineering against the Backdrop of Intelligent Manufacturing. *International Journal of Frontiers in Engineering*, 6(1), 9–14. https://doi.org/10.25236/IJFET.2024.060102.

17. Jiménez López, E., and Cuenca Jiménez, F. (2022). Technical Considerations for the Conformation of Specific Competences in Mechatronic Engineers in the Context of Industry 4.0 and 5.0. *Processes*, 10(8), 1445. https://doi.org/10.3390/pr10081445.
18. Ali, M. A., Balamurugan, B., and Sharma, V. (2022). IoT and Blockchain Based Intelligence Security System for Human Detection Using an Improved ACO and Heap Algorithm. In: *2022 2nd International Conference on Advance Computing and Innovative Technologies in Engineering (ICACITE), Greater Noida, India*, pp. 1792–1795. IEEE. https://doi.org/10.1109/ICACITE53722.2022.9823827
19. Adebisi, O., Afolayan, A., Ayoade, I., Adejumobi, P., and Adejumobi, I. (2024). Integration of Deep Learning Techniques in Mechatronic Devices and Systems: Advancement, Challenges, and Opportunities. In: *2024 International Conference on Science, Engineering and Business for Driving Sustainable Development Goals (SEB4SDG), Omu-Aran, Nigeria*, pp. 1–6. https://doi.org/10.1109/SEB4SDG60871.2024.10630414
20. Zaitceva, I., and Andrievsky, B. (2022). Methods of Intelligent Control in Mechatronics and Robotic Engineering: A Survey. *Electronics*, 11(15), art. 2443. https://doi.org/10.3390/electronics11152443.
21. Channi, H. K. (2024). *Role of Data Acquisition, Sensors, and Actuators in Mechatronics Industry*, Intelligent Techniques in Mechatronics. John Wiley & Sons.

5 Fault Diagnosis and Predictive Maintenance Using AI in the Mechatronics Industry

Shadia Yahya Baroud

5.1 INTRODUCTION

In the vanguard of Industry 4.0 (IR 4.0), mechatronics industry (MI) confronts a crucial challenge: equipment reliability and optimization of maintenance. Predictive maintenance (PdM) serves as the backbone of this advanced manufacturing era, to reduce the operational disruptions and prolong the equipment's life (Elahi et al., 2023). In this chapter, we review an application of artificial intelligence (AI) in fault diagnosis and PdM of mechatronics or smart maintenance systems.

Since advanced mechatronic systems have become widespread in many industries, maintenance strategies should be efficient, proactive, and feasible. AI technologies, such as machine learning (ML)/deep learning (DL), lead this irreversible downturn into unprecedented precision in fault diagnosis, enabling PdM (Arinez et al., 2020). In this chapter, we unravel the complexities of integration of AI within the mechatronics sphere and its critical role in the augmentation of diagnostic accuracy and predictive capability.

Data collection forms the foundation to the digital journey of Industry 4.0 beginning with data harvest followed by pouring it into intelligent analysis using AI. The applications of Industrial Internet of Things (IIoT) in smart factories enable complete real-time monitoring of manufacturing processes as part of the connected infrastructure of technologies and resulting in a closed-loop system.

5.2 INDUSTRY 4.0

The last ten years can be characterized as the era of manufacturing digital transformation, also known as Industry 4.0. As illustrated in Figure 5.1, this evolution includes cloud computing, AI, ML, cyber–physical systems (CPS), sensors and software technology, enhanced connectivity technology like "5G and 'Wi-Fi6,' IoT, digital twins, blockchain, and robotics with the impact of Industry 4.0 on the digitization of manufacturing" (Baroud et al., 2024; Rüßmann et al., 2015). These innovations are revolutionizing classical manufacturing strategies, improving performance, flexibility, and agility, and contributing to the realization of the interconnected and smart factories.

DOI: 10.1201/9781032615790-5

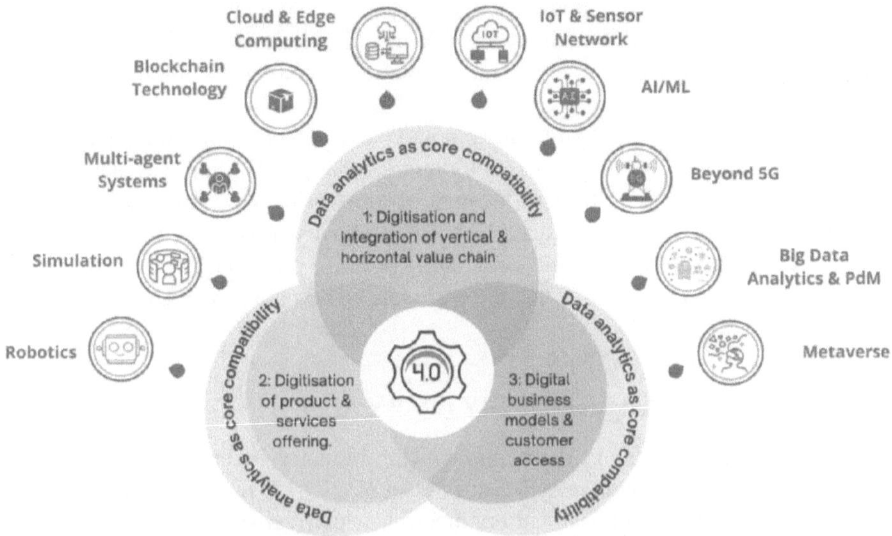

FIGURE 5.1 The 10th crucial pillars of Industry 4.0. (Adapted from Oztemel and Gursev, 2020.)

In the light of Industry 4.0, it is important to discuss the manufacturing strategies adopted by manufacturers, integration issues and trends affecting the manufacturing sector (Baroud et al., 2023; George et al., 2023; Manufacturing and Trends, 2021).

Industry 4.0 propels the manufacturing sector into an era marked by autonomous information exchange, advanced machine control, and interoperable production systems (Lemstra and de Mesquita, 2023). One of its fundamental objectives is to establish connectivity and integration within production environments (DIN and DKE, 2023), facilitating the creation of extensive data footprints via sensors and machine monitoring. However, challenges are inherent in the existing manufacturing frameworks, particularly due to centralized control systems and data heterogeneity arising from diverse manufacturing vendor products (Haricha et al., 2023; Lu et al., n.d.). Centralized control structures may impede the seamless integration of Industry 4.0 technologies, leading to compatibility issues with legacy systems (Lattanzi et al., 2021). Moreover, the mixed nature of data produced by various vendors complicates the standardization of communication and data interoperability within the digital framework, and thus the successful execution of Industry 4.0 endeavors (Tao et al., 2019).

5.3 PREDICTIVE MAINTENANCE

For industries or processes that could not afford unscheduled downtime, like utilities, transportation, power generation, public and emergency services, etc., there is now much more traffic of PdM (Baroud et al., 2024; Pech et al., 2021). In the area of production, maintenance, and inventory, it is planning strategy that is essential for strategic planning. Still, it is not always possible everywhere because of technological and logistical factors (Nguyen and Medjaher, 2019). PdM plays a crucial role in the production environment due to unpredictable machine failures causing schedule displacement, customer delivery delay, and employee overworking (Elahi et al., 2023; Pech et al., 2021).

Through continuous system monitoring and early detection of faults using historical datasets and ML, along with visual indicators of wear (Guillen Rosaperez, 2020; Paolanti et al., 2018), PdM utilizes a range of tools prioritized for timely maintenance. PdM has been designed to meet the current Industry 4.0 requirements, aiming to reduce costs, eliminate waste production, and minimize major failure rate (Ayvaz and Alpay, 2021). Coon et al. also divide the experience and data antecedents of maintenance (Farooq et al., 2020).

5.4 EVOLUTION OF INDUSTRY 4.0 BY ADAPTING MI

Mechatronics is a keystone of the manufacturing industry, meaningfully influencing its advancements and processes (Riesener et al., 2023). It represents an integrated approach that combines mechanical engineering, electrical systems, electronics and intelligent computer control (Adegbite et al., 2023; Shetty and Kolk, 2011). This integration aims to design and manufacture robust, cost-effective, reliable, adaptable, and versatile industrial products and processes (Nnodim et al., 2021). Since its establishment by Japanese engineer T. Mori of Yaskawa Electric Corporation in 1969, mechatronics has seen significant evolution. The name "mechatronics," combining "mecha" for mechanical and "tronics" for electronics, represents the integration of these fields with information technology (IT) (Habib, 2007).

The mechanical innovations in fields like machine tools and power generation seek to diminish or, preferably, eradicate physical strain on humans (Kuru and Yetgin, 2019). Conversely, advancements in electronics and IT focus on reducing mental strain. Mechatronics, thus, offers relief from both physical and mental tasks in various life aspects. Its philosophy lies in emerging technologies from traditional interdisciplinary domains, including computer science and engineering, to create comprehensive, standalone systems within new products. As mechanical systems began incorporating microelectronics and IT, new potentials emerged for mechanical design and automated functions (Alciatorc and Histand, 2019).

General Motors' development of Programmable Logic Controllers (PLCs) in 1968 marked a turning point in industrial automation (Bolton, 2006). Together with control systems as well as with other algorithms, for instance, the proportional-integral-derivative (PID), supervisory control and data acquisition (SCADA),it catalyzed a significant transformation in the automation landscape of the industrial sector (Jerker, 2017).

Mechatronics, a field that intertwines various disciplines, is progressing rapidly, significantly influencing different industries' futures. In their study, Nnodim et al. (2021) thoroughly examine emerging trends in mechatronics, underscoring the growing complexity and interdisciplinary nature of engineering system development. These revolutionize mechatronics with their focus on significant features in computer engineering, simulation, modeling, power electronics, electromechanical motion devices, micro-electro-mechanical systems (MEMS), microprocessors, and distributive system platforms (DSPs). This work stresses modeling, analysis, simulation, virtual prototyping, and visualization in the development of advanced mechatronic systems. Additionally, we have seen this trend in sectors like automotive, robotics, aerospace, and consumer goods where there is a need for versatile and scalable products.

Integration of software, electronics, and mechanical components in mechatronics has made it possible to develop innovative and interdisciplinary systems that are effective and more adaptable than ever before to address the needs of modern industries (Adegbite et al., 2023).

5.5 CHAPTER FLOW

Since then, engineers and researchers have identified significant advancement in computer engineering, simulation and modeling, power electronics, electromechanical motion devices, MEMS, microprocessors, and DSPs that they have reinvented mechatronics. The study is a call for the use of models, analysis, simulations, and unique forms of prototyping and visualization in designing complex mechatronics systems. This is most apparent in those industries that have a high incidence of requiring a large number of jobs, especially those that are both very flexible and very scalable following the rules and expectations of the greater economy: automotive, robotics/aerospace, even the consumer marketplace, for instance.

The ability to create innovative, interdisciplinary systems, which are more effective, adaptable, and can meet the diverse needs of the modern industries arises from the need to integrate software, electronics, and mechanical components within mechatronics (Adegbite et al., 2023).

The applications of fault diagnosis algorithms will be explored in the next sections along with the evolution of AI algorithms that utilize data from system behavior, sensor feedback, and historical maintenance records, therefore accelerating this fault diagnosis process. AI-powered tools have reduced total downtime and improved system reliability and productivity (Aldrini et al., 2023; Aliyu et al., 2022; Contreras-Valdes et al., 2020; Elahi et al., 2023). Moreover, in this chapter, how to implement the AI-based PdM outlined in Chapter 1 will also be shown. When applied for accurate prediction of various remaining useful life (RUL) of the mechatronic components involved in a system, it redefines PdM, the scheduling of maintenance, and the optimal utilization of resources. We try to investigate using an example derived from practice whether AI can improve the resilience or productivity of mechatronic systems and talk about key issues and trends of AI development in this promising and rapidly developing field.

5.6 ADVANCEMENTS IN AI FOR FAULT DIAGNOSIS

The current phase of manufacturing has been triggered by Industry 4.0 that has disrupted the flow of production, which is based on PdM powered by AI and ML (Minh et al., 2022). The evolution of the PdM is particularly important for its ability to predict equipment failures in advance of their occurrence. The process here consists of collecting and assessing data from a number of sensors to determine patterns which suggest potential misbehavior of the equipment. This paper contends that PdM can lead to improvements in production quality, system control, cost management, and machinery or systems downtime reduction (Arafat et al., 2024).

With the infusion of AI into the mechatronics sector, fault diagnosis underwent a revolution. Fitted to the field, ML/DL models now accurately and rapidly predict equipment failures in the PdM space, enabling an era of PdM that brings higher uptime and efficiency (Aliyu et al., 2022; Yu and Zhang, 2023).

5.7 ML ENHANCEMENTS

The later advancement or simply advancements rely at the center on ML algorithms that have shown to be capable of identifying failure patterns and potential malfunctioning before they lead to system breakdowns (Pandey, 2021). In developing supervised learning models including Random Forests for the prediction of a machine failure and learning algorithms such as Support Vector Machines, large datasets have been analyzed to identify tiny flaws in the operation of the machine that may lead to the later failure. Unsupervised learning approaches such as clustering and neural networks (NNs) are also being increasingly used to detect new fault types with no labeled data, expanding the diagnostic capabilities beyond known issues (Fernandes et al., 2022). A comprehensive categorization of ML categories and models applied to maintenance-oriented projects (Nacchia et al., 2021) is shown in Figure 5.2.

The effectiveness of current diagnosis algorithms of faults in ML systems is, however, much improved at this time. In particular, Coelho et al. (2022) proposed a novel approach to quantify the duration of the stoppages of staking press lines using (a) the

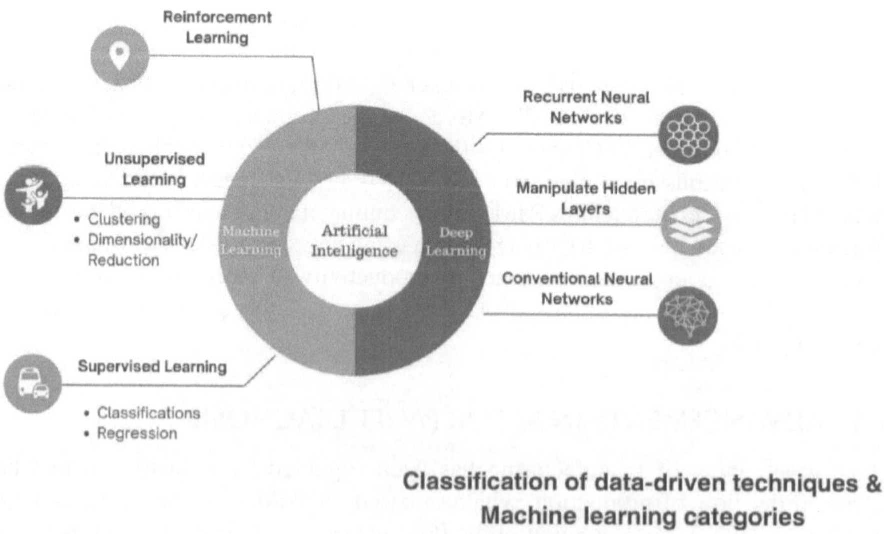

FIGURE 5.2 ML and DL algorithms. (Achouch et al., 2022.)

time slices and feature extraction, accompanied by anomaly detection, and (b) ML classifier algorithms. Random Decision Trees (RDT) was also the best among the 12 classifiers with the Receiver Operating Characteristic (ROC) Area Under the Curve (AUC) index of (96%) increasing the macro F1-score by (22.971%) from the individual classification method (b). The analysis was conducted on a dataset of (13,568) real-world cases categorized into two binary classes: normal state of operation and the faulty state.

5.8 DL BREAKTHROUGHS

Hailed as a subclass of ML, DL uses multilayered NNs to deliver a more robust fault diagnosis system.

To this end, convolutional neural networks (CNNs) are very commonly used to process and interpret involved sensor data into actionable insights. As demonstrated by these models, they are particularly successful at feature extraction and autonomously uncovering minute details within high dimension data that cannot be visualized or dealt with by traditional computational methods (Yu and Zhang, 2023).

Hadi et al. (2023) explored AutoML-based enhancing techniques of fault classification using AutoML techniques in the IIoT for PdM. The study introduced two models, PyCaret and Auto Deep Neural Network (AutoDNN), which were implemented to find ball bearing faults, done with Auto Keras. The models were used under the Case Western Reserve University bearing faults dataset and resulted in high accuracy, recall, precision, and F1 score. This research demonstrates extraordinary exploitation by AI in IIoT applications, especially in the areas of manufacturing and energy, of the process optimization for PdM and minimizing time and associated costs.

5.9 INTEGRATION WITH MECHATRONIC – REAL-WORLD APPLICATION IN MECHATRONICS SYSTEMS

Mechatronic system is integrated with AI-based model based on advancements in sensor technology and data processing power (Ryalat et al., 2024). The necessary data granularity is provided by high-fidelity sensors and provides the ability to process and analyze real-time data or the CPS. By virtue of working together, this process provided a more global and profound look at health and behavior of the system, creating more dependable and faster fault diagnosis. Figure 5.3 represents a schematic of the CPS manufacturing structure.

For "spacecraft fault detection, isolation and recovery" (FDIR), Hülsmann and Förstner (2020) investigated the use of AI and ML. The research study identified existing FDIR strategies and has developed an original predictive technique, using modeling, explainable AI, and multiple ML techniques, to assess telemetry analysis and detect anomalies. The study focused on AI's possibility to enhance systems, as well as the crucial importance of interpreting AI to improve the efficiency and autonomy of spaceships with the help of predictive FDIR. In addition, the study addressed these challenges – restricted data availability and difficulty of applying DL models – to propose data-driven and expert knowledge-based strategies for FDIR implementation.

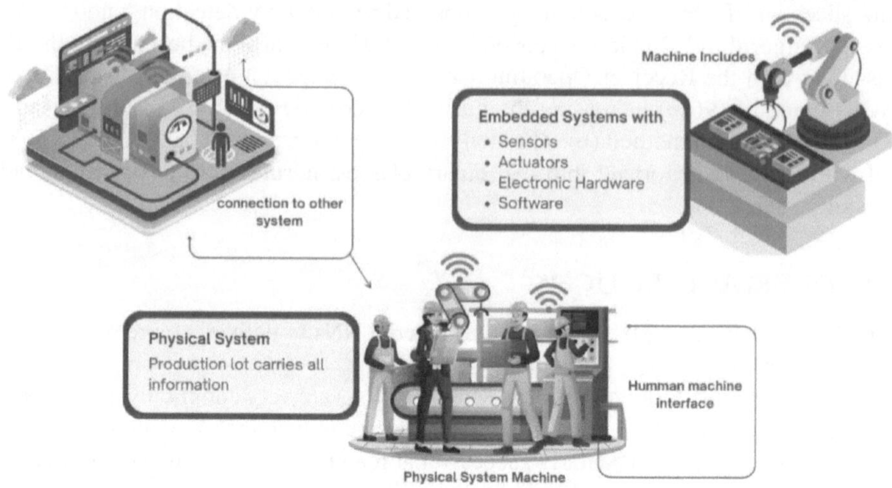

FIGURE 5.3 CPS's manufacturing structure. (Adapted from Alcácer and Cruz-Machado, 2019.)

The work of Zaghdoudi et al. (2022) is based on the development of a predictive analysis methodology which has been developed mainly for industrial systems, and in particular to the forecast of supplier delays. It is a technique of collecting and analyzing reliable industrial data aiming at producing a model that could help in decision-making and increase efficiency. They argue their work proves that the analysis and prediction are applicable in other than conventional systems of mechatronics industrial sectors. In this regard, they employ the ML techniques to sketch a roadmap of industrial improvement and flexibility, thus emphasizing the large-scale applicability and fungibility of the very predictive approaches in the industrial context.

5.10 DATA PROCESSING AND SENSORS TECHNOLOGY

Pivotal in supporting AI-driven fault diagnosis (Lee, 2023) are advances in sensor technology and data processing used.

In the study of Ozkat et al. (2023), they developed a novel feature, mean peak frequency (MPF) derived from vibration signals spectrograms, and adapted a ML method to predict RUL of machines. For calculating RUL, they predicted MPF values by using a long short-term memory (LSTM) and a set threshold was also used. Their method was validated under three experimental conditions: varying throttle levels, blades condition, and produced RUL estimates of "4s," "10s" and "10s", and the corresponding root mean square error (RMSE) results of "3.7142 Hz," "1.4831 Hz," and "1.3455 Hz." The study also pointed out that their only focus was on one flight operation: hover, and one feature, MPF. Consequently, they suggested that other features or some modeling technique be employed to compute the forecasting range in subsequent analyses, for example, real-life applications or operation (like manipulating turns or false signals).

The introduction of sensors interfaces introduces enormous complexity to mechatronic systems. In Ali et al. (2022), the challenges of designing and deploying adaptable sensor interfaces for industrial use are explored. The challenge is to establish circuits to carry high voltage and acquire and process data reliably, also with compatibility and scalability. The issues are addressed with a cross-disciplinary strategy that requires knowledge from electronics, software engineering, and industrial design.

5.11 CHALLENGES IN AI ADOPTION

Overall, trial and error still represent the most common approach for implementing AI in fault diagnosis (Liagkou et al., 2021). However, relatively large amounts of high-quality data is still a significant barrier to data quality and availability (Zhao et al., 2021). Additionally, the interpretability of AI decisions is also an issue that stakeholders need to put their trust in the AI's "black box" outputs but misled by its underlying decision-making process (Lemstra and de Mesquita, 2023).

Nevertheless, there are still challenges on the road to AI integration with fault diagnosis (Zhao et al., 2021). Fang et al. (2023) in their study on autonomous vehicles found that data quality, volume requirements, and the interpretability of AI decisions are hurdles for that technology. In particular, the work presents an interpretable fault diagnosis approach with an Autoencoder-based Latent Dynamics for Sensor Anomaly Estimation (ALDSAE) model for anomaly detection and a residual explainer for post-diagnosis interpretation. It shows how environmental effects influence sensor data and how this is connected to the results of anomalous behavior measurement. From accuracy, the ALDSAE outperforms traditional detectors, and the explainer is as fast, practical, and important as feature analysis can be.

Effective human robot interaction integration in industrial settings is a major challenge within the field of mechatronics (Mukherjee et al., 2022; Vemuri and Thaneeru, 2023). Rodriguez-Guerra et al. (2021) highlight the challenges in building these robotic operating at the same time as workers interact in safe and efficient ways on the shop floor. The problem is to design collaborative systems that are secure and efficient. This therefore calls for proper understanding of the dynamic constraints of human–robot interactions and the development of systems that may easily adapt to human presence and conduct.

5.12 CONCLUSION

In particular, recent advances in AI have redefined the capabilities of fault diagnosis in MI. The change toward intelligent, data-driven maintenance practice is a paradigm shift of sorts in manufacturing mimicking Industry 4.0 goals. However, ongoing research, development and worldwide industry adoption of these AI applications, and the resulting advances are crucial to realizing the full and maximum potential of smart maintenance strategies.

Shetty et al. (2022) analyzes the impact of recent technological advancements and their contributions in cost reduction, shorter product development time, and reduced prototyping time and feed forward manufacturing in various fields. For instance, development of unmanned smart aircraft leads to a reduction of time as well as costs besides minimizing errors and reduction of effects on environment. Furthermore, the application of the capabilities of AI in the security systems, for example, perimeter defense and threat detection, has greatly increased defense capabilities. Consequently, these advancements are a clear testament to the critical role mechatronics plays in modern warfare and aerospace technology, as precision, efficiency, and innovation play a decisive role.

5.13 PDM TECHNIQUES BASED ON AI

The utilization of AI within the MI differs from the conventional maintenance in the ways that it represents a radical transformation (Chomklin et al., 2023). This section examines how AI-driven predictive analytics and predictive prognostics are changing maintenance strategies by optimizing resource utilization as well as extending equipment lifecycle (Rojek et al., 2023).

5.13.1 AI-Driven Prognostics for PdM

In prognostics, AI's positioning is a prediction of the RUL of equipment, the cornerstone of a robust PdM. In the present work, Rojek et al. (2023) investigated the application of AI for PdM in industrial conditions. The method was developed from a computational analysis and simulation on several sets of real-world industrial data. The results support the use of preventive maintenance (PM) with large-scale documented sensor data and well-trained ML algorithms. The study shows that the most basic of AI solutions and applications can generate high productivity at low implementation cost, optimizing materials, energy, and water usage. For various industries, this approach is an increasingly important one, as it will improve maintenance efficiency and decrease costs.

One such example is a work of Orlowska-Kowalska and Wolkiewicz (2022) to apply ML to PdM on hydroelectric power plants specifically for turbine load cycle management. Researchers developed a model specific to the relationship between the turbine's load cycle and four different ML algorithms. The research confirmed that PdM in hydroelectric plants is promising when we apply ML.

5.13.2 Maintenance Schedules Optimization

Further, PdM empowers the optimum maintenance schedules using the power of AI. Through accurate forecasting of when maintenance is required, AI moves companies from fixed interval maintenance to condition-based schedules. Not only does this preserve sources, but it also prevents disruption to operations. Examples such as Azab et al. (2021) indicate that ML algorithms or models can solve maintenance tasks while balancing the workload, solve balancing the workload and reducing the number of excessive interventions in flow-shop scheduling.

5.13.3 Enhancing Equipment Life Cycle

An important benefit of an AI in PdM is that it can extend equipment life cycles. Timely maintenance interventions, enabled by AI, reduce the wear and tear to such extent that the durability as well as the operational life period of machinery is extended (Elahi et al., 2023).

In the case of the hydroelectric power plants, Vallim filo et al. (2022) propose an ML application in a PdM framework for equipment load cycle, being applied to the turbine. To industrialize the predictive model, the researchers-built variables related to the turbine load cycle. For four ML algorithms, they explored what the most important variables were for each. In this case, the framework was applied and the ratio on maintenance prediction has an accuracy of up to 98%. Overall, they stabilize the use of ML in PdM inside industrial environments with this approach.

5.13.4 Obstacles to AI Integration in PdM

AI is very helpful in extending equipment life cycles. Thus, AI is used only for the mitigation of excessive wear and tear of machinery through timely maintenance interventions such that they can be made to be more durable and prolong machinery operational life (Elahi et al., 2023).

According to Vallim Filho et al. (2022), the turbine's load cycle is proposed as a PdM context using ML for equipment in hydroelectric power plants. With variables interrelated to the turbine load cycle, the researchers industrialized a predictive model. Wherein they explored four ML algorithms and identified the motivation variables as the most important for each. The framework is applied to a real-world case and high accuracy rates around 98% are achieved for maintenance prediction. This approach implies the efficiency of utilizing ML in PdM of industrial systems.

Despite the benefit it can provide for PdM, implementing AI is difficult. Notable issues and challenges are performance, or the need for expert personnel to understand AI outcomes, data quality, as well as integration with existing systems. This is dovetailed by Raza (2023) as well, that such challenges are to be overcome for the adoption of AI in PdM to be successful.

5.13.5 Conclusion

The most important virtue of AI in PdM is the prolongation of equipment life cycles. AI contributes through timely maintenance interventions, to reduce excessive wear to extend the machinery durability and operational life (Contreras-Valdes et al., 2020). Using AI to do the advance analyzing, companies and factories not only predict failures, but they could also optimize schedules of machinery maintenances, increasing its life cycle in the overall process. Further refinements and improvements of these techniques seem to be in the hands of continued advances in AI technologies (Schmidt, 2023).

5.14 CASE STUDY: IMPLEMENTING PREDICTIVE ANALYTICS IN "BENTELER AUTOMOTIVE'S LIGHTHOUSE FACTORY"

5.14.1 Introduction

The trends and popularization of predictive analytics is the subject of this case analysis in the framework of the "BENTELER Automotive" and the BOOST 4.0 project;

von Enzberg et al. (2020) on smart maintenance. This is because BENTELER Automotive is specialized in the automotive, energy, and mechanical engineering business through delivering safety critical parts. The biggest aim here was to go ahead and start applying PdM techniques so as to improve the efficiency and dependability of some critical applications such as hydraulics and conveyors.

5.14.2 PdM in Hydraulic Press Systems

- *Challenge and Solution:* The hot forming line had problems of oil leakage from the hydraulic press that was essential in shaping sheet metal. A PdM solution, relying on advanced analytics, is implemented to react and predict possible leakages.
- *Outcomes:* As a result, this smart maintenance strategy enabled faster, more effective maintenance interventions and greatly reduced dependency on manual diagnosis.

5.14.3 Enhancing Conveyor Belt Operations

- *Operational Focus:* With scrap metal moving only along the scrap belt, any pile-up would lead to operational halts. Predictive analytics were used to identify early signs of potential disruptions.
- *Results:* These diagnostic algorithms had been successfully implemented preventing stoppages and increasing response times in maintenance.

5.14.4 Analyzing the Impact and Outcomes

- *Performance Improvements:* The integration of these PdM solutions notably improved overall production efficiency.
- *Quantitative Benefits:* The effectiveness for the implementation of this method is evidenced by means like the mean time to repair rates, the mean time between failures, and a proper utilization of the available resources in maintenance. These outcomes resulted in significant enhancements in operational efficiency, cost savings, and prolonged equipment lifespan.

5.14.5 Lessons Learned and Best Practices

- *Knowledge Sharing:* The case study showed the benefit of having domain expert on hand and good communication between maintenance engineer, its specialist, and data scientists.
- *Methodology:* The adoption of semi-formalized methods such as CONSENS was very helpful when it came to structured knowledge transfer.
- *Holistic Approach:* The study emphasized the importance of a holistic approach – combining domain knowledge with data analysis and practical application of PdM systems – to make PdM systems more effective.

Consequently, this is a detailed analysis of "BENTELER Automotive's lighthouse factory" for practical evidence of how AI and predictive analytics are changing maintenance practices in MI. Case study outcomes documented in this case study show that smart maintenance solutions are able to significantly enhance the efficiency, reduce costs by significant amounts, and ensure greater reliability of equipment.

5.15 CHALLENGES AND EMERGING RESEARCH DIRECTIONS

As AI and the Industry 4.0 are constantly transforming MI, several problems, challenges, and opportunities of future development have emerged (Bousdekis and Mentzas, 2021). This section focuses on the main issues related to employment of AI technologies in computer technologies for PdM with the notable lines of the development in the sphere of MI.

5.16 CHALLENGES IN AI IMPLEMENTATION

1. *Data Quality, Volume, and Availability:* A major issue when using AI for fault diagnosis and PdM is the requirement for large volumes of high-quality data (Pandey, 2021). Accurate, well-structured data is a major factor that the performance of AI models relies on. Lack of complete and coherent data can significantly degrade predictive performance that underpins AI-enabled maintenance solutions to reduce their efficiency and reliability.
2. *Existing Systems and Advanced Technologies Integration:* Another significant challenge is integrating AI technologies with legacy systems in manufacturing. Many mechatronic systems in operation today were not designed with AI compatibility in mind. As a result, retrofitting these systems to accommodate AI can be complex, time-consuming, and costly (Schmidt, 2023).
3. *Skill Gap and Workforce Development:* AI integration into PdM also calls for professional human resources capable of making, deploying, and analyzing AI models. Such approaches entail a demand for training programs and educational activities required to produce maintenance engineers and technicians, as well as the required AI knowledge (Suganob et al., 2022).
4. *Interpretability and Trust:* The problems with the particular AI models are as follows: dilemma of complexity, controllability, and their black-box feature (Zhu et al., 2023). More often, to receive end-users' and stakeholders' confidence, trustworthy AI solutions based on explainable and interpretable models should be used to understand the reasoning of AI models (Falcone et al., 2022).
5. *Data Privacy and Cybersecurity:* Advanced sensors and actuators, coupled with ever-increasing demands for operational integrity and operator safety, necessitate robust data privacy and cybersecurity. As such, security concerns arise; therefore, manufacturers tend to avoid integrating open-source platforms with their current systems and outsourcing data processing. Advanced technologies must be used on these platforms to keep the availability of data secure, as well as a close cooperation between the enterprise's sensors and systems is a prerequisite to ensure successful integration, thus securing the process (Bousdekis and Mentzas, 2021).

5.17 FUTURE RESEARCH TRENDS

1. *Emerging Innovations:* Some current trends include application of edge and cloud computing (with real-time analytics) to improve PdM performance in remote areas and Explainable AI (XAI) for a better understanding of AI to enhance PdM in various ways, and the use of AI to interconnect with other even more developed autonomous platforms and robotics towards the achievement of maximum maintenance productivity is current (Brusa et al. 2023).
2. *Advancements in AI Algorithms:* Continued research in AI is expected to yield more advanced and efficient algorithms. Future advancements may include enhanced DL models and real-time adaptive learning proficiencies, with algorithms requiring minimal data for accurate predictions. Integrating AI-powered PdM with advanced robotics and autonomous systems promises to create even more efficient and responsive maintenance solutions.
3. *Cross-Industry Applications:* Exploring the application of AI in PdM across various sectors within the MI can lead to the discovery of universal best practices and innovative solutions. Cross-industry collaborations may yield insights that drive the domain forward.
4. *Integration and Interoperability:* Breakdowns always exist in every manufacturing system. Multi Agent Systems (MAS) are used to manage breakdown probability reduction in the manufacturing industries by coordinating and being smart in performing PdM (Baroud et al., 2024; Singh et al., 2023).
5. *Ethical and Responsible AI:* It is expected that as AI permeates more phases of manufacturing, particularly in its applications of robotics, ethical considerations and how AI is being used ethically will become more significant. This research will ensure that AI applications will be transparent, fair, social, and the values aligned.

5.18 CONCLUSION AND FINAL REMARKS

The integration of AI technologies will be essential to how the MI can execute its journey into smarter, AI-driven manufacturing. AI technologies keep evolving and can introduce PdM applications to the MI resulting in a more efficient, reliable, and sustainable manufacturing situation.

In the following chapter, a closer examination of the PdM applications of AI and ML/DL across numerous industries is made. Other research findings demonstrate that these technologies can obtain high system efficiency, raise reliability, and decrease reliance on costs.

Significantly, the chapter helps in understanding how AI-driven PdM can transform industrial operations. The publication underlines the significance of data quality, cybersecurity, and systems integration of complex systems, providing practical advice and innovative solutions to multiple fields.

This chapter is rather practical and has important consequences for numerous types of industries. Last but not the least, this chapter also demonstrates how AI and ML/DL have the capability to enhance the maintenance approaches and, in the process, minimize system unavailability with maximum overall performance. These insights can help industries to use these insights to adopt more sustainable, efficient, and smart maintenance techniques to adapt to the changing need of the contemporary industrial landscape.

REFERENCES

Achouch, M., Dimitrova, M., Ziane, K., Sattarpanah Karganroudi, S., Dhouib, R., Ibrahim, H., and Adda, M. (2022). On Predictive Maintenance in Industry 4.0: Overview, Models, and Challenges. *Applied Sciences (Switzerland)*, *12*(16), 1–22. https://doi.org/10.3390/app12168081

Adegbite, A. O., Nwasike, C. N., Nwaobia, N. K., Gidiagba, J. O., Enabor, O. T., Dawodu, S. O., Adefemi, A., and Ezeigweneme, C. A. (2023). Mechatronics in Modern Industrial Applications: Delving into the Integration of Electronics, Mechanics, and Informatics. *World Journal of Advanced Research and Reviews*, *20*(3), 1179–1197. https://doi.org/10.30574/wjarr.2023.20.3.2561

Alcácer, V., and Cruz-Machado, V. (2019). Scanning the Industry 4.0: A Literature Review on Technologies for Manufacturing Systems. *Engineering Science and Technology, an International Journal*, *22*(3), 899–919. https://doi.org/10.1016/j.jestch.2019.01.006

Alciatore, D. G., and Histand, M. B. (2019). *Introduction to Mechatronics and Measurement Systems* (5th ed.). McGraw-Hill Education.

Aldrini, J., Chihi, I., and Sidhom, L. (2023). Fault Diagnosis and Self-healing for Smart Manufacturing: A Review. *Journal of Intelligent Manufacturing*, 35, 2441–2473. https://doi.org/10.1007/s10845-023-02165-6

Ali, M., Hassan, A., Honarparvar, M., Nabavi, M., Audet, Y., Sawan, M., and Savaria, Y. (2022). A Versatile SoC/SiP Sensor Interface for Industrial Applications: Implementation Challenges. *IEEE Access*, *10*, 24540–24555. https://doi.org/10.1109/ACCESS.2022.3152379

Aliyu, R., Mokhtar, A. A., and Hussin, H. (2022). Prognostic Health Management of Pumps Using Artificial Intelligence in the Oil and Gas Sector: A Review. *Applied Sciences (Switzerland)*, *12*(22), 11691. https://doi.org/10.3390/app122211691

Arafat, M. Y., Hossain, M. J., and Alam, M. M. (2024). Machine Learning Scopes on Microgrid Predictive Maintenance: Potential Frameworks, Challenges, and Prospects. *Renewable and Sustainable Energy Reviews*, *190*(Part A), 114088. https://doi.org/10.1016/j.rser.2023.114088

Arinez, J. F., Chang, Q., Gao, R. X., Xu, C., and Zhang, J. (2020). Artificial Intelligence in Advanced Manufacturing: Current Status and Future Outlook. *Journal of Manufacturing Science and Engineering, Transactions of the ASME*, *142*(11), 1–16. https://doi.org/10.1115/1.4047855

Ayvaz, S., and Alpay, K. (2021). Predictive Maintenance System for Production Lines in Manufacturing: A Machine Learning Approach Using IoT Data in Real-time. *Expert Systems with Applications*, *173*(November 2020), 114598. https://doi.org/10.1016/j.eswa.2021.114598

Azab, E., Nafea, M., Shihata, L. A., and Mashaly, M. (2021). A Machine-Learning-Assisted Simulation Approach for Incorporating Predictive Maintenance in Dynamic Flow-Shop Scheduling. *Applied Science*, *11*(24), art. 11725. https://doi.org/10.3390/app112411725

Baroud, S. Y., Yahaya, N. A., and Elzamly, A. M. (2023). Spare Parts Management Analytics for Smart Maintenance: A Review on Dealing with Challenges to Warranty Profitability. *AIP Conference Proceedings*, 2808. https://doi.org/10.1063/5.0133121

Baroud, S. Y., Yahaya, N. A., and Elzamly, A. M. (2024). Cutting-Edge AI Approaches with MAS for PdM in Industry 4.0: Challenges and Future Directions. *Journal of Applied Data Sciences*, *5*(2), 455–473. https://doi.org/10.47738/jads.v5i2.196

Bolton, W. (2006). Programmable Logic Controllers. *Programmable Logic Controllers* (4th ed.), 1–292. Newnes. https://doi.org/10.1016/B978-0-7506-8112-4.X5018-9

Bousdekis, A., and Mentzas, G. (2021). Enterprise Integration and Interoperability for Big Data-Driven Processes in the Frame of Industry 4.0. *Frontiers in Big Data*, *4*(June), 1–18. https://doi.org/10.3389/fdata.2021.644651

Brusa, E., Cibrario, L., Delprete, C., and Di Maggio, L. G. (2023). Explainable AI for Machine Fault Diagnosis: Understanding Features' Contribution in Machine Learning Models for Industrial Condition Monitoring. *Applied Sciences (Switzerland)*, *13*(4), 2038. https://doi.org/10.3390/app13042038

Chomklin, A., Jaiyen, S., and Wattanakitrungroj, N. (2023). A Survey of AI Techniques Based on Predictive Maintenance in Lean Manufacturing. *Science, Technology, and Social Sciences Procedia*, *2023*(4), 1–14.

Coelho, D., Costa, D., Rocha, E. M., Almeida, D., and Santos, J. P. (2022). Predictive Maintenance on Sensorized Stamping Presses by Time Series Segmentation, Anomaly Detection, and Classification Algorithms. *Procedia Computer Science*, *200*(2019), 1184–1193. https://doi.org/10.1016/j.procs.2022.01.318

Contreras-Valdes, A., Amezquita-Sanchez, J. P., Granados-Lieberman, D., and Valtierra-Rodriguez, M. (2020). Predictive Data Mining Techniques for Fault Diagnosis of Electric Equipment: A Review. *Applied Sciences (Switzerland)*, *10*(3), 1–24. https://doi.org/10.3390/app10030950

DIN and DKE, German Standardization Roadmap Industrie 4.0, 5th ed., Berlin/Frankfurt, Germany: DIN e. V. and DKE, Apr. 2023. [Online]. Available: www.dke.de/en/areas-of-work/industry/standardization-roadmap-industry-40

Elahi, M., Afolaranmi, S. O., Martinez Lastra, J. L., and Perez Garcia, J. A. (2023). A Comprehensive Literature Review of the Applications of AI Techniques through the Lifecycle of Industrial Equipment. *Discover Artificial Intelligence*, *3*(1), 43. https://doi.org/10.1007/s44163-023-00089-x

Falcone, Y., Faqrizal, I., and Salaün, G. (2022). Probabilistic Analysis of Industrial IoT Applications. *Proceedings of the 12th International Conference on the Internet of Things (IoT)*, pp. 41–48. [Online]. Available: https://doi.org/10.1145/3567445.3567461

Fang, Y., Min, H., Wu, X., Lei, X., Chen, S., Teixeira, R., and Zhao, X. (2023). Toward Interpretability in Fault Diagnosis for Autonomous Vehicles: Interpretation of Sensor Data Anomalies. *IEEE Sensors Journal*, *23*(5), 5014–5027. https://doi.org/10.1109/JSEN.2023.3236838

Fanibhare, S., Sarkar, N. I., and Al-Anbuky, A. (2021). Smart Manufacturing and Tactile Internet Based on 5G in Industry 4.0: Challenges, Applications and New Trends. *Electronics*, *10*(24), 3175. https://doi.org/10.3390/electronics10243175

Farooq, B., Bao, J., Li, J., Liu, T., and Yin, S. (2020). Data-Driven Predictive Maintenance Approach for Spinning Cyber-Physical Production System. *Journal of Shanghai Jiaotong University (Science)*, *25*(4), 453–462. https://doi.org/10.1007/s12204-020-2178-z

Fernandes, M., Corchado, J. M., and Marreiros, G. (2022). Machine Learning Techniques Applied to Mechanical Fault Diagnosis and Fault Prognosis in the Context of Real Industrial Manufacturing Use-cases: A Systematic Literature Review. *Applied Intelligence*, *52*(12), 14246–14280. https://doi.org/10.1007/s10489-022-03344-3

George, A. S., Hovan George, A. S., and Baskar, T. (2023). Edge Computing and the Future of Cloud Computing: A Survey of Industry Perspectives and. *Partners Universal International Research Journal*, *02*(02), 19–44. https://doi.org/10.5281/zenodo.8020101

Guillen Rosaperez, D. A. (2020). Self-Learning Methodology for Failure Detection in an Oil-Hydraulic Press: Predictive maintenance. M.S. thesis, School of Electrical Engineering and Computer Science (EECS), KTH Royal Institute of Technology, Stockholm, Sweden, 2020. [Online]. Available: https://kth.diva-portal.org/smash/record.jsf?pid=diva2:1523089

Habib, M. K. (2007). Mechatronics – A Unifying Interdisciplinary and Intelligent Engineering Science Paradigm. *IEEE Industrial Electronics Magazine*, *1*(2), 12–24. https://doi.org/10.1109/MIE.2007.901480

Hadi, R. H., Hady, H. N., Hasan, A. M., Al-Jodah, A., and Humaidi, A. J. (2023). Improved Fault Classification for Predictive Maintenance in Industrial IoT Based on AutoML: A Case Study of Ball-Bearing Faults. *Processes*, *11*(5), 1507. https://doi.org/10.3390/pr11051507

Haricha, K., Khiat, A., Issaoui, Y., Bahnasse, A., and Ouajji, H. (2023). Recent Technological Progress to Empower Smart Manufacturing: Review and Potential Guidelines. *IEEE Access*, *11*, 77929–77951. https://doi.org/10.1109/ACCESS.2023.3246029

Hülsmann, M., and Förstner, R. (2020). Predictive Maintenance as Enhancement of Fault Detection, Isolation and Recovery Strategies of Spacecraft. *Proceedings of the International Astronautical Congress, IAC*, 2020(October), 12–14.

Jerker, D. (2017). Local Cloud Internet of Things Automation: Technology and Business Model Features of Distributed Internet of Things Automation Solutions. *IEEE Industrial Electronics Magazine*, *December*, 8–21.

Kuru, K., and Yetgin, H. (2019). Transformation to Advanced Mechatronics Systems within New Industrial Revolution: A Novel Framework in Automation of Everything (AoE). *IEEE Access*, *7*, 41395–41415. https://doi.org/10.1109/ACCESS.2019.2907809

Lattanzi, L., Raffaeli, R., Peruzzini, M., and Pellicciari, M. (2021). Digital Twin for Smart Manufacturing: A Review of Concepts Towards a Practical Industrial Implementation. *International Journal of Computer Integrated Manufacturing*, *34*(6), 567–597. https://doi.org/10.1080/0951192X.2021.1911003

Lee, M. R. (2023). A Review on Intelligent Control Theory and Applications in Process Optimization and Smart Manufacturing. *Processes*, *11*(11), 3171. https://doi.org/10.3390/pr11113171

Lemstra, M. A. M. S., and de Mesquita, M. A. (2023). Industry 4.0: A Tertiary Literature Review. *Technological Forecasting and Social Change*, *186*(February 2022), 1–11. https://doi.org/10.1016/j.techfore.2022.122204

Liagkou, V., Stylios, C., Lamprini, P., and Petunin, A. (2021). Challenges and Opportunities in Industry 4.0 for Mechatronics, Artificial Intelligence and Cybernetics. *Electronics*, *10*(2001), 1–23. https://doi.org/10.3390/electronics10162001

Lu, Y., Liu, C., Wang, K. I.-K., Huang, H., and Xu, X. (2020). Digital Twin-driven Smart Manufacturing: Connotation, Reference Model, Applications and Research Issues. *Robotics and Computer-Integrated Manufacturing*, 61, art. 101837. https://doi.org/10.1016/j.rcim.2019.101837

Minh, D., Wang, H. X., Li, Y. F., and Nguyen, T. N. (2022). Explainable Artificial Intelligence: A Comprehensive Review. *Artificial Intelligence Review*, *55*(5), 3503–3568. https://doi.org/10.1007/s10462-021-10088-y

Mukherjee, D., Gupta, K., Chang, L. H., and Najjaran, H. (2022). A Survey of Robot Learning Strategies for Human–Robot Collaboration in Industrial Settings. *Robotics and Computer-Integrated Manufacturing*, *73*(June 2021), 102231. https://doi.org/10.1016/j.rcim.2021.102231

Nacchia, M., Fruggiero, F., Lambiase, A., and Bruton, K. (2021). A Systematic Mapping of the Advancing Use of Machine Learning Techniques for Predictive Maintenance in the

Manufacturing Sector. *Applied Sciences (Switzerland), 11*(6), 1–34. https://doi.org/10.3390/app11062546

Nguyen, K. T. P., and Medjaher, K. (2019). A New Dynamic Predictive Maintenance Framework Using Deep Learning for Failure Prognostics. *Reliability Engineering and System Safety, 188*, 251–262. https://doi.org/10.1016/j.ress.2019.03.018

Nnodim, C. T., Arowolo, M. O., Agboola, B. D., Ogundokun, R. O., and Abiodun, M. K. (2021). Future Trends in Mechatronics. *IAES International Journal of Robotics and Automation (IJRA), 10*(1), 24. https://doi.org/10.11591/ijra.v10i1.pp24-31

Orlowska-Kowalska, T., and Wolkiewicz, M. (2022). Fault Diagnosis and Prognosis of Mechatronic Systems Using Artificial Intelligence and Estimation Theory. *Electronics (Switzerland), 11*(21), 3528. https://doi.org/10.3390/electronics11213528

Ozkat, E. C., Bektas, O., Nielsen, M. J., and la Cour-Harbo, A. (2023). A Data-driven Predictive Maintenance Model to Estimate RUL in a Multi-rotor UAS. *International Journal of Micro Air Vehicles, 15*, 1–14. https://doi.org/10.1177/17568293221150171

Oztemel, E., and Gursev, S. (2020). Literature Review of Industry 4.0 and Related Technologies. *Journal of Intelligent Manufacturing, 31*(1), 127–182. https://doi.org/10.1007/s10845-018-1433-8

Pandey, R. (2021). Integration of Machine Learning Algorithms in Mechatronic Systems for Predictive Maintenance. *Mathematical Statistician and Engineering Applications, 70*(2), 1822–1829. https://doi.org/10.17762/msea.v70i2.2475

Paolanti, M., Romeo, L., Felicetti, A., Mancini, A., Frontoni, E., and Loncarski, J. (2018). Machine Learning Approach for Predictive Maintenance in Industry 4.0. *2018 14th IEEE/ASME International Conference on Mechatronic and Embedded Systems and Applications, MESA 2018, January 2020*, 1–6. https://doi.org/10.1109/MESA.2018.8449150

Pech, M., Vrchota, J., and Bednář, J. (2021). Predictive Maintenance and Intelligent Sensors in Smart Factory: Review. *Sensors, 21*(4), 1–39. https://doi.org/10.3390/s21041470

Raza, F. (2023). AI for Predictive Maintenance in Industrial Systems. *Cosmic Bulletin of Business Management, 2*(1), 167–183. https://doi.org/10.13140/RG.2.2.27313.35688

Riesener, M., Kuhn, M., Boßmann, C., and Schuh, G. (2023). Methodology for the Design of Interdisciplinary Modules for Mechatronic Modular Product Platforms. *Procedia CIRP, 119*, 675–680. https://doi.org/10.1016/j.procir.2023.03.119

Rodriguez-Guerra, D., Sorrosal, G., Cabanes, I., and Calleja, C. (2021). Human–Robot Interaction Review: Challenges and Solutions for Modern Industrial Environments. *IEEE Access, 9*, 108557–108578. https://doi.org/10.1109/ACCESS.2021.3099287

Rojek, I., Jasiulewicz-Kaczmarek, M., Piechowski, M., and Mikołajewski, D. (2023). An Artificial Intelligence Approach for Improving Maintenance to Supervise Machine Failures and Support Their Repair. *Applied Sciences (Switzerland), 13*(8), 4971. https://doi.org/10.3390/app13084971

Rüßmann, M., Lorenz, M., Gerbert, P., Waldner, M., Justus, J., and Harnisch, M. (2015). *Industry 4.0: The Future of Productivity and Growth in Manufacturing Industries*. Boston Consulting Group [Online]. Available: www.bcg.com/publications/2015/engineered_products_project_business_industry_4_future_productivity_growth_manufacturing_industries

Ryalat, M., Franco, E., Elmoaqet, H., Almtireen, N., and Al-Refai, G. (2024). The Integration of Advanced Mechatronic Systems into Industry 4.0 for Smart Manufacturing. *Sustainability (Switzerland), 16*(19), 1–39. https://doi.org/10.3390/su16198504

Schmidt, M. (2023). Integrating Machine Learning Techniques for Advancing Industry 4.0: Opportunities, Challenges, and Future Directions. *International Journal of Engineering Sciences and Technologies, 1*(2), 1–11. https://doi.org/10.58531/ijest/1/2/3

Shetty, D., and Kolk, R. A. (2011). *Mechatronics System Design* (2nd ed.). Cengage Learning.

Shetty, D. K., Prerepa, G., Naik, N., Bhat, R., Sharma, J., and Mehrotra, P. (2022). Revolutionizing Aerospace and Defense: The Impact of AI and Robotics on Modern Warfare. *Proceedings of the 4th International Conference on Information Management & Machine Intelligence*, 1–8. Association for Computing Machinery. https://doi.org/10.1145/3590837.3590856

Singh, V., Sharma, N., and Sharma, S. K. (2023). Analyzing the Role of Multi-Agent Technology in Minimizing Breakdown Probabilities in Manufacturing Industries. *Journal of Quality in Maintenance Engineering*, *30*(1), 221–247. https://doi.org/10.1108/jqme-04-2023-0040

Suganob, N. J., Joseph Alagar, J., Arestotiles, R., Tagaan, J. G., Jane Gloria, J., Cabilogan, J. I., Jun Arcay, M., Macadato, M., Baldonado, F., Marajas, E., Salaan, C. J., Concepcion, R., and Dadios, E. (2022). Design and Development of Mechatronics Trainer for Skills Competency Assessment. *2022 IEEE 14th International Conference on Humanoid, Nanotechnology, Information Technology, Communication and Control, Environment, and Management, HNICEM 2022*, 1–6. https://doi.org/10.1109/HNICEM57413.2022.10109540

Tao, F., Zhang, H., Liu, A., and Nee, A. Y. C. (2019). Digital Twin in Industry: State-of-the-Art. *IEEE Transactions on Industrial Informatics*, *15*(4), 2405–2415. https://doi.org/10.1109/TII.2018.2873186

Vallim Filho, A. R. de A., Farina Moraes, D., Bhering de Aguiar Vallim, M. V., da Silva, L. S., and da Silva, L. A. (2022). A Machine Learning Modeling Framework for Predictive Maintenance Based on Equipment Load Cycle: An Application in a Real World Case. *Energies*, *15*(10), 1–41. https://doi.org/10.3390/en15103724

Vemuri, N., and Thaneeru, N. (2023). Enhancing Human-Robot Collaboration in Industry 4.0 with AI-driven HRI. *Power System Technology*, *47*(4), 341–358.

von Enzberg, S., Naskos, A., Metaxa, I., Köchling, D., and Kühn, A. (2020). Implementation and Transfer of Predictive Analytics for Smart Maintenance: A Case Study. *Frontiers in Computer Science*, *2*(November), 578469. https://doi.org/10.3389/fcomp.2020.578469

Yu, J., and Zhang, Y. (2023). Challenges and Opportunities of Deep Learning-based Process Fault Detection and Diagnosis: A Review. *Neural Computing and Applications*, *35*(1), 211–252. London: Springer. https://doi.org/10.1007/s00521-022-08017-3

Zaghdoudi, M. A., Hajri-Gabouj, S., Varnier, C., Zerhouni, N., and Ghezail, F. (2022). Predictive Analysis Methodology for Industrial Systems: Application in Supplier Delays Prediction. *2022 IEEE Information Technologies and Smart Industrial Systems (ITSIS)*, 1–6. https://doi.org/10.1109/ITSIS56166.2022.10118391

Zhao, Z., Wu, J., Li, T., Sun, C., Yan, R., and Chen, X. (2021). Challenges and Opportunities of AI-Enabled Monitoring, Diagnosis & Prognosis: A Review. *Chinese Journal of Mechanical Engineering (English Edition)*, *34*(1), 1–29. https://doi.org/10.1186/s10033-021-00570-7

Zhu, Z., Lei, Y., Qi, G., Chai, Y., Mazur, N., An, Y., and Huang, X. (2023). A Review of the Application of Deep Learning in Intelligent Fault Diagnosis of Rotating Machinery. *Measurement: Journal of the International Measurement Confederation*, *206*(June 2022), 112346. https://doi.org/10.1016/j.measurement.2022.112346

6 Control Systems Optimization Using AI in Mechatronics Engineering

Senthil Kumar R., Saravanan S., Pandiyan P., and Balakumar P.

6.1 INTRODUCTION

The efficiency and protection of mechatronic systems are generally dependent on their control systems, which are frequently complex and dynamic in nature. The intelligent controlling approach can be construed rather freely from the perspective of automation. The way the concept is implemented in real-time, swiftly, and while the system fulfills its intended functions, distinguishes it from conventional feedback systems. Therefore, if the control law is continuously updated, traditional adaptive automation can be regarded as intelligent. Such systems can be categorized as borderline from a categorization standpoint [1]. When examining the influences on intelligent control strategies for mechatronic, it is easy to identify the evolution of strategies from mathematical computation that later moved into the arena of computer user interface design, processed, and examined big data, and, currently, have been proactively used in industrial systems management [2]. The tremendous expansion of literature begins to slow in the early 2010s and does not stop.

Numerous methods of applying AI to control system optimization exist. The ranges of the control settings that minimize a preset deviation metric, like the average squared deviation between the intended and actual outcomes of the system, could be found using AI methods [3]. For a specific system or application, new control principles that prove more efficient can be created using AI algorithms. These algorithms can be used, for instance, to create adaptive control rules that can adapt to modifications in the system's characteristics or operating system. The resilience of control systems to shocks and ambiguity can be increased using AI algorithms. For instance, fault-tolerant control (FTC) mechanisms that can keep running even if some of the system's elements fail can be created using AI algorithms.

Different artificial intelligence (AI) techniques and fields, including machine learning (ML), Fuzzy Logic (FL), artificial neural networks (ANNs), evolutionary computing, and genetic algorithms (GA), are employed in robotics and mechatronics.

DOI: 10.1201/9781032615790-6

ML techniques include supervised and unsupervised learning, ensemble, and reinforcement learning (RL). Intelligent control algorithms analyze big data sets and discover useful patterns from actions made using a variety of probabilistic, statistical, and optimization techniques [4]. As opposed to neural networks (NNs) and RL, which are useful in the fields of autonomous control, scheduling, and logistics, traditional knowledge and ensembles are mostly employed for processing and categorizing data sets [5]. NNs, however, perform best when dealing with unidentified features and complicated data. The dependability and security of employing smart control methods in self-governing machines interfacing with the real environment is a significant concern.

Numerous mechatronic applications have successfully exploited AI-based control system optimization. Robotic manipulators, arms, and other devices have all been controlled to the best of their abilities using AI techniques. Robotic systems now operate more accurately, quickly, and efficiently in terms of energy. Machine tool controls, such as those for Computer Numerical Control (CNC) machine tools and milling equipment, have been improved with the assistance of AI algorithms. As a result, machine tools have become more effective and efficient. The management of automobile systems, including engine controls and brake assist systems, has been improved using intelligence algorithms. Consequently, automobiles are now more efficient, safer, and consume less fuel. AI has the potential to revolutionize the design and implementation of control mechanisms for mechatronic devices. The performance and effectiveness of mechatronic systems can be significantly enhanced through AI-based control system optimization.

Robotic controllers are being trained using RL [1]. By communicating with the environment and earning incentives for good actions, RL agents might learn to operate robots to carry out complicated tasks, such as running, walking, and grabbing things. Adaptive control rules are being developed for mechatronic systems using ML. The best control settings for a system during different conditions of use may be found using ML algorithms. As a result, the system may continue to operate well, irrespective of the atmosphere changes. For mechatronic platforms, fault-tolerant systems of control are being created using deep learning (DL). DL algorithms are able to detect and correct faults in system parts. It makes it possible for the system to function regardless of whether some of the parts break. We could expect increasingly creative and efficient methods to apply AI for optimizing automation in mechatronics engineering as the field of AI continues to advance.

6.2 INTELLIGENT CONTROLLERS FOR CONTROL SYSTEMS OPTIMIZATION IN MECHATRONICS ENGINEERING

Intelligent control blends traditional automatic control theory with cutting-edge concepts like NNs. It leverages mathematical methods to enhance performance through learning from experience. Various modern control methods are available; however, this section primarily focuses on the pivotal role of ML. ML enables adaptive control, predictive modeling, anomaly detection, optimization, and RL, playing a pivotal role in the evolution of intelligent control [1].

6.3 ADAPTIVE CONTROL METHODS

Adaptive control is a fundamental component of intelligent control systems. It is a method with a strong theoretical foundation that allows a system to preserve its effectiveness even when the parameters of the system and the characteristics of its environment are uncertain or constantly changing. This adaptability is essential in real-world applications, where conditions are rarely static.

The "Stripe Algorithm" (SA), created by Yakubovich, is well known for its efficiency in online ML tasks. It continuously performs better than gradient descent learning algorithms and conventional linear techniques like logistic and linear regression [6]. It is crucial to acknowledge that although SA has benefits like interpretability and simplicity of use, it might not always perform as well as sophisticated nonlinear models. The particular application and its needs will determine whether to use SA or more complicated models. Furthermore, as demonstrated in ref. [7], a strong theoretical foundation does not ensure smooth practical implementation. Theory frequently desires to be adapted and improved to provide the intended outcomes in the intricate and dynamic contexts of real-world control systems.

6.4 OPTIMIZATION METHODS

Prior to ML, optimization techniques were traditionally developed to find a function's extremum, as described in ref. [8]. This optimality theory is an organizing principle for a wide range of ML issues, which are typically expressed as the minimization of a functional J with respect to a parameter θ.

The precise form of the objective to be reduced is contingent upon the ML approach. The objective of tasks such as regression/classification is to reduce the number of forecasted errors on the available dataset. The goal of RL is to optimize the gain from plant activities. This objective can be learned through the utilization of various search algorithms.

As a result, mathematical optimization encompasses a diverse range of types, methods, and applications, making it a versatile and essential tool in both optimization problems and ML algorithms.

6.5 ARTIFICIAL NEURAL NETWORKS

The historical trajectory of ANNs has been characterized by fluctuations until the early 1990s, when their fundamental principles were carefully determined. These networks are portrayed by the function Rn [1]. A neuron possesses n inputs (x_i) with associated weights (w_i), which scale the signals transmitted through connections. These weighted signals ($x_i w_i$) are then aggregated in an adder, resulting in the sum of the weighted neuron outputs: $\Sigma(x_i w_i)$. This summation is subsequently processed through an activation function $\phi(\text{net})$, which transforms it into an output value for the neuron. A simplified diagram illustrating the neuron's operation is depicted in Figure 6.1. The essence of NN learning lies in identifying the optimal architecture and weights that minimize the training error across all potential training samples.

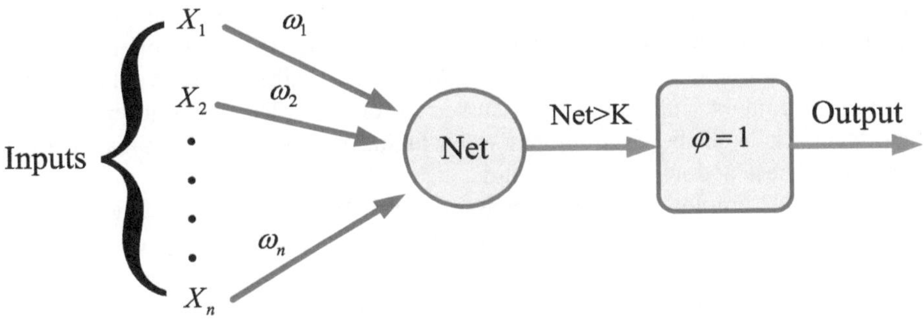

FIGURE 6.1 Structure of NN.

ANN proves particularly efficient in controlling nonlinear and noisy systems. The algorithm offers adaptability and can function in real time once trained. Recent advancements in NN structures and properties aim to address existing limitations. Traditional heuristic approaches to NN training can lead to uncertain solutions and bottlenecks, often requiring a preparatory training dataset [9]. There are drawbacks to employing NNs in mechatronics system (robotics), including lengthy training times, increased unpredictability in training outcomes, the potential for unsuitable control of expensive equipment, and the current feasibility of implementing NNs primarily in the method of Ultra Large Scale Integration (ULSI).

6.6 FL APPROACH

The FL framework, initially suggested by Zadeh et al. [10], centers around the notion of a fuzzy set, characterized by an "element membership function" that assigns values within the interval [0, 1], as outlined in ref. [1]. FL perceptive can be effectively represented through NNs. Today, there exist numerous "neuro-fuzzy networks" that can be approximately characterized as universal approximators, as per the framework detailed in ref. [11]. These networks are noteworthy for their capacity to emulate human-like reasoning patterns and their NN structure.

6.7 REINFORCEMENT LEARNING

Within the framework of RL, the fundamental concepts involve the agent and its interaction with the environment. The agent takes actions (at) while existing in a specific state (st) and is rewarded (rt) accordingly. The agent's subsequent actions (at + 1) in the same state (st) are influenced by the received reward (rt + 1). Through this ongoing interaction with a dynamic and uncertain environment, the agent aims to learn an "optimal decision policy," which encapsulates the ideal sequence of actions for maximizing its cumulative state.

Kurrek et al. [12] outlined a versatile method for developing control policies applicable to an extensive range of robots, environments, and manipulative tasks, including both physical and virtual environments, implemented through computer programs.

Kim et al. [5] discussed the effectiveness of RL by using an inverted pendulum as an example, controlled by a Proportional-Integral-Derivative (PID) controller, although in a virtual environment. A pivotal aspect of this process is the real-time collaboration among environment with agent, as discussed in ref. [13]. The agent accepts immediate feedback from the environment when taking random or suboptimal actions, quantifying these actions through rewards.

Werbos [14] employed the encompassing term "approximate dynamic RL programming." In a broad sense, approximate dynamic programming encompasses all computational tools aimed at solving the maximum accurate value to Bellman expression, as needed for optimal programming solutions. From a control theory perspective, RL can be viewed as a form of optimal adaptive control achieved through an advanced controller which converges toward optimality.

6.8 MULTI-OBJECTIVE OPTIMIZATION FOR CONTROL SYSTEMS IN MECHATRONICS ENGINEERING

For the designer, multiple goals and function are crucial in the majority of engineering challenges. Usually, the designer should simultaneously optimize a few competing objectives. For instance, the designer will aim to balance two opposing goals: engine power and fuel conservation. These issues have a collection of optimal design vectors known as the Pareto front, as opposed to single-objective optimization challenges, where there exists a single extremum indication for an issue [15]. These solutions' most crucial quality is the fact that the others control neither of these. One of these options is picked by the designer as the best fit for their needs.

Creating a vector of choice variables that satisfies constraints to provide acceptable levels to all of the objective functions has been described as multi-objective optimization [16]. However, it is expected that all functions should have minimal error without losing generality. These issues can be described numerically as:

Find the vector $X^* = \left[X_1^*, X_2^*, \ldots\ldots\ldots, X_n^*\right]^T$ to optimize

$$F(X) = [f_1(X), f_2(X), \ldots, f_k(X)]^T \tag{6.1}$$

6.9 FTC USING AI IN MECHATRONICS ENGINEERING

A control approach called FTC allows a system to function even when there are concerns. In mechatronics systems that are frequently complicated and operate in difficult settings, this is essential [17]. There are several ways that AI might be applied to mechatronics engineering to enhance the effectiveness of FTC systems. The block diagram of fault-tolerant controller in mechatronics system is shown in Figure 6.2. The system's health and performance are routinely assessed through a range of tools and monitoring techniques. AI and ML techniques like SVM and NNs can be used to observe sensor values and identify leavings from expected behavior [18]. Finding out the exact kind and location of the fault once it has been found is important. By

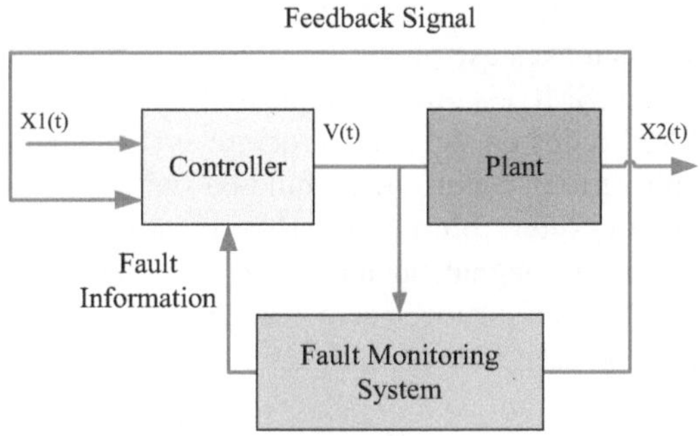

FIGURE 6.2 Block diagram of fault-tolerant controller in mechatronics systems.

analyzing sensor data and system behavior, AI algorithms could be used to separate faults and identify problematic parts or subsystems.

Accurate fault isolation can be carried out by integrating model-based techniques and data-driven tactics. Once the defect has been determined, AI-based strategies could be used to identify its severity and effects. Making decisions concerning whether to run the system indefinitely, move to a backup system, or start a safe shutdown is made easier with the use of this data. The control system is modified using AI-based control methods to take the faulty state into account while ensuring reliable and secure operation. It might involve adopting alternative control methods, employing redundant components, or altering the control settings. Real-time control operation optimization can be achieved via adaptive monitoring and RL methods.

The performance of the system along with past data can be used to train AI algorithms, which can then consistently improve the fault isolation, supervision, and detection methods. Adaptable regulation and RL techniques allow the system to generate fault behaviors and adjust to changing conditions. The efficacy and security of FTC systems must be ensured by rigorous testing and validation procedures. AI can assist in creating realistic fault scenarios for testing purposes and to confirm the efficacy of the control strategies during various fault scenarios.

A potential means to apply AI in FTC is to develop fault detection and diagnostic (FDD) systems that can accurately identify and distinguish flaws in real time. Errors that the FTC system is ignorant of cannot be corrected. Huge past information sets could be used to educate AI-based FDD algorithms to recognize patterns that point to defects. They may also be used to create system models that can be utilized to find and fix problems even when there is noise and ambiguity.

Making control algorithms, which may correct errors, is another method to apply AI in FTC. This is problematic because errors can affect the system in a number of ways, and it might be tricky to build control. Adaptive control algorithms can update these variables in real time to account for modifications in the system's dynamics brought on by faults that might be created using them. FDD systems are being created using NNs to precisely identify and diagnose defects in robots that manipulate objects. The development of FTC systems, which can correct engine flaws in aero planes, uses fuzzy logic. The development of FTC algorithms that are able to learn how to correct flaws in self-driving cars makes use of RL.

Sliding-mode control (SMC) is one of the most often used techniques for controlling robotic arms. SMC relies on dynamic structure systems, which are made up of several structures with various features that could be switched between. In SMC, the entire system is forced to "slide" through the cross section of the standard features by use of a discontinuity control signal, such as +k and k. A system is considered to be in the process of sliding when it moves across a curved or sliding surface [19]. The following formula may be used to explain the most basic SMC strategy:

$$u = k \cdot \text{sign}(e), \tag{6.2}$$

where e is the error in the control signal, k is the constant that represents the size of the change, and sign () is the sign function.

The employment of a sliding-mode observer (SMO) for identifying faults represents one of the SMC fault-tolerant approaches [20]. The creation of an error estimation technique is enabled by an estimate of the indescribable state and the modeling of uncertainty made feasible by the employment of a nonlinear observer. AI can often significantly boost the effectiveness of FTC systems in mechatronics engineering. Defects could be promptly found and fixed with AI-based FDD systems, and faults could be compensated for in a variety of scenarios using AI-based control algorithms.

6.10 CONTROL SYSTEMS OPTIMIZATION FOR ROBOTICS USING AI IN MECHATRONICS ENGINEERING

AI algorithms could be applied to find optimal controller settings, create controllers that are resistant to disturbances, and make controllers that are capable of handling sophisticated nonlinear systems. Although the use of AI in robot control system optimization is still in its infancy, there are a lot of potential benefits. AI-enabled devices are certainly capable of generating intelligent, effective, and resilient robots, which might have a significant influence on a variety of sectors including manufacturing, medicine, and mobility.

Self-driving cars install AI-enabled governs to manage their brakes, steering, and speed. These controllers use data gathered from Light Detecting and Ranging (LiDAR), laser and radar detectors, or recording devices to assess the environment and determine the safest course of action. The industrial robots that perform operations such as welding, drawing, and assembling are controlled by AI-powered processors. These types of AI controls might be used to improve the robots' efficiency and precision. Healthcare robots that conduct functions including surgeries and physiotherapy are controlled by AI-enabled controls. These controls might be used to increase these robots' precision as well as safety.

ANNs are effective at controlling noisy, nonlinear systems. The continuous improvements to ANN's structures and attributes are intended to address the current drawbacks. Heuristic ANN training processes demand the creation of an example for training and might result in confusing results and deadlocks. Stabilization and program control tasks necessitate that feedback is present in the loop. The control approach depends on the feasible observed output as the system's state vectors are often not accessible for observation. Tasks like tracking the trajectory and speed stabilization are commonplace in mechatronics and robots. At the system output, these factors are simple to measure [21].

A block diagram of the plan of each block in the fault identification process is presented in Figure 6.3. A self-learning trajectory monitoring loop and an adaptive NN-based speed control loop for an autonomously wheel mobile robot were proposed in ref. [22]. A performer, an opponent, and a shadowy critic are all hidden neurons in a three-layer NN with a hyperbolic tangent activated function, and it can be written as,

$$\overline{Y_i} = a\tanh(b\sum_{j=0}^{J} \overline{w_{ij}}(n)\overline{x_j}(n)), \overline{x_0}(n) = 1, (a,b) > 0 \tag{6.3}$$

The linear activation function for output neuron is

$$\overline{Z_k}(n) = c\sum_{i=0}^{I} w_{ki}(n)\overline{Y_i}(n)), \overline{Y_0}(n) = 1, C > 0 \tag{6.4}$$

To create the self-learning process, a unique inverse kinematics and rotational model of the robot is identified. The weights used in the reverse velocity models are modified after computing the inversion error. As a result, the robot was given command without the need of a controller, training data, or logical control principles. The automated operation of human and unmanned airliners using dynamic flight control techniques is the subject of several studies utilizing ANN [23]. This is particularly true for autonomy under challenging climatic conditions and during system malfunctions. ML techniques enhance the precision of the control and react to unidentified non-linear flying situations.

NNs could be used to create a trainable flight control system, as demonstrated in ref. [24]. Based on the information collected, huge sets of neural controllers are taught to

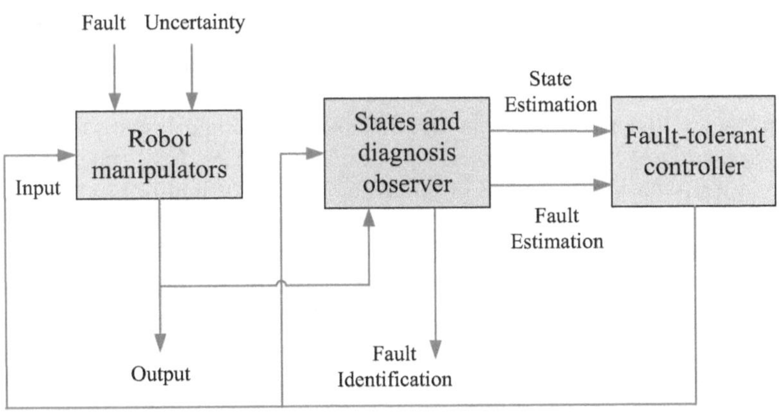

FIGURE 6.3 Fault-tolerant process in robot manipulators.

exhibit the desired performance. Then, using methods of statistics, the efficacy of the training could be evaluated. NNs are highly suited for handling challenging nonlinear dynamics and fault tolerance issues because of their learning capabilities. Spiking NN controllers have the capacity for adaptability and online learning while transitioning from simulations to the real world. In comparison with earlier generations of NN, these NN structures include fewer neurons and neural connections. In particular, ML techniques enhance control precision and enable the system to respond effectively to unidentified nonlinear flight situations, even during system failures or autonomous operation in challenging atmospheric conditions. ML techniques enhance the precision of the control and react to unidentified nonlinear flying situations.

For walking bipedal robots, an effective control strategy based on a support vector algorithm is proposed in ref. [25] to reduce power consumption under unidentified system dynamics model parameters and short data sample sizes. With restrictions on robot articulating perspectives, the novel controller, which is included in the best controller, seeks to reduce the energy-associated cost function. ML techniques were used to evaluate the downhole pump's health and conduct further diagnostics, as demonstrated in ref. [26]. Data was gathered using a dynamometer to gauge the load on the sucker rod, and it was then processed using particular software designed for creating and assuring the real-time functioning of systems for collecting and analyzing data about the observing component. To minimize production mishaps, save expenditures on repairs, and enhance every aspect of manufacturing, prediction implementation according to ML is a cutting-edge and efficient method.

6.11 CONTROL SYSTEMS OPTIMIZATION FOR ENERGY SYSTEMS USING AI IN MECHATRONICS ENGINEERING

Energy efficiency represents one of the most significant issues facing cities today when it comes to sustainable energy sources. By identifying the best option, energy efficiency should be raised to satisfy demand and limits [27]. A smart city's inhabitants will have high-quality lives through effective utilization of resources if sustainable energy is given. To satisfy their energy needs, smart cities must advance their complete renewable energy policies by using other forms of energy such as Photovoltaic (PV), wind turbine, thermal collectors, biomass, and geothermal energy [28]. It is difficult to give the city safe, affordable power. It is always possible to utilize modelling and optimization approaches for interconnected systems.

Using an intelligent management tool that uses fuzzy logic systems to improve the battery system for scheduling lowers operational expenses. Real-time management requires a well-considered choice of unit size. The use of multi-objective optimization in real-time grid management has been explored by researchers. To defeat multi-objective optimization and achieve Pareto sets, common boundary approaches are applied. Several researchers present instances of how fuzzy logic is being effectively applied to address the optimization problem or build appropriate grid optimization plans for sources of alternative energy.

The doubly fed induction generator has been managed using AI control techniques including Proportional-Integral (PI) and fuzzy. The mathematical modeling and evaluation of control systems were made easier by vector control approaches. Additionally, the fuzzy–neuro hybrid controller performed better than the PI online tuning approach. Methods using fuzzy logic and NNs have several limitations. Several of these restrictions are determined by their parameters. This integration of fuzzy with neuro algorithm is employed throughout the Maximum Power Point Tracking (MPPT) optimization process as the NN will be dependent on duration, errors of training, and nonlinear task estimate if input data are not included within the training set [29].

AI algorithms have already been applied to defect diagnostics and weather forecasting, which aid in the smooth operation of the renewable energy system. Variability in the solar power producing system is caused by solar irradiance, changes in the weather, and the presence of clouds. The amount of solar radiation has an impact on the temperature. Temperature prediction makes it feasible to forecast solar energy. The temperature forecast can be solved more effectively with fuzzy logic. The AI algorithms Radial Basis Function (RBF) and Multi-Layer Perceptron (MLP) are utilized to predict solar radiation [30].

A PV module's effectiveness is not linear as it is influenced by the surrounding environment. Therefore, we employ a few techniques that can always be fully monitored by a PV device. The MPPT tactics range in complexity from simple to quite sophisticated. Though there are ways that monitor temperature and/or irradiation requirements to accomplish the MPP, most solutions assess voltage/current or electricity. The P–V and the I–V curves are two properties of the PV module that specify its function points under various operating situations.

Figure 6.4 depicts an Internet of Things (IoT)-based PV monitoring system. There are three layers to the autonomously tracking PV system utilizing IoT. The layers

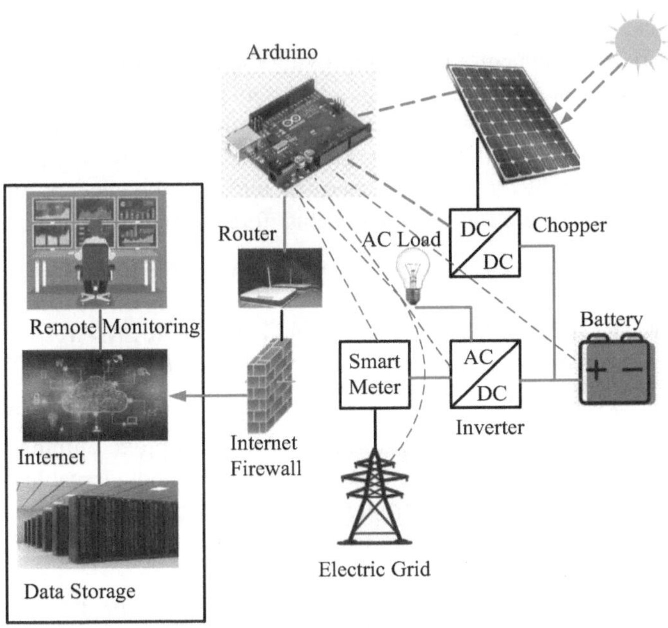

FIGURE 6.4 Observing of PV system using IoT [31].

are the PV system installation, the link to the internet entry point for associations, and finally the autonomous control and monitoring layer [31]. To track MPPT points from the PV system, FL and ANN have been employed. For MPPT forecasting using wind speed, a variety of NN frameworks including multi-layer perceptrons, back propagation NNs, multi-layer adaptable linear networks, and probabilistic NNs are utilized. When creating a specific integration system, the MPPT method is typically chosen. Flexibility, affordability, the time it takes to manage the most power points, and the efficiency of its operation are the considerations. Although the back propagation approach has a few drawbacks, including the absence of persuasive convergence, its straightforward and effective mathematical methodology demonstrates the fundamental idea of neuro-computing. This probabilistic approach has the limitation that it cannot accurately reflect the rapid shift in the hybrid model's outputs [32].

As wind farms largely rely on the wind's variable character, the ongoing expense of the energy network under evaluation is increased by the unpredictability of wind power output. The relationship between wind power and the speed of the wind is largely linear. For wind speed estimate in renewable power plants, an NN model has been utilized. The relative humidity (RH), wind speed, temperature, and pressure are estimated and predicted using a different technique that employs genetic expression programming (GEP). Short-term forecast of less than 24 hours is insufficient for solar power plant optimization and operation. Multi-layer perceptron could therefore enhance the efficacy 24 hours ahead of solar irradiation. The wavelet-network approach is additionally employed to solve the data shortage problem. To train ANN, the Newton technique Levenberg–Marquardt (LM) algorithm was employed. Input statistics from time series, including slope, average, and standard deviation, are generated and utilized as inputs to fuzzy and artificial neural predictors rather than using the direct wind speed [33].

Researchers have concentrated on improving the quality of the electricity system in response to current customer demand. The implementation of load forecasts for consumer demand is taken into account in the future perspective. To optimize artificial algorithms, hybrid soft computing approaches should be applied, and a suitable transfer function has to be researched. To integrate the concept of smart grids and cities into next-generation energy systems, forecasting load techniques need be created [34, 35].

6.12 CONTROL SYSTEMS OPTIMIZATION FOR SMART HOMES AND BUILDINGS USING AI IN MECHATRONICS ENGINEERING

In mechatronics engineering, AI can also be used to improve automation systems for smart buildings and homes [36]. More sophisticated and adaptable controls that are able to acquire from data and progressively enhance their performance may be created using AI algorithms. This can end up in smart structures that become more sustainable, pleasant, and effective [37].

To forecast and optimize energy use, controllers can be trained using ML algorithms. An ML system, for instance, may be trained to regulate a thermostat to reduce energy usage while keeping an appropriate temperature. To increase safety and comfort, DL algorithms could be utilized to create regulators which can learn from highly dimensional sensor information. A DL algorithm could be used to develop a smart lighting system controller that uses occupancy and light intensity sensors to optimize room illumination for satisfaction and energy efficiency. Through imparting knowledge, controllers can learn from errors, and RL algorithms could improve the performance of intelligent home and building systems. RL algorithms could be used, for example, to educate a controller for an intelligent irrigation system when to water plants in the most efficient manner based on weather data and soil moisture data.

IoT technology's rapid data generation could include inaccurate information that is not appropriate for in-depth examinations. IoT data statistical analysis requires the application of industry-accepted techniques, as well as a comprehensive evaluation of the data to ensure correctness and applicability. The initial review of the data is essential for verifying the variables, detecting outliers, cleansing the data, finding values that are missing, and other tasks. Determining the precise method of data collection is essential. A preliminary evaluation of the raw data is necessary to achieve objectives including emphasizing data aspects, summarizing the data in an approachable way, visualizing data, and preparing the data for in-depth research.

AI-based Smart Building Automation Controller is shown in Figure 6.5. The hybrid location of items is determined using the method of IoT-based localization in the home. A hybrid algorithm is a problem-solving abilities algorithm that integrates two or more separate algorithms. Sensor-based monitors and deploying are the primary functions of IoT applications, and knowing the peculiarities of the internal surroundings is essential to enhancing efficiency. Exploratory data analysis (EDA) is a method for examining and evaluating datasets with the aim of identifying their main characteristics. To identify significant trends or recurrent patterns, it involves

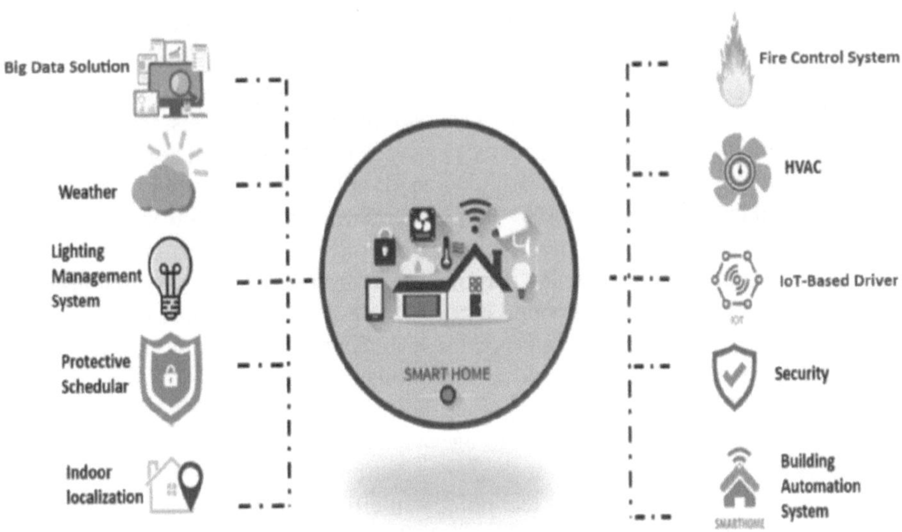

FIGURE 6.5 Smart Building [36].

utilizing a number of data visualization tools. EDA looks at what the data could show outside of formal modeling and gives an understandable overview of dataset variables and how they relate to one another. The easiest way to assess data on energy use is to create separate plots for each variable. The seasonal effect wave and other factors that determine how the data is distributed are shown in the line plots, which provide a high-level representation of the energy information gathered across 4 years.

Based on specified conditions and the related brightness levels supplied by the consumer, the system's controller had the ability to choose the required amount of illumination for the control premises as it assessed the illumination system. By moving the electrically controlled curtains, the controller was able to optimize the natural light utilization for the appropriate illuminance settings. When there is insufficient natural light, the controller was able to run an artificial lighting source to add more light. This function makes it possible to keep the living space's lighting at a higher level of comfort. Mechatronics engineering is at the forefront of providing sustainable, effective, and user-friendly living and working places, employing control systems optimization for smart buildings and homes. These systems will grow more complex, adaptable, and ecologically mindful as technology develops.

6.13 CHALLENGES IN CONTROL SYSTEMS OPTIMIZATION USING AI IN MECHATRONICS ENGINEERING

Despite substantial investments from both academia and industry in the development of autonomous driving (AD)—(mechatronics system) technology, numerous challenges continue to pose significant hurdles in various aspects of these systems:

Complex Nature of AD Systems: AD systems involve a series of interconnected decision-making processes, where one decision influences another. While improvements have been made in certain aspects, many components within these systems remain interdependent. As an example, A multi-criteria decision-making method was introduced in selecting the most appropriate driving maneuvers. This decision-making process involves several sequential stages, with the initial phase emphasizing safety-critical factors much like human decision-making, and involves taking into account various considerations. Consequently, an efficient motion planning system for AD may synergize effectively with a high-energy feedback controller. In contrast, simple controllers might consume less energy but necessitate a more complex approach to motion planning to maintain robustness [38, 39]. Thus, the imperative need for developing intelligent frameworks becomes evident, as it plays a crucial role in balancing these conflicting metrics and dynamically generating optimal solutions.

Dynamicity of Road Environment: Nowadays, there is a shared belief that cities are becoming busier and undergoing substantial changes. This is because of things like colorful road advertisements and better lighting, which are part of the move to use more digital technology in cities. Despite researchers introducing a variety of sensor-based solutions, such as vision, radar, diverse modalities, and lasers, attaining high accuracy levels, particularly in dynamic road scenarios, remains a significant challenge. Compounding this complexity, there is a growing trend among individuals to retain personal luxury vehicles, which contributes to delicate traffic congestion. The complexity of the AD environment is greatly increased by these trends, making it more difficult to precisely identify, track, and recognize different tasks in the field of AD.

Big Data and Real-Time Processing: The autonomous car uses a variety of sensing tools to keep a close eye on its environment. These tools include cameras, sensors, LiDAR, and other comparable technologies, all of which are constantly gathering a lot of data. This data encompasses high-quality content, like higher-resolution videos, owing to the paramount importance of AD. As a result, processing this extensive data in real time presents a remarkable challenge, demanding careful consideration of factors such as accuracy, power usage, and cost [40, 41].

Intelligent Data Prioritization: As previously mentioned, a significant volume of diverse data is collected, leading to big data. Existing literature emphasizes the impracticality of processing the entirety of this data within an autonomous vehicle. Hence, the deployment of data prioritization mechanisms becomes imperative. These mechanisms serve the purpose of selecting only the most vital information for further processing while discarding the remainder. This prioritization process should have the capability to make decisions based on a diverse range of data acquired under varying environmental conditions [42].

Robustness and Adaptability: Research studies suggest that collecting and processing data for autonomous vehicles in specific environments is relatively straightforward. However, many widely used AI techniques in AD are trained on data collected in these specific environments, making them less dependable in unfavorable weather conditions. Google's team recently addressed this challenge by proposing the concept of an all-weather autonomously driven vehicle. Nevertheless, the difficulty escalates when the environment becomes uncertain due to factors like snow, rain, and fog, which can impact the quality of captured data. Consequently, the AD system must demonstrate robustness and adaptability to handle diverse environmental conditions and execute various tasks associated with mobile autonomous entities.

Integration/Fusion of Sensory Data: Real-world applications pose a challenge when it comes to achieving optimal performance and precise accuracy using a single sensor. Consequently, decision-making algorithms are frequently employed to process combined data obtained from various sensors. Autonomous vehicles typically rely on two primary sensor types: environmental perception sensors, which detect objects in the vehicle's vicinity, and localization sensors, which track the vehicle's position. Fusion algorithms fall into two subcategories: ML methods, such as deep NNs, and multi-sensor information fusion for state assessment, exemplified by the Kalman filter (KF). Many research studies have introduced sensor fusion-based models that incorporate a variety of sensors and fusion algorithms. These models primarily emphasize accuracy enhancement, with limited exploration of the practical feasibility of implementing these techniques. The primary challenges in AD encompass sensing, real-time processing, communication, and the implementation of learning-based control strategies. There is substantial potential for developing robust, lightweight, and efficient fusion-based pipelines customized specifically for autonomous vehicles.

DL for Autonomous Drive: Recent studies have emphasized the critical importance of explaining the decisions made by AI models, especially in situations where these decisions have significant impacts on human lives, such as in healthcare or legal contexts. AD utilizing DL approaches [43] is no exception to this requirement for model transparency and it becomes possible not only to illuminate the inner workings and knowledge acquired by DL models [44].

DL models used in AD are not influenced by imbalances of training data ensure better overall performance, resulting in a more dependable understanding of the vehicle's surroundings and compliance with potential regulatory requirements. The interpretation of the model that has learned and the construction of possible alternative scenarios using Explainable Artificial Intelligence (XAI) techniques can define the boundaries of model performance, reveal potential sources of bias, and analyze the decision-making process to identify any weaknesses in the model. Without advancements in model explainability, the practicality of AD functions that leverage the potent modeling capabilities of DL will remain limited. Challenges like these are crucial to address the continued advancement and safe deployment of AD technology.

6.14 CONCLUSION

Optimization of AI-based control systems is a fast-growing subject with several potential uses in mechatronics engineering. This chapter emphasizes the synergistic relationship between AI and control theory as well as the necessity for more modern approaches to the difficult problems of technical system regulation. Using AI techniques, it is now feasible to operate mechatronic systems in ways that previously appeared inconceivable, with higher computing efficiency and simplicity. The seamless fusion of mechanical, electrical, and computer engineering with AI offers a potent method for creating and optimizing complex systems. The chapter also examines the field's possibilities and issues, such as the demand for greater data and computing power and the difficulties in assuring the security and dependability of AI-based systems. Also, the potential and challenges of AI-based control system optimization and forecasts how this discipline will impact future mechatronics engineering are discussed.

REFERENCES

1. Zaitceva, I., and Andrievsky, B. (2022). Methods of intelligent control in mechatronics and robotic engineering: A survey. *Electronics*, 11(15), 2443.
2. Dittrich, M. A., Uhlich, F., and Denkena, B. (2019). Self-optimizing tool path generation for 5-axis machining processes. *CIRP Journal of Manufacturing Science and Technology*, 24, 49–54.
3. Ramu, S. K., Irudayaraj, G. C. R., Subramani, S., and Subramaniam, U. (2020). Broken rotor bar fault detection using Hilbert transform and neural networks applied to direct torque control of induction motor drive. *IET Power Electronics*, 13(15), 3328–3338.

4. Annaswamy, A. M., and Fradkov, A. L. (2021). A historical perspective of adaptive control and learning. *Annual Reviews in Control*, 52, 18–41.
5. Kim, J. B., Lim, H. K., Kim, C. M., Kim, M. S., Hong, Y. G., and Han, Y. H. (2019). Imitation reinforcement learning-based remote rotary inverted pendulum control in openflow network. *IEEE Access*, 7, 36682–36690.
6. Lipkovich, M. (2022). Yakubovich's method of recursive objective inequalities in machine learning. *IFAC-PapersOnLine*, 55(12), 138–143.
7. Perel'man, I. I. (1991). Analysis of modern adaptive control methods from the standpoint of application to automatization of technological processes. *Автоматика и телемеханика*, (7), 3–32.
8. Andrievsky, B. R., and Fradkov, A. L. (2021). Speed gradient method and its applications. *Automation and Remote Control*, 82, 1463–1518.
9. Momeni, E., Samaei, M., Hashemi, A., and Dowlatshahi, M. B. (2023). A review on the application of soft computing techniques in foundation engineering. In: *Artificial Intelligence in Mechatronics and Civil Engineering: Bridging the Gap* (pp. 111–133). Springer Nature.
10. Zadeh, L. A., Klir, G. J., and Yuan, B. (1996). *Fuzzy Sets, Fuzzy Logic, and Fuzzy Systems: Selected Papers* (Vol. 6). World Scientific.
11. Cybenko, G. (1989). Approximation by superpositions of a sigmoidal function. *Mathematics of Control, Signals and Systems*, 2(4), 303–314.
12. Kurrek, P., Jocas, M., Zoghlami, F., Stoelen, M., and Salehi, V. (2019, July). AI motion control—a generic approach to develop control policies for robotic manipulation tasks. In: *Proceedings of the Design Society: International Conference on Engineering Design* (Vol. 1, No. 1, pp. 3561–3570). Cambridge University Press.
13. Menghal, P. M., and Laxmi, A. J. (2011, April). Real time simulation: A novel approach in engineering education. In: *2011 3rd International Conference on Electronics Computer Technology* (Vol. 1, pp. 215–219). IEEE.
14. Werbos, P. J. (2012). Reinforcement learning and approximate dynamic programming (RLADP)—foundations, common misconceptions, and the challenges ahead. In: *Reinforcement Learning and Approximate Dynamic Programming for Feedback Control*, pp. 1–30. Wiley-IEEE Press.
15. Sutton, R. S., and Barto, A. G. (2018). *Reinforcement Learning: An Introduction*. MIT Press.
16. Su, Y., Zheng, C., and Mercorelli, P. (2016). Nonlinear PD fault-tolerant control for dynamic positioning of ships with actuator constraints. *IEEE/ASME Transactions on Mechatronics*, 22(3), 1132–1142.
17. Khajavi, M. N., Notghi, B., and Paygane, G. (2010). A multi objective optimization approach to optimize vehicle ride and handling characteristics. *World Academy of Science, Engineering and Technology*, 4, 502–506.
18. Ali, K., Mehmood, A., and Iqbal, J. (2021). Fault-tolerant scheme for robotic manipulator—nonlinear robust back-stepping control with friction compensation. *PLoS One*, 16(8), e0256491.
19. Milecki, A., and Nowak, P. (2023). Review of fault-tolerant control systems used in robotic manipulators. *Applied Sciences*, 13(4), 2675.
20. Jerbi, H., Abassi, R., VijiPriya, J., Hmidi, F., and Kouzou, A. (2021). Passivity-based asynchronous fault-tolerant control for nonlinear discrete-time singular Markovian jump systems: A sliding-mode approach. *European Journal of Control*, 60, 95–113.
21. Yu, X., Feng, Y., and Man, Z. (2020). Terminal sliding mode control—an overview. *IEEE Open Journal of the Industrial Electronics Society*, 2, 36–52.

22. Van, M., Franciosa, P., and Ceglarek, D. (2016). Fault diagnosis and fault-tolerant control of uncertain robot manipulators using high-order sliding mode. *Mathematical Problems in Engineering*, Article ID 7926280, 14 pages. https://doi.org/10.1155/2016/7926280.
23. Lin, W. S., and Yang, P. C. (2008). Adaptive critic motion control design of autonomous wheeled mobile robot by dual heuristic programming. *Automatica*, 44(11), 2716–2723.
24. Kumar, R. S., Raj, I. G. C., Alhamrouni, I., Saravanan, S., Prabaharan, N., Ishwarya, S., Kumar, R. S., Raj, I. G. C., Alhamrouni, I., Saravanan, S., Prabaharan, N., Ishwarya, S., Gokdag, M., and Salem, M. (2023). A combined HT and ANN based early broken bar fault diagnosis approach for IFOC fed induction motor drive. *Alexandria Engineering Journal*, 66, 15–30.
25. Kruger, T., Mossner, M., Axmann, J., Vorsmann, P., and Kuhn, A. (2010). Adaptive flight control for unmanned aircraft using a stable neural network observer. In: *Proceedings of the AIAA Infotech at Aerospace 2010, Atlanta, GA, USA, April 20–22, 2010*. American Institute of Aeronautics and Astronautics (AIAA).
26. Wang, L., Liu, Z., Chen, C. L. P., Zhang, Y., and Lee, S. (2012). Support vector machine based optimal control for minimizing energy consumption of biped walking motions. *International Journal of Precision Engineering and Manufacturing*, 13, 1975–1981.
27. Sharaf, S. A., Bangert, P., Fardan, M., Alqassab, K., Abubakr, M., and Ahmed, M. (2019, March). Beam pump dynamometer card classification using machine learning. In: *SPE Middle East Oil and Gas Show and Conference* (p. D032S066R003). SPE.
28. Kanase-Patil, A. B., Kaldate, A. P., Lokhande, S. D., Panchal, H., Suresh, M., and Priya, V. (2020). A review of artificial intelligence-based optimization techniques for the sizing of integrated renewable energy systems in smart cities. *Environmental Technology Reviews*, 9(1), 111–136.
29. Zekić-Sušac, M., Mitrović, S., and Has, A. (2021). Machine learning based system for managing energy efficiency of public sector as an approach towards smart cities. *International Journal of Information Management*, 58, 102074.
30. Ullah, S., Qureshi, I. M., and Zohaib, M. M. (2016, January). Design of an Artificial-Intelligence based controller for doubly fed induction generator based wind energy conversion system. In: *2016 International Conference on Intelligent Systems Engineering (ICISE)* (pp. 318–325). IEEE.
31. Saravanan, S., and Babu, N. R. (2016). Maximum power point tracking algorithms for photovoltaic system—a review. *Renewable and Sustainable Energy Reviews*, 57, 192–204.
32. Ramu, S. K., Irudayaraj, G. C. R., and Elango, R. (2021). An IoT-based smart monitoring scheme for solar PV applications. In: *Electrical and Electronic Devices, Circuits, and Materials: Technological Challenges and Solutions* (pp. 211–233). Wiley-Scrivener.
33. Al-Othman, A., Tawalbeh, M., Martis, R., Dhou, S., Orhan, M., Qasim, M., and Olabi, A. G. (2022). Artificial intelligence and numerical models in hybrid renewable energy systems with fuel cells: Advances and prospects. *Energy Conversion and Management*, 253, 115154.
34. Chàvez-Ramirez, A. U., Vallejo-Becerra, V., Cruz, J. C., Ornelas, R., Orozco, G., Munoz-Guerrero, R., and Arriaga, L. G. (2013). A hybrid power plant (solar–wind–hydrogen) model based in artificial intelligence for a remote-housing application in Mexico. *International Journal of Hydrogen Energy*, 38(6), 2641–2655.
35. Chui, K. T., Lytras, M. D., and Visvizi, A. (2018). Energy sustainability in smart cities: Artificial intelligence, smart monitoring, and optimization of energy consumption. *Energies*, 11(11), 2869.

36. Kumar, R. S., Pugalhanthi, P. L., Dhanyaa, N., Kaviyanjali, V., Dinesh, K., and Kathirvel, K. (2024). IoT-based smart monitoring topologies for energy-efficient smart buildings. In: *Modern Computational Techniques for Engineering Applications* (pp. 79–106). CRC Press.
37. Satav, S. D., Lamani, D., Harsha, K. G., Kumar, N. M. G., Manikandan, S., and Sampath, B. (2023). Energy and battery management in the era of cloud computing: Sustainable wireless systems and networks. In: *Sustainable Science and Intelligent Technologies for Societal Development* (pp. 141–166). IGI Global.
38. Muhammad, K., Ullah, A., Lloret, J., Del Ser, J., and de Albuquerque, V. H. C. (2020). Deep learning for safe autonomous driving: Current challenges and future directions. *IEEE Transactions on Intelligent Transportation Systems*, 22(7), 4316–4336.
39. Sangapu, S. C., Prasad, K. N., Kannan, R. J., Chen, T. M., and Sathiyanarayanan, M. (2023). Impact of class imbalance in VeReMi dataset for misbehavior detection in autonomous vehicles. *Soft Computing*, 1–11.
40. Valliappan, S., Bagavathi Sivakumar, P., and Ananthanarayanan, V. (2019). Efficient real-time decision making using streaming data analytics in IoT environment. In: *International Conference on Advanced Computing Networking and Informatics: ICANI-2018* (pp. 165–173). Springer Singapore.
41. Rithani, M., Kumar, R. P., and Doss, S. (2023). A review on big data based on deep neural network approaches. *Artificial Intelligence Review*, 56(12), 1–37.
42. Muhammad, K., Rodrigues, J. J., Kozlov, S., Piccialli, F., and de Albuquerque, V. H. C. (2020). Energy-efficient monitoring of fire scenes for intelligent networks. *IEEE Network*, 34(3), 108–115.
43. Adadi, A., and Berrada, M. (2018). Peeking inside the black-box: A survey on explainable artificial intelligence (XAI). *IEEE Access*, 6, 52138–52160.
44. Saiharsha, B., Diwakar, B., Karthika, R., and Ganesan, M. (2020, June). Evaluating performance of deep learning architectures for image classification. In: *2020 5th International Conference on Communication and Electronics Systems (ICCES)* (pp. 917–922). IEEE.

7 Robotics and AI in Mechatronics Industry

Geetha M., Suresh Seetharaman, Parimala Gandhi A., and Gomathi E.

7.1 INTRODUCTION TO ROBOTICS AND AI IN MECHATRONICS INDUSTRY

Industrial robotics and mobility generally involves robots, cobots, and autonomous systems performing various tasks in industrial warehouse settings [1]. Industrial robotics engineers use them to automate repetitive, dangerous, and high-precision tasks. Examples include welding, painting, assembly, packaging, sorting, and quality control in various industries. E-commerce growth has fueled the need for efficient logistics making mobility crucial. Mobility enables the seamless movement of people and goods from one place to another. This can be seen in the demand and innovation around autonomous vehicles, cranes, shuttles, and autonomous storage and retrieval systems (ASRS). There has also been significant growth in the software systems and programs required to operate these solutions. These technologies enable injury-free work environments with increased productivity.

Although cobots do not require a distinct place to work, they are far less expensive to work than conventional industrial robots. This shows that they do not need cages for support and shielding from people to function without risk on a typical plant floor. They can select parts, carry out production tasks like sanding, polishing, and screwing, and run traditional manufacturing works like stamping and injection molding presses. They can also utilize cameras equipped with computer vision technology to carry out quality control inspections.

Automotive manufacturers, such as BMW and Ford, frequently employ cobots for gluing and welding, lubricating camshafts, putting oil into engines, and doing quality control checks.

7.2 FUNDAMENTALS OF ROBOTICS AND AI IN MECHATRONICS INDUSTRY

Industrial artificial intelligence (AI) stands out in these ways:

- *Infrastructures:* Capable of processing in real time, which guarantees industrial-grade reliability with strict security requirements and interconnection, and are highly valued in hardware and software.

 DOI: 10.1201/9781032615790-7

- *Data:* Large-volume, high-velocity data in a certain range of units, products, regimes, etc. are necessary for industrial AI.
- *Algorithms:* They necessitate the combination of digital, heuristic, and physical knowledge. Highly complex because of governance, deployment, and model management.
- *Making Decisions:* In an industrial setting, there is typically very little room for error, thus managing ambiguity is difficult. Efficiency is especially crucial for solving large-scale optimization issues.

The primary goal of industrial AI is to provide tangible value by reducing waste, enhancing quality, improving operator performance

7.3 ROBOTIC MANIPULATORS: KINEMATICS, DYNAMICS, AND CONTROL

Telemanipulators and the capacity for machine numerical control [2]. Telemanipulators are remotely operated devices which have a gripper and an arm. The commands the person gives through his control device are followed by the gripper and arm. Radioactive material was handled by telemanipulators first.

Machines can be accurately controlled with respect to a stated coordinate system. It was first applied in 1952 at MIT, where it was issued in the formation of the first machine programming language, called APT (Automatic Programmed Tools). The first programmable telemanipulator was developed by integrating these two methods. In 1961, the first industrial robot was employed on these principles familiarizing with the industrial settings, such as auto factories. The aim of spontaneous mobility in autonomous networks and production mechanism propelled the formation of mobile robots. The former resulted in the development of driverless transport systems that were made use of in the late 1970s on factory floors to transfer items to various stages of the production process.

Robots are solid devices that use physical manipulation to perform tasks. They possess effectors to apply physical forces to their surroundings and sensors to sense environment. As mentioned, there are three primary types of robots: manipulators, mobile robots, and humanoid robots.

7.3.1 AI and Robotics

The robot can be controlled by a software that interprets sensor data, makes decisions about what to do next, and then commands the effectors to take action in a realistic manner.

7.3.2 Hardware

7.3.2.1 Sensors

Robots and their environment are perceptually interfaced by sensors. Passive sensors, such as cameras, are capable of detecting signals caused by external sources in the surrounding environment. However, there are also active sensors that release energy into the surroundings, such as sonar, radar, and lasers. Environment-based items reflect this energy. The necessary information can then be collected using these reflections.

In general, active sensors yield more data than passive ones. However, they also use more electricity. This may cause issues for mobile robots that depend on batteries to carry their energy with them. Regardless of whether they are active or passive, we have three different types of sensors.

Record the distances to objects

- Produce an entire image of the environment
- Measure a property of the robot itself.

Range finders are used by many mobile robots to determine how far away items are from them.

Radar and laser are effective for long-distance measurements, while certain specialized sensors can measure extremely large distances. For close-range sensing, options include touch-sensitive skin, bump panels, and whiskers.

Imaging sensors constitute the second significant class of sensors. These cameras capture photos of the surroundings, which can subsequently be subjected to image recognition and computer vision algorithms for analysis.

Proprioceptive sensors are the third major class. These provide the robot with self-awareness. Shaft decoders, which count motor revolutions in tiny increments, are frequently installed on motors in robotic joints to capture the configuration of the joint. Using torque and force sensors is important when handling breakable items or when positions are unclear. Imagine a robot manipulator weighing a ton putting in a lightbulb.

7.3.2.2 Effectors

Robots move, alter their body shape, and interact with their surroundings through the use of effectors. To understand how a robot interacts with its environment, the concept of degree of freedom (DOF) can be used as an abstract notion.

Each independent direction in which a robot can move is counted as one degree of freedom (DOF), as illustrated by the example of an autonomous underwater vehicle (AUV). DOFs can be categorized into three spatial directions: x, y, and z, as well as three angular orientations: pitch, roll, and yaw. Additionally, a dynamic state may be included, which represents the rate of change for each kinematic dimension. Non-rigid robot bodies may have more DOFs due to their ability to deform. For example, the human wrist has three DOFs: rotation, side-to-side movement, and up-and-down motion.

Based on the placing an object, like a hand, at a specific location and orientation requires six DOFs, as shown in Figure 7.1.

The manipulator consists of one prismatic joint (P) and five revolute joints (R), for a total of precisely six DOFs. Prismatic joints produce sliding motion, whereas revolving joints produce rotational motion. You can see that your arm has more than six DOFs by using it as an example. You can still rotate your elbow even if you place your hand on the table. It would be very difficult for robots to operate with only the minimum number of DOFs. Manipulators with more DOFs are needed.

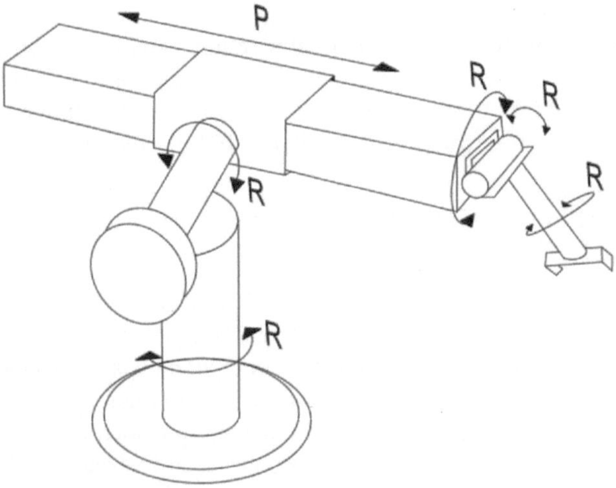

FIGURE 7.1 Arm robot.

Robots that move are a unique breed. It is not necessary to have actuated elements that correspond to the number of DOFs. Consider an automobile. It has two DOFs and can move in both directions. However, you will discover that the car is three-dimensional if you explain its kinematic configuration. The user can drive their automobile in any direction and to any (x, y) position on a level surface, such as a parking lot. As you can see, the vehicle has two controlled DOFs and three effective DOFs. If a robot has more effective DOFs than controllable DOFs, we call it nonholonomic; if the two numbers are equal, we call it holonomic.

7.3.2.3 Movement

A unique class of effectors for mobile robots are the locomotion systems, such as wheels, tracks, and legs. Two independently actuated wheels, one on each side, make up the differential drive. The robot travels at the same speed in a straight path when both wheels turn in opposition directions. The synchro drive is essential for enabling each wheel to rotate independently. Chaos might easily result from this. However, your robot is safe if you guarantee that every wheel points in the same direction and moves at the same pace.

If the controllable DOF is less than total DOF, nonholonomic drives are selected. Holonomic drives, which typically consist of three or more wheels and allow for independent movement and orientation, are used by advanced robots.

7.3.2.4 Power Sources

To operate their effectors, robots require power. The electric motor is the most often used mechanism for both manipulator actuation and mobility. Other methods that could be used are hydraulic actuation with pressurized fluids and pneumatic actuation with compressed gas. They are not commonly utilized, although they do have some application niches.

7.3.2.5 Bits and Pieces

A majority of robots have digital communication capabilities, such as wireless networks. Particularly now, such modules have become more affordable. They can serve as a back link to the robots' home station or a means of communication between robots. Lastly, a body frame is required to hang all the components

7.4 ROBOT PERCEPTION AND SENSING USING AI IN MECHATRONICS INDUSTRY

Robotic perception [3] is a fast-growing field that allows robots to perceive, understand, and communicate with their surroundings. Because of advancements in sensor technology, machine learning algorithms, and computer capacity, robots are now able to perform tasks in industrial, medical, and other environments.

Numerous advantages come with robotic perception, such as lower costs, higher efficiency, and enhanced safety. But there are obstacles to overcome, like limitations in sensor technology, processing complicated sensory input, and combining perception with other robotic skills.

It will take ongoing research and development funding as well as cooperation between subject-matter specialists to address these issues. However, there are a lot of potential advantages to robotic perception, and if it develops further, it will surely change how humans use and interact with robotics.

7.4.1 The Key Technologies in Robotic Perception

A number of essential technologies are essential to robot perception. Robots can effectively observe and interact with their environment because of the technologies discussed below.

The AI field of robot is to empower computers to understand and evaluate pictures and videos. Robots detect and identify objects, people, and other items in their environment with the help of computer vision technologies.

7.4.1.1 LiDAR (Light Detection and Ranging)

LiDAR sensors locate and gauge an object's distance from its surroundings using laser light. LiDAR sensors are widely used by organizations in robots for obstacle recognition and 3D mapping.

7.4.1.2 Radar

Another tool for ranging and object detection is radar. Beyond LiDAR's detection range, items can be seen using radar devices, which employ radio waves to detect objects.

7.4.1.3 Ultrasonic Sensors

High-frequency sound waves are used by ultrasonic sensors to measure distance and identify the presence of objects. Robotic engineers incorporate these sensors to enable objects' detection and obstacle avoidance.

7.4.1.4 Inertial Measurement Units (IMUs)

IMUs measure the robot's orientation, acceleration, and magnetic field by combining gyroscopes, magnetometers, and accelerometers. IMUs are used by robots for gesture detection, motion tracking, and localization.

7.4.1.5 Global Positioning System (GPS)

The robot can navigate and track its precise location/movement—all because of innovative technology. This technology is frequently used by robotic system manufacturers in outdoor robots, like drones and autonomous cars.

7.4.1.6 Microphones and Speech Recognition

Robots are equipped with microphones and speech recognition technology to facilitate human–robot communication, enabling users to instruct the robot in normal language.

7.4.1.7 Need for Robotic Perception

A robot's power to engage with its environment in a purposeful way relies heavily on its capacity for perception and sensing. These abilities are used by robots for path finding, object handling, user interface, and ambient awareness. The robots can analyze environment, identify barriers, and prevent impacts.

Robots have the ability to sense an object's position, size, and shape to process the operations that need to assembly, sorting, and packaging. In situations where robots and humans communicate, robots must be able to identify and interpret human movements and words. Robot designers equip robots by enabling them to assess and measure several environment parameters for efficient observation.

Technologies for perception and sensing would unlock robots' full potential, transforming industries through enhanced automation and precision.

7.4.1.8 Limitations of Robotic Perception/Sensing

In recent years, although robotic sensing and perception have improved, manufacturers still need to solve these technologies' limitations. One of the primary obstacles is to identify robots' poor context awareness, which can make it challenging for them to understand the significance of certain things.

Furthermore, the limited range of certain sensing technologies makes it challenging for robots to identify distant objects or dimly illuminated situations. Another drawback is that some sensor systems are susceptible to variations in weather or lighting, which may affect the accuracy of their data.

Additional obstacles hinder the efficacy and precision of robotic perception. Despite these challenges, ongoing research in this area helps mitigate them, thereby improving the functionality and usability of robotic sensing and perception technologies.

7.5 ROBOT PLANNING AND NAVIGATION USING AI IN MECHATRONICS INDUSTRY

The path planning of mobile robot control has made use of reinforcement learning techniques along with integrated representations of the environment.

While the algorithm traverses the nonholonomic model, sensor measurements provide it with up-to-date information on the environment and its dynamic components. It is not necessary to estimate the forms of the barriers or even know their velocities when using the approach that is described. We have used extensive computer simulations to verify our path planning solutions' effectiveness [4].

The path planning algorithm based on an integrated representation of the environment data has been applied to the presented technique for the purposes of validation and comparison. An integrated representation of the environmental data serves as the foundation for the planning algorithm.

7.6 ROBOT LEARNING AND ADAPTATION USING AI IN MECHATRONICS INDUSTRY

7.6.1 Explaining the Significance and Circumstances of Robotic System Adaptability

A number of aspects are for comparing and evaluating adaptive robotic systems based on the needs, technological structure, and accomplishments of the robotic systems. From an applied and conceptual standpoint, we try to offer a nomenclature that is both general and distinctive. Our goals in this regard do not include a comprehensive and all-encompassing assessment of the genuine applications that are now in use.

7.6.2 Examining and Classifying the Relevant Adaptive Robotics Work

Early studies in robotics, control, and machine learning research groups may be cited or correlated with the idea of adaptive robotics. A brief assessment of the relevant research and an abstract taxonomy are offered to help identify the most crucial practical goals. The primary objective of this taxonomy is to pinpoint the methodological bounds that will be applied in the following contribution: the mathematical characterization of adaptability for robotic systems. It is not meant to classify the accomplishments based on their technical complexity or application.

7.6.3 Redefinition of Adaptive Robotics' Standard Mathematical Framework

Framing the challenge as an adaptive robotic decision-making system draws maximum effort. Next is formulation based on mathematical models that draw inspiration from Markov Decision Processes and Control theory, then developing robust and flexible robot programs for uncertain environments, and, ultimately, providing hope for the next step of automation, whether it be for social services or industrial manufacturing [5].

7.7 HUMAN–ROBOT INTERACTION USING AI IN MECHATRONICS INDUSTRY

Figure 7.2 shows the human–robot interaction (HRI), which can be done in various terms such as coexistence, cooperative, and collaborative. In coexistence, the workplace

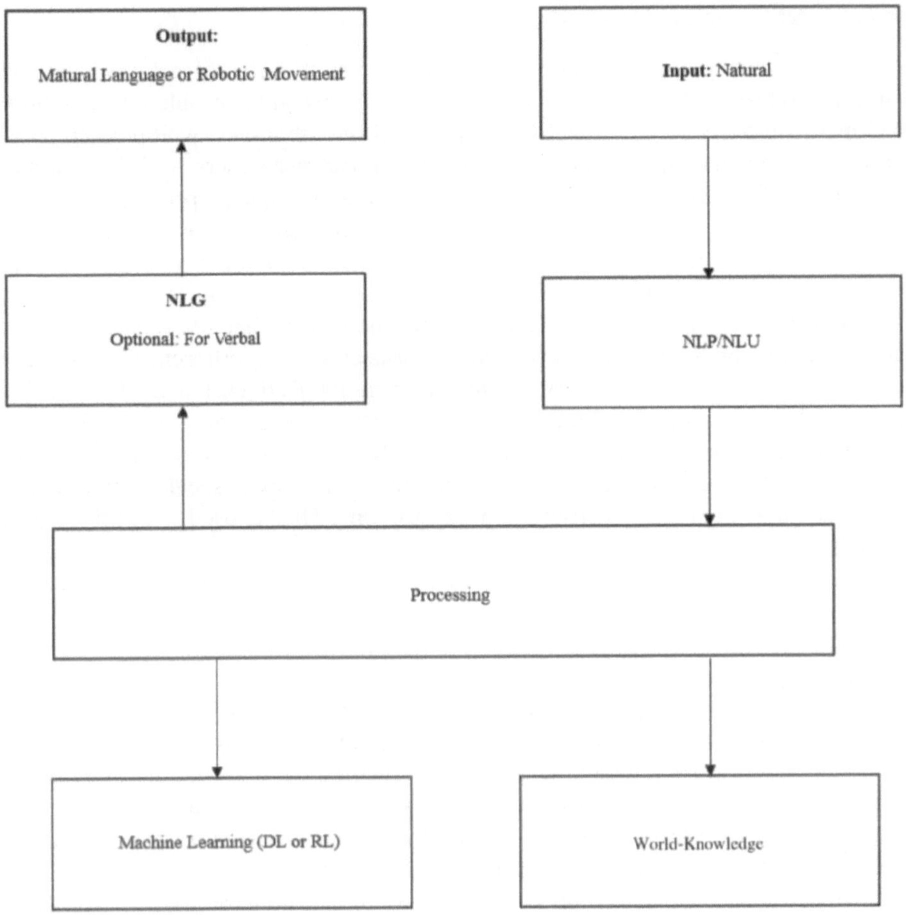

FIGURE 7.2 Human–robot interaction.

is shared between people and live robots side by side, without any boundaries. Both humans and robots operate in the same workspace, but they do not work on the same component at the same time. Instead, their movements are sequential. Cooperation method involves a person and a robot on the work component and at the same time they both will be in motion. Collaborative robotics refer to robots that respond immediately to the movements of human workers. This model is based on various HRI systems currently available. Verbal HRI input is given in natural language which is pre-processed by natural language processing (NLP) and natural language understanding (NLU) system. Then it is processed by machine learning (ML) technologies which use either deep learning or reinforcement learning, or sometimes both. We need World-Knowledge (for language understanding), i.e., real-world mapping for correlating input to actual objects in surroundings. We will review different AI technologies which are in practice for verbal HRI, their advantages and pitfalls, and further advancements [6]. NLP and ML, two AI technologies, have changed HRI field drastically and so it is crucial to review them.

7.7.1 NLP and NLU in HRI

NLP is the technology that understands the process in terms of natural language. NLU is a subset of NLP, but there are differences between the problem they address and methodologies they follow. NLP and NLU differentiation is well pointed out in article. There is a confusion of understanding among researchers, so it is important to point out the differences. NLP is basically "natural language processing" which includes part-of-speech tagging, text categorization, segmentation, machine translation, and syntactic parsing, whereas NLU is prominently about understanding the natural language which constitutes different techniques such as semantic parsing, sentiment analysis, summarization, working mechanism of chatbots, dialogue agents, and paraphrasing. So, we can say that NLU works on gaining different meanings and context of the language by understanding the grammar. Both NLP and NLU then use ML and deep learning techniques for post-processing and learning mechanisms. As mentioned in Section 7.1, the HRI differs from the applications of robots. Command to a robot for movement is a one-sided dialogue where human speaks, and output is a robotic movement for an activity. In that case, only NLP is used with ML or deep learning.

There are other types of HRI systems such as humanoid robots, social robots where robots also reply verbally, in which case NLU and natural language generation play an important role. For pure one-to-one verbal communication, a robot needs to understand what a human is saying. So, to understand that a robot needs to understand the semantics, grammar, and meaning of the language. An implemented system based on embodied construction grammar and simulation semantics is described. While considering HRI, today the research is ongoing in NLU, but the irony is that most of the HRI systems today are not pure natural systems but based on basic NLP techniques like test processing and pattern recognition. NLU is primarily based on capturing syntax and grammar construction from the first stage and using World-Knowledge, semantics, and context of verbal communication. So, NLU first focuses on the semantics of the language. Semantics are derived from two models: literal models and interpreted models. Literal Models use part-of-speech tagging techniques of NLP, word dependencies and syntax structures to analyze the meaning. Therefore, NLU uses basic NLP techniques first. Interpreted models help to extract situation-related information which are developed from multi-modalities. So, literal models crack the semantics and grammar with the help of interpreted models ready for further execution plan generation and Knowledge-World mapping in NLU; execution plan generation is used to plan robot output execution based on task knowledge which is already extracted from natural text or command as explained above. Execution plan generation uses three different models: probabilistic model, a model based on Hidden Markov Models, and Naive Bayes model.

7.7.2 Logic Models and Cognitive Model

Knowledge-World mappings are developed for the real-world situation to act or respond for input given, which contains real-world referents. It is crucial for robots to continuously conceive and acquire new information by performing interaction with the environment and humans. To understand the semantics of the language, Combinatory Categorial Grammar lexicons are effectively used for HRI systems. CCG is the state-of-the-art system for mapping natural language to meaning representations using extensive calculus methods.

Another novel processing of natural text is Language Grounding which comes after pre-processing. Commanding robots with language require both semantic understanding and subsequent result generation with reference to the real-world referents mentioned in the command. Referring to an object "the red ball" or "the tall man" and mapping them to the surrounding world is an example of Symbol Grounding Problem. Symbol Grounding involves a technique for connecting semantics interpretations and information to real-world data from its sensory perceptron. Language can be better learned when it is presented and interpreted in the context of real world. So, Grounded Language acquisition is a bridge which joins the meaning of communication and real-world data or objects to apply to the physical world. Natural Language Generation is a sub-area of NLP in which a lot of research has been done recently. NLG is used for those interactions where an output from the robot is expected in verbal format or in form of natural speech.

HRI is a much-needed technology in industrial robotics because of its advantages in implementing command-based interaction resulting in the movement of a robot. It also discussed different AI technologies, especially NLP and NLU, and deep learning methodologies involved in HRI. The important argument is about natural language. Natural language is ambiguous and there are various factors that must be considered when we process or analyze verbal input. Few aspects are these: first is semantics, the meaning of natural language is a basic significant part. In language, two words having the same length, characters, and pronunciation can have different meanings (e.g., the word "bat"). The second aspect is context of the language. To interpret correctly, contextual information is essential. Computers sometimes are not able to remember contextual information and then it becomes a troublesome situation for them to analyze. Semantics and context of the natural language go hand in hand. When it comes to syntax, computers are great in processing or manipulating them, but when it comes to meanings and understanding rules or concepts, then they struggle. So, when you go with the words other than those on which HRI system is trained, your system is likely to encounter difficulties. It is very easy for us to learn a new language, new words, verbs, but it is not the same when it comes to computers. After reviewing many research papers and work, and different robot HRI systems, we infer that all major technologies of current HRI systems are based on big data and basic NLP methods. We believe that the task of language understanding and teaching language to robots is among the most significant things to be included in HRI methodologies. The field of language understanding with deep learning is expected to make a large impact in the coming few years, while Reinforcement Learning with NLU and deep learning will also advance in technology. Currently it is somewhat difficult to teach robots natural language. We expect that RNN and MT should play a more important role in language understanding along with Reinforcement Learning. Reinforcement Learning is in frequent use for teaching movements to robots and in Game Playing, but we believe that it can be also used for language understanding with semantic parsing.

7.8 COBOTS USING AI IN THE MECHATRONICS INDUSTRY

Designs with reliability and safety in mind, this can also improve workers' performance and working conditions, while increasing human trust in collaboration. During HRC, people and robots collaborate in the same space. Cobots are made to halt before they cause injury to a human teammate inadvertently. Cobots should also be lightweight to minimize their inertia and facilitate sudden stops. Even now, certain cobots can be trained to carry out logistical duties by having other people simply point their arms once.

By integrating AI-powered robots, we can bridge the gap between humans and technology, enhancing business models to better align with consumer expectations. In industrial workplaces, these robots help ensure worker safety and perform difficult tasks such as cutting, grinding, welding, and inspection autonomously. The ability of AI robots to complete tasks on the machine learning platform further enhances their capabilities.

7.8.1 Cobots

The goal of contemporary cobot research is to make cobots as intelligent and capable as humans in terms of vision, cognizance, learning, adaptability, and manipulation. For tasks like welding, assembly, safety inspections, material handling, polishing, etc., ergonomic HRC must be improved by building robot motors, perception, control planning, and learning in a secure way. The following are research issues that researchers are working on to find cognitive solutions in the industry. Cobots learn and develop skills without specific coding, mimicking human perception and movement to complete a physical activity. It has the ability to process a challenging task in a large open space. Research is progressing on the human–robot collaborative work environment at psychological empowerment with international standards ISO 10218-1:2011 and ISO/TS 15066:2016. These standards can offer substitutes for physical barriers that maintain safety and well-being. Expanding the scope of true collaboration requires operators to interact safely with an industrial robot with little to no training throughout the full production cycle.

By applying creative interaction strategies that make use of motion control tools, augmented reality, and multi-sensory interfaces, robots operating in manufacturing tasks can be made safer and easier to operate. These methods also result in less expensive training and quicker prototyping. Additionally, new technology can be employed to enhance the safety of collaborative robots, making them more efficient and safer to work with.

Industrial robots help automate tasks, such as helping people lift heavy parts or working on assembly lines, and these supply chain robots transport products within buildings. Given that both the cobot and the robot are capable of doing comparable tasks, the distinctions between the two methodologies were exhibited to illustrate their respective applications and to indicate which is superior in specific situations. The benefits and drawbacks of cobots were talked about. The cobot's performance can be influenced by a number of metrics, such as control strategies, pre-processing methods, and sensing approaches.

An overview of the different types of robots and cobots, tools for detecting or simulating, the task and its possible locations, and the various kinds of AI-based control technology are presented in this work. ML and deep learning approaches were used in numerous evaluated studies to manage the cobot task. This evaluation also covered the results of the chosen papers, including topics such as robot ability, training procedure, safety concern, accuracy, and time delay. Ultimately, recommendations and future directions for the cobot–AI domain interaction were given by this systematic review.

7.9 SWARM ROBOTICS USING AI IN MECHATRONICS INDUSTRY

A swarm is typically made up of several members that are identical and move forward asynchronously.

Compared to the group as a whole, these people only possess one skill: they lack intelligence and are unable to accomplish the swarm's objectives alone. Furthermore, it has been demonstrated that complex collective actions can be replicated by group members.

Self-organization is important for decentralization. Combining positive feedback, negative feedback, randomness, and many interactions is essential for self-organization. Over time, the residues of these pheromones will vanish, lowering their appealing potency. This evaporation process takes less time to reach shorter trails.

The ants therefore most likely go through them more frequently than lengthy ones. In this way, nature provides an answer to the conundrum of determining the quickest route between a colony and food source (see Figure 7.3).

7.9.1 Features

Swarm robots possess three main features.

- *Robustness:* Though some components may fail, the system must still be able to complete the intended mission. The main benefit of having autonomous agents in a distributed system is that the system as a whole is robust even when certain agents fail.

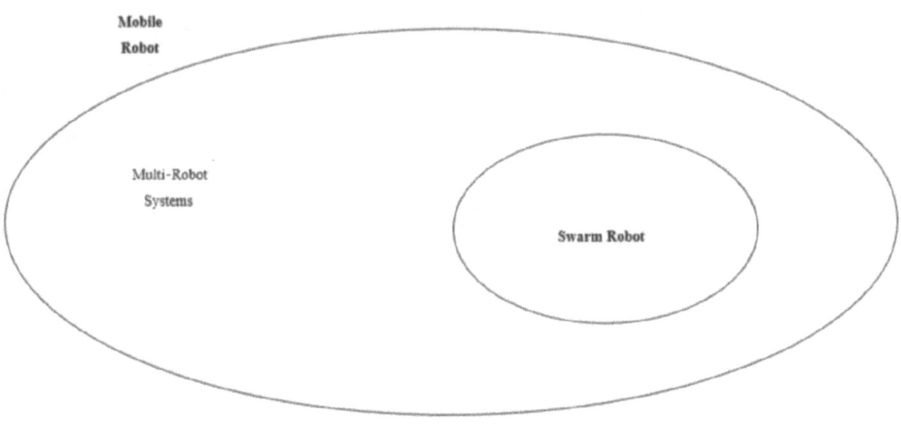

FIGURE 7.3 Delimitation of fields: mobile robotics, multi-robot systems, swarm robotics.

- *Flexibility:* Systems have to be adaptable. Despite having limited resources for observation and communication, they must manage a variety of roles and responsibilities.
- *Scalability:* The system has to work even with a smaller number of components. The performance of the mission cannot be changed by the addition or removal of personnel.

7.9.2 Comparison of Swarm Robotics and Multi-Robot Systems

A collection of multi-agent systems, or multi-robot systems (MRSs), has robots as its individual agents. These systems incorporate numerous self-governing robots that communicate with one another while possessing specified objectives and sensory data. The robots are programmed to take on a specific shape. Additionally, swarm robotics generally refers to devices with limited resources. MRS is not limited to this scope; it frequently manages autonomous robots in complex supply chains or in the industrial sector. The acting capabilities are a key point of differentiation between MRS and swarm robotics. Swarms are often homogeneous groups of agents with limited individual capabilities. Individual acting capability is seen by MRS as a crucial component in many appliances. Additionally, these systems have a significant degree of heterogeneity, which is uncommon in swarm robotics.

7.10 AUTONOMOUS ROBOTS USING AI IN MECHATRONICS INDUSTRY

Autonomous robots are designed to carry out activities without humans, such as drone aircraft. They vary with size, cost, AI, mobility, dexterity, and utility, from robotic process automation and data capture capabilities. An increasing number of autonomous robots are equipped with AI, allowing them to perceive and comprehend.

Autonomous robots have the ability to improve routine operations speed and accuracy, especially in manufacturing and warehousing, and decrease the possibility of worker harm under hazardous conditions.

Autonomous robots are expected to play a significant role in the supply chain, particularly in manufacturing, final assembly, and warehousing. In the future, these robots are anticipated to take over various segments of the supply chain, freeing up human labor for higher-value, less hazardous, and more strategic tasks.

Future supply chains will use autonomous robots more frequently to function more like humans.

For instance, advancements in haptic sensors provide the sensation of touch, which will enable robots to grab objects without requiring modifications to programming or mechanical elements.

With increasing sophistication, autonomous robots are able to operate alongside humans, with shorter setup times, and reduced need for supervision. Autonomous robots improve the ability to function autonomously with increasing quality and productivity.

End-to-end supply chain process will become more seamless as the market for autonomous robots enlarge. Nowadays, a number of businesses employ autonomous robots for certain tasks, using different robots to test improvements. Robots that make robots may become the standard in supply chains in the future as creative businesses develop and grow, helping to optimize manufacturing processes.

The main ways that autonomous robots collaborate with supply chain innovation are by lowering operational expenses and by raising potential revenue. In particular, self-governing robots can improve the productivity and efficiency, decrease the rate of error, rework, and risk, as well as enhance worker safety.

7.11 INDUSTRIAL ROBOTICS USING AI IN MECHATRONICS INDUSTRY

Industrial robots are autonomous robotic devices that have the ability to swiftly and accurately execute various manufacturing operations. For repetitive tasks, they can operate entirely on their own, or they can be partially supervised by humans, AI, machine vision components, visual sensors, or other guidance systems. They are largely motivated by positional data and predefined tasks, and they are programmed to function in sync with the other machines in a manufacturing run. Industrial robots come in a wide variety of forms. The following list offers a quick overview of their variations:

- Collaborative robots,
- Cartesian "X-Y-Z" robots,
- Articulated "robot arm" robots
- "Wearable" mobile robots
- Assembly of Selective Compliance Robot Arm, also known as "SCARA,"
- Robots-Delta;
- Cylindrical-coordinate;
- Spherical "rotary";
- Parallel connection; and serial manipulators

7.12 MEDICAL ROBOTICS USING AI IN MECHATRONICS INDUSTRY

AI is increasingly being used in healthcare for diagnostics and white-collar automation. However, the field of medical robots might be a little underdeveloped. This is probably due to laws pertaining to mechanized surgery [7]. AI software is gradually making its way into the medical robotics industry, and it is expected to play a much larger role in the future, driven by increased funding and a growing pool of AI talent in medical robotics companies. In particular, it investigates: What AI can do for medical robotics and how is it already being used by healthcare clients? Very few case studies that demonstrated a hospital or health network's effectiveness with AI-based medical robotics were available to us.

The current state of AI in medical robotics vendors includes an assessment of the AI expertise within these firms and methods to verify if a vendor's program is actually utilizing AI.

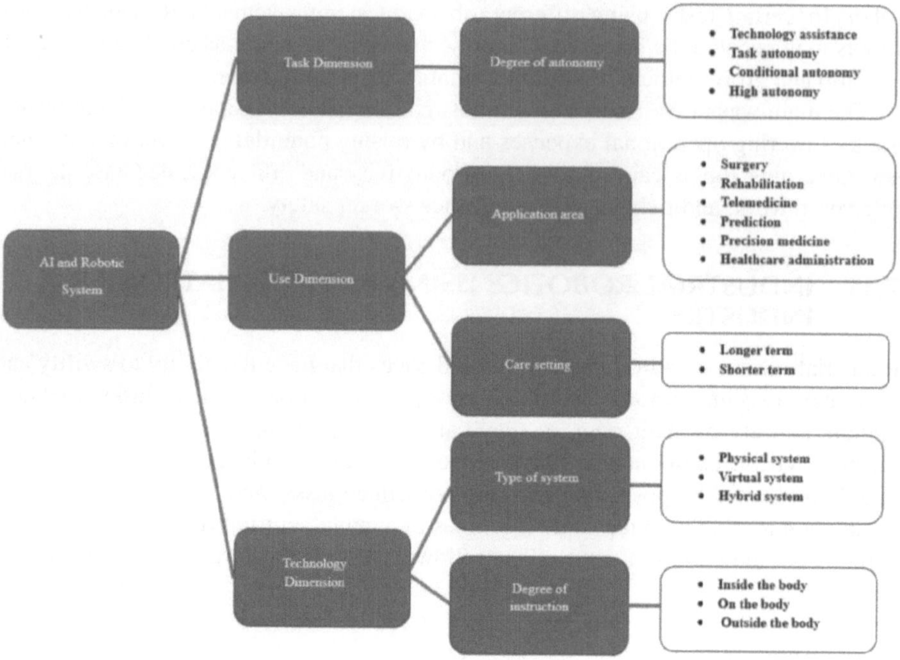

FIGURE 7.4 Classification based on application area.

7.12.1 AI and Robotic Systems Classification in Medicine

Healthcare robot is categorized by three factors: technology, task, and use. The level of autonomy of the system defines the "task" dimension. Figure 7.4 shows the various application areas of AI in robotic systems [8].

7.12.2 AI and Robotics in Surgery

Robotics-assisted surgery, which involves the use of a mechanical device to assist surgery instead of a human being, is increasingly impacting common surgical procedures, particularly minimally invasive surgery. In surgery, three types of robotic systems are employed: Master–slave systems have no autonomous components; they are fully dependent on the actions of the operating surgeon. Active systems carry out pre-programmed functions while staying under the supervision of the surgeon. Semi-active systems enable a surgeon to supplement the pre-programmed component of the system. The hand movements are sent to surgical instruments during teleoperation or laparoscopic surgery, and the devices replicate those movements.

7.12.3 Robotics and AI for Rehabilitation

Rehabilitation functions like risk avoidance, treatment, and monitoring are supported by AI robotic system. Fall detection systems are the example of a system that uses wearable technology or smart sensors installed in an area to automatically notify family members, emergency services, or medical personnel when help is needed. With time, AI enables these systems to learn and recognize people's typical behavioral patterns and traits. Additionally, systems are able to evaluate environmental concerns, such as the presence of fall hazards (like stairwells) or off-light home objects. Physical systems can perform monitoring, therapeutic social functions, and physical aid (lifting objects, opening doors, etc.). With the use of exoskeletons, robots can assist stroke patients in regaining their ability to move and in replacing lost function.

7.12.4 Telemedicine AI Robotics

Telemedicine system functions include triage, diagnosis, non-surgical treatment, surgical treatment, monitoring, consultation, and specialty care.

To ascertain the severity of a patient's illness and the priority of therapy, medical triage evaluates the patient's present symptoms, indicators, and test findings. AI-powered mobile health apps are becoming increasingly common for diagnosis and therapy optimization.

Internet-of-Things (IoT) technologies let "smart homes" incorporate wearables and smart mobile devices. They can gather contextual and patient data, help people with daily living, track their progress toward personalized care plans and rehabilitation objectives, send out reminders, and notify caregivers when help is needed.

In addition, physical systems exist that are capable of providing specialized care: Robot DE NIRO can intelligently retrieve or move objects, autonomously traverse through areas on demand, and engage in natural, dependable, and safe HRI.

7.12.5 AI Robotics for Precision Medicine

Precision medicine customizes interventions based on the unique patient, taking into account their genetic variations and other relevant characteristics such as age, gender, and ethnicity. Applications for digital health can also include information about food intake, activity level, emotional state, and so forth.

- Adoption Obstacles for Robotics and AI in Healthcare
- Opposition to Technology
- Openness and Explicitness
- Liability, Accountability, and Responsibility
- Data Integrity, Protection, and Quality
- Security and Safety

7.13 AGRICULTURAL ROBOTICS USING AI IN MECHATRONICS INDUSTRY

Mechatronics and robotics are becoming increasingly significant in agricultural production and administration. The robots are capable of multitasking, having high sensory acuity, being consistently operational, and being able to work in challenging environments. For many years, agricultural practices have employed electromechanical

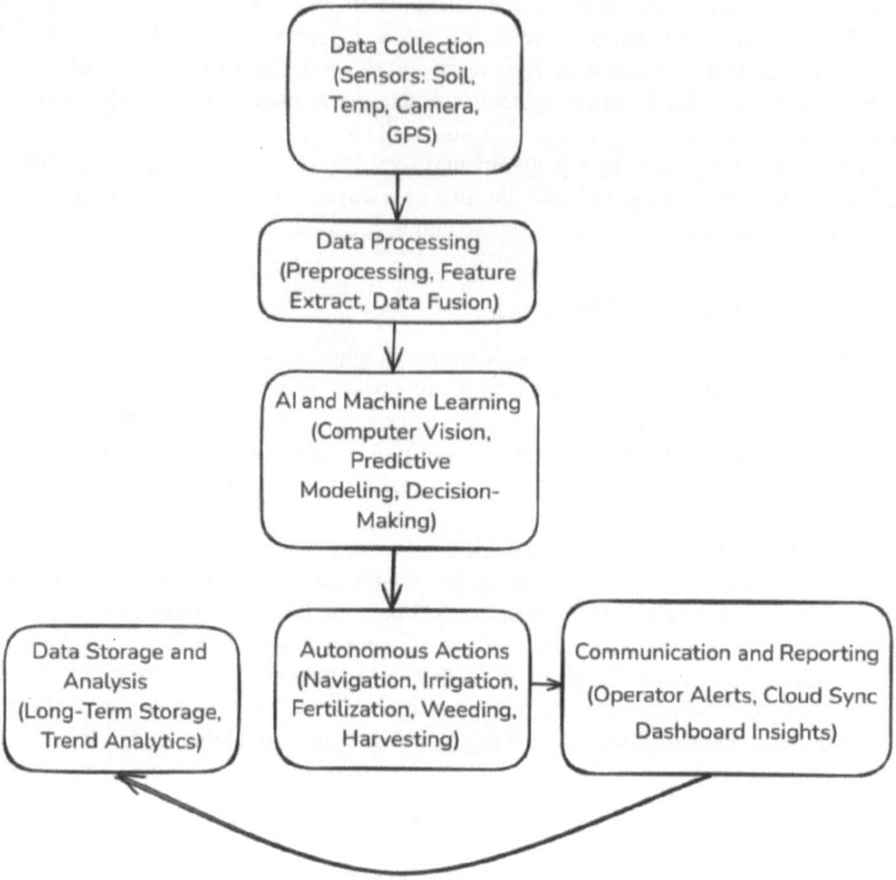

FIGURE 7.5 Agricultural robot.

systems. Research on agricultural robotic systems using various precision farming tools is ongoing on a global scale. This study discusses the features and shortcomings of robots that have been developed globally. Indian Agricultural Research Institute (IARI) has additionally developed robotic planters. It has been discussed how far autonomous vehicles can go [9].

Agricultural robots, shown in Figure 7.5, can help offset the shortage of agricultural laborers and growing global population. A "farm robot" that combines machine vision, image processing, and mechatronics can provide the best answer for autonomous agricultural operations in terms of accuracy, cost, less labor-intensive work for humans, and environmental preservation.

7.13.1 MF-Scamp Robots

The creation of MF-Scamp robots was reported by Pedersen et al. [10]. This robot is intended for harvesting, weeding, and scouting. Either four or six wheels are driven by it. The robot's intelligent hoe tools identify the crops in rows and columns using a vision sensor. It then precisely guides itself after that, and herbicide use has decreased as a result of this. Moreover, weeds between the crops are identified using a color sensor. By using fewer tractors, this type of design lowers the cost of spraying. The increased expenses for small farmers are one of the system's drawbacks (see Figure 7.6).

Madsen and Jakobsen had already designed this platform back in 2001. After that, Aalborg University in Denmark developed it as a third-generation autonomous research platform for precision farming. With two motors for each wheel, one for traction and the other for steering, it boasts four-wheel drive and four-wheel steering, allowing the robot to move with great mobility [11]. The robot has a track width of

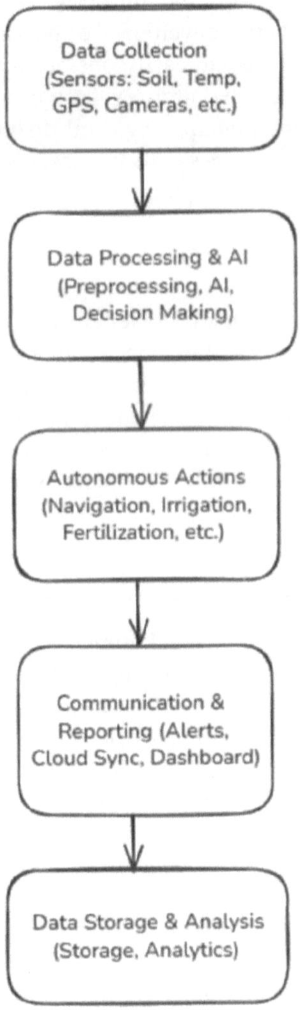

FIGURE 7.6 MF-Scamp robots.

1 m and a high clearance of 60 cm. The platform has a Real-Time Kinematic Global Positioning System installed. For functions like weeding tools, sensors, or spraying devices, there is a unit above the head of the frame.

Farm management software has an anti-collision system and a navigation system that calculates route plans in the field [12]. The orientation, shape, and size of the weeds can be used to generate weed patches throughout the entire field. GP Sand GIS systems are used to create weed maps of fields. The API platform has a capacity of 4.32 hectares per hour and a speed of 3.6 km per hour. This makes the weed mapping system possible and makes it easier to precisely spray patches with the appropriate herbicide combination. In this manner, it lowers the use of herbicides from 30% to 75%. However, there are two difficulties that need to be addressed with regard to gathering data about weeds and physically pulling weeds.

The inter-row area between crops, the intra-row area among plants, and the near-crop area within the leaf and root envelope are the various working spaces found in agricultural land. The machine can easily operate in the direction of the robot in the inter-row area, but managing the intra-row area is quite challenging as the space is defined by the uneven crop plant spacing. Soil disturbance near crops should be avoided as it can result in soil wetness. To use laser weeding or micro spraying, a system with an intelligent sensor is required.

7.13.2 AgBot II

AgBot II was created by Queensland University of Technology (QUT), Australia, using sensor networks, drones, satellite, and weather. A suitable statistical program has been used to create mathematical models. This is to support farmers in making decisions about irrigation, fertilizer, herbicide, and pesticide applications. A prototype with software-based operation and sensors may operate alone or in groups, with vision sensors enabling robots to "see" if a plant is a rose or a weed and respond chemically or mechanically.

7.13.3 Generalized Architecture of Agricultural Robots

Agricultural (farm) robots use various technologies such as GPS, vision processing, and LiDAR to navigate in the field. An IC engine, battery, and solar power could be the agricultural robot's primary power sources (see Figure 7.7). The fundamental idea behind the agricultural robot is its stability in motion, which is dictated by the motor's speed and revolutions per minute. For the robot to be stable, obstacle recognition and linear and angular movement control are crucial. Effective field operations include the best possible seed placement, weed control, spraying, harvesting, crop monitoring, and so forth. The advancements in autonomous farming robots have the potential to benefit precision farming. This will support ongoing crop status parameter monitoring using various sensing technologies, such as micronutrient availability, biomass index, pest and disease status, water stress, temperature stress, etc., to enable more effective crop remediation. The timely completion of numerous tasks by the farm robot can make up for the shortage of agricultural laborers. A single platform known as a "farm robot" or "agricultural robot" can combine multiple technologies

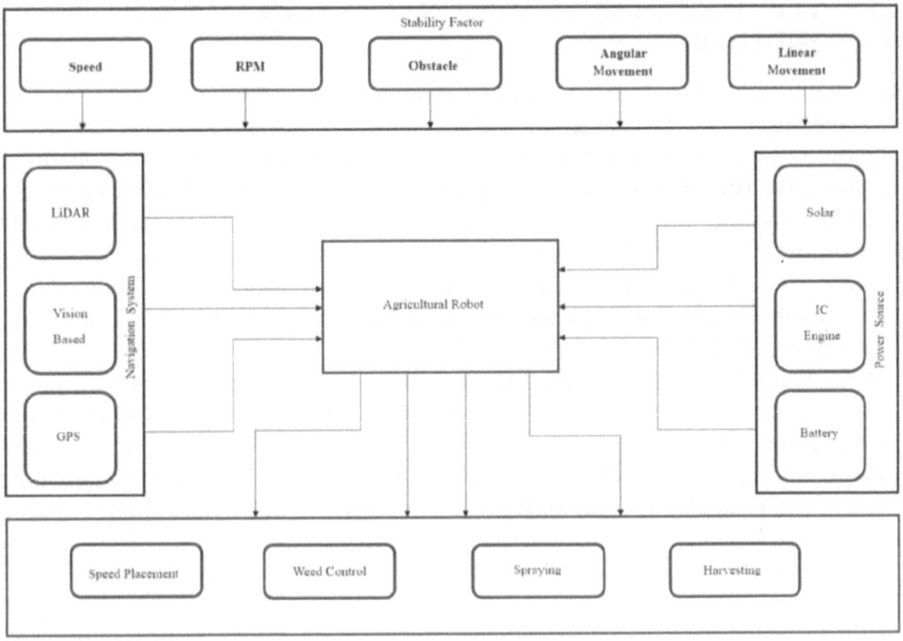

FIGURE 7.7 Architecture of agricultural robot.

such as machine vision, image processing, and mechatronics to provide the best possible solution for autonomous agricultural operations in terms of accuracy, economy, fewer human labor-intensive tasks, and environmental preservation.

The robot's modeling intelligence enables it to comprehend the real world and predict possible outcomes. The following high-level elements constitute the concept of "intelligence": language, reasoning, analogy, planning, common sense, and perception. *Hard generalization and large datasets*: These days, machines are able to translate speech and recognize images with remarkable accuracy after undergoing intensive training on large datasets. These skills are attained by using statistical approximations that are derived from the available data. However, the model frequently fails when the system needs to handle novel scenarios and there is a lack of training data.

Humans can abstract concepts and norms and apply them to a wide range of situations, and we know that humans are capable of performing recognition tasks even with limited input. This degree of abstraction and generalizability is still lacking in the AI systems. *Lack of interpretation and a black box*: The absence of interpretation in the current AI system is another problem. Deep neural networks, for instance, comprise millions of parameters, making it impossible to determine why the network produces positive or negative results. The trained models created from a convolution neural network are frequently uninterpretable, despite some recent work on visualizing high-level features utilizing the weight filters in the network. As a result, the majority of researchers employ contemporary AI techniques as a mystery.

7.14 ROBOTICS IN SPACE USING AI IN MECHATRONICS INDUSTRY

Robots are utilized to carry out manual tasks such as gathering samples for analysis, measuring, and moving around the astronaut [13]. The most common robot now employed in space is the rover. In 2018, NASA created an AI system using photos of the moon. By using a neural network algorithm, astronauts were able to locate planets like Mars and the moon with a high degree of success. Because robots are more likely to survive in space than humans, sending them there has far more benefits. The challenges faced by humans are radiation, distance from earth, gravity, isolation, and hostile environments. The languages that NASA uses are Python, ML, linear regression, C/C++, MATLAB, and Fortan.

NASA employed neural networks in addition to these languages to produce a functional model of the solar system. Various websites serve as a kind of central store for NASA data. NASA and ESA launched probes to Jupiter in 2000, followed by Saturn in 2004. George Devol invented the first intelligent humanoid robot in 1954, and Unimation built it. It was a fully digitally controlled programmable robot.

The exploration of the cosmos has always been a fascinating endeavor, pushing the boundaries of human knowledge and technology. The AI role in advancing space robotics becomes increasingly significant. AI has the potential to transform space exploration, increasing its efficiency, economy, and human safety.

One of the most significant challenges in space exploration is the vast distances and harsh environments that spacecraft and robots must traverse. These variables make it challenging for human operators to control the machines. AI systems are being developed to allow space robots to process and evaluate environmental data and make choices without human assistance. For vast space missions, when communication lag may prevent real-time control, this degree of autonomy is vital.

Another area where AI plays a vital role is in the development of advanced navigation systems for space robots. Traditional navigation methods, such as GPS, are not available in deep space, making it challenging for robots to find their position and navigate accurately. Data from various sensors can be collected using AI algorithms to enable space robots.

7.15 CHALLENGES AND FUTURE DIRECTIONS IN ROBOTICS AND AI IN MECHATRONICS INDUSTRY

Robustness of AI: Almost all ML algorithms suffer from the fact that the majority of current AI systems are easily tricked. AI will undoubtedly have a significant impact on our lives in the future. Humans will become increasingly dependent on AI systems for entertainment, employment, and survival as the amount of knowledge available to us increases.

They can work alongside or in tandem with humans to attain higher service standards in some of these areas, where they can replace labor-intensive human labor and generate new potential applications [14-17]. The age of augmented intelligence is thought to be the next phase of AI. Wearable technology and ubiquitous sensing systems are moving toward intelligent embedded systems, which will essentially be a physical extension of humans. In contrast to AI systems, human senses, information retrieval, and physical capabilities are limited. AI algorithms and cutting-edge sensing technologies could keep an eye on our surroundings and interpret our intentions, enabling smooth communication. Technological developments in sensing and computation technology will make it possible to connect human behavior and brain function to the point where AI self-awareness and emotions can be more realistically observed and replicated. Recently, academic institutions as well as IT companies have shown a renewed interest in quantum computing. The computational capacity it promises is considerably beyond our imagination, possibly important to the field of AI, even if the field is still in its infancy and there are significant obstacles to overcome.

REFERENCES

1. Estifanos Tilahun Mihret. (2020). Robotics and artificial intelligence. *International Journal of Artificial Intelligence and Machine Learning*, 10(2), 22. https://doi.org/10.4018/IJAIML.2020070104
2. Kawasaki, H. (n.d.). Robot kinematics and dynamics. *Eolss.net*. Retrieved January 6, 2024, from www.eolss.net/sample-chapters/c18/e6-43-38-01
3. Premebida, C., Ambrus, R., and Marton, Z.-C. (2019). Intelligent robotic perception systems. In E. G. Hurtado (Ed.), *Applications of Mobile Robots*. IntechOpen.
4. Borboni, A., Reddy, K. V. V., Elamvazuthi, I., AL-Quraishi, M. S., Natarajan, E., and Azhar Ali, S. S. (2023). The expanding role of artificial intelligence in collaborative robots for industrial applications: A systematic review of recent works. *Machines*, 11(1), 111. https://doi.org/10.3390/machines11010111
5. Dias, P. G. F., Silva, M. C., Rocha Filho, G. P., Vargas, P. A., Cota, L. P., and Pessin, G. (2021). Swarm robotics: A perspective on the latest reviewed concepts and applications. *Sensors (Basel, Switzerland)*, 21(6), 2062. https://doi.org/10.3390/s21062062
6. Hemantkumar, G. K., and Giriraj, P. (2023). Mechatronics and robotics for several applications. *International Journal of Research Publication and Reviews*, 4(7), 1868–1875. https://doi.org/10.55248/gengpi.4.723.48570
7. Denecke, K., and Baudoin, C. R. (2022). A review of artificial intelligence and robotics in transformed health ecosystems. *Frontiers in Medicine*, 9, 795957. https://doi.org/10.3389/fmed.2022.795957
8. Deo, N., and Anjankar, A. (2023). Artificial intelligence with robotics in healthcare: A narrative review of its viability in India. *Cureus*, 15(5), e39416. https://doi.org/10.7759/cureus.39416
9. Elbasi, E., Mostafa, N., and Alarnaout, Z. (2023). Artificial intelligence technology in the agricultural sector: A systematic literature review. *IEEE Access*, 11, 171–202.
10. Pedersen, S., et al. (2008). Development of the MF-Scamp robot for precision agriculture. *Journal of Field Robotics*, 25(5), 402–410.
11. Bisgaard, T., et al. (2004). Development of an autonomous platform for precision farming. *Proceedings of the International Conference on Field Robotics*, 256–263.
12. Bak, B., & Jakobsen, T. (2004). Development of farm management software for precision farming applications. *Journal of Agricultural Engineering*, 22(3), 135–142.
13. Yoshida, K. (2009). Achievements in space robotics. *IEEE Robotics and Automation Magazine*, 16(4), 20–28.

14. Foust, J. (2020). InSight mole making slow progress into Martian surface. *Space News* [online]. Available: https://spacenews.com/insight-mole-making-slow-progress-into-martian-surface/
15. Geetha, M., Suresh, S., Jothinathan, K., Lekshmy, P. L., Anish, J. P. M., Swamy, S., and Valarmathi, I. R. (2024). A comparative study of long short-term memory based long-term electrical load forecasting techniques with hyper parameter optimization. *International Journal of Electrical and Computer Engineering (IJECE)*, 14(6), 7080–7089. http://doi.org/10.11591/ijece.v14i6.pp7080-7089
16. Natarajan, C., Janorious Hermia, J., Amutha, J., and Geetha, M. (2024). Enhancing performance of power allocation for VLC networks by non-orthogonal multiple access-MIMO. *International Journal of Electrical and Electronics Research (IJEER)*, 12(2), 421–427. https://doi.org/10.37391/IJEER.120213
17. Amosedinakaran, S., Mano Raja Paul, M., Kannan, S., and Geetha, M. (2023). Performance analysis of battery E-vehicle system using open loop and closed loop application. *Measurement: Sensors*, 31, 100963. https://doi.org/10.1016/j.measen.2023.100963

8 Arm Controlled Waste Picking Robot Rover Utilizing Image Recognition Model Server, iOS, and Microcontroller Application over Bluetooth

Bhaswanth Reddy Isani, Tharun S., Viswa Brahmana Nanda Kishore, Sumit Kumar Jindal, Shubham Mangesh Lad, and Hemprasad Yashwant Patil

8.1 INTRODUCTION

Nowadays we can observe that people are throwing waste everywhere irresponsibly. We do not follow the instructions. We are creating tons of waste every day. Waste includes every waste material such as food, plastic, metal waste, and chemical waste. Even though we are instructed to follow certain rules and regulations about treating the waste by keeping it in bins, we are too lazy and irresponsible to follow those rules. We can observe rampant violations of rules on roads, beaches, hospitals, hotels, schools, colleges, etc. This waste cannot easily break down and dissolve in the environment, especially when it comes to plastic. A man generates at least 800 g of waste every day [1]. So, we can estimate the total waste that can be generated by every person every day. This is not constant, but it has been increasing rapidly day by day which leads to many adverse consequences and threaten our environment and ecosystem. However, we can say that we are not doing that intentionally because we are doing that in a hurry to focus on our work and to save time.

Waste has been classified into many types; the basic subgroups of waste are liquid waste and solid waste. Solid waste is further classified into many types. They are:

DOI: 10.1201/9781032615790-8

- Medical waste
- Reclaimable waste
- Precarious waste
- E-waste
- Agricultural waste
- Household waste
- Industrial waste

This will lead to many problems such as water pollution and air pollution. We produce plastic waste which is extremely dangerous and cause air pollution as well as water pollution. When plastic is exposed to sunlight, the carbon particles in the plastic get dissolved in the atmosphere and this pollutes the air and leads to the formation of increasingly deadly gases like methane which causes air pollution. These gases will cause diseases related to the heart, respiratory problems, lung cancer, etc. and they may cause even death. The metal waste and plastic which are thrown near beach and river areas will dissolve in water and pollute the same. Many people drink this polluted water without filtering, which also leads to deadly diseases [2].

Waste not only includes plastic and metal waste but also includes paper waste; every person contributes to such waste at least once or twice a day. However, paper easily dissolves in the soil [3]. Paper waste might not create problems like pollution but wasting paper is also a reason for global warming. We can easily recycle and properly reuse paper, but we are just throwing it away once we use it instead of throwing it in a bin. This causes the production of more paper, and for the production of paper, we have to cut down the trees. Likewise, it is the reason for air pollution and global warming. Medical waste is one of the most threatening issues in our world. Medical waste like tablets contains chemicals. When they are thrown away, they get exposed to sunlight, release chemicals into the earth's atmosphere, and pollute air. Another form of medical waste is saline needles. When it comes to maintaining proper trash storage bins, we have to do that for this medical waste, because it is extremely dangerous to everyone. We should not throw the saline needles away. When we step on it or hold it accidentally, it may cause transmission of disease from one person to another.

Likewise, we have many issues with waste. It also has been increasing rapidly in our daily life. Additionally, we are also producing more waste day by day. The amount of waste we produce daily has been increasing day by day. As per the surveys conducted by different organizations, we are generating at least 15 tons of waste every day and approximately 5.5 million tons per year [4]. This rate will surely increase in the coming years. If things go on at this pace, we can soon see waste everywhere. Our towns and cities will be filled with waste. It will lead to air pollution and water pollution and threaten our lives. So, this pollution has to be monitored, and we have to take appropriate measures and follow the techniques which help in reducing the waste generation at least by 1% or 2%. It can bring enormous changes in the earth's atmosphere. So, we have to monitor the generation of waste so that we can achieve sustainable development [5].

Therefore, the main objective of our project is to collect and study the waste that has been thrown everywhere without human contact. We know that there are many hazardous chemicals and sensitive metals included in waste. Also, by keeping an eye on the recent global issues like coronavirus disease 2019 (COVID-19) pandemic, we have been prompted to reconsider touching anything. During that period, we had to wear masks and body shields, and use proper hand washers and sanitizers, which led to creating a huge waste throughout the world. This became a huge problem as no one dared to touch other's belongings. So even though we have the intention to collect that waste we would not do that. According to reports, at least 10,000 to 12,0000 tons of excess waste was produced only through medical waste, so we can estimate the waste generated by the other sources. Also, there are many hazardous materials that cause many diseases when we touch them accidentally. For example, chemical waste containing many deadly chemicals have been thrown everywhere. Furthermore, extremely sensitive materials like saline needles are the reason for transferring the diseases.

By observing these issues, we came up with the idea of creating a device to collet waste as part of waste management efforts. The advantage of this device is that there is no need for making human contact to collect the waste. The robotic hand collects the waste and places it in a bin. It will collect the waste from the targeted places and store it in appropriate places. So, this project provides a solution for many issues like collecting sensitive items and medical waste. As there is no human contact required in this collecting process, we can put an end to transferring the diseases. We can operate in places where human entry is restricted like industrial areas for several reasons, and make our work possible in a simple manner.

8.2 IMAGE RECOGNITION

Image recognition can be viewed in accordance with computer vision in the range of our project. In our project, computer vision is used to recognize the waste in the vicinity of the bot and dispose it. To understand and implement this feature, we must first try to understand computer vision and its branches which might be helpful to integrate this feature in our project.

8.2.1 Computer Vision

Computer vision is a field that consists of various disciplines which at their core depend upon acquiring images or a series of images, processing them, analyzing them, and transforming them into decisions or symbolic information [6].

Computer vision uses image processing to import the required information from a source, employing concepts like learning theory and statistics to come to a conclusion. In nonprofessional terms, computer vision acts to copy the human vision system and its decisions as shown in Figure 8.1.

There is a plethora of branches to computer vision, and each serves a different purpose and uses a different implementation, such as scene reconstruction, object detection, video tracking, motion estimation, image restoration, and learning indexing. We are interested in the fields of object detection and object recognition to implement in the navigation of our bot.

Before we try to understand concepts like object detection and recognition, let us take a look at the history of computer vision first.

FIGURE 8.1 Image recognition simpler output.

8.2.2 History

Universities pioneering in artificial intelligence started working on computer vision in the late 1960s. It was created to mimic the human visual system, and to be a steppingstone to provide robots the ability of intelligent behavior. In 1966, it was believed that adding a computer to a camera and having it to describe or say what it captured could be a summer project [7].

During that time, one key differentiation between digital image processing and computer vision was the aspiration to achieve comprehensive scene understanding by extracting three-dimensional model structures from images. Many fundamental aspects of today's computer vision algorithms were established in the 1970s, encompassing the labeling of lines, extraction of non-polyhedral shapes and edges from captured images, estimation of optical flow, polyhedral modeling, depiction of objects as connections between smaller structures, and motion estimation. This emphasis on developing techniques for analyzing image content laid the groundwork for the advancements witnessed in the field of computer vision [8, 9].

During the 1970s, several studies focused on more rigorous quantitative and mathematical aspects of computer vision. These include the deduction of shape from multiple cues such as contour models, the concept of scale-space, shading, and texture and focus acknowledged as snakes [10]. Researchers also realized that most of these mathematical conceptions could be satisfied within the same optimized framework as Markov random fields and regularization. Some of the past research concepts became more productive than others in the 1990s. Research in projective three-dimensional reconstruction or remodeling led to better learning of camera calibration. With the advent of methods for optimization in camera calibration, it was realized that a lot of the ideas were already explored in the field of photogrammetry. This led to the development of methods to get sparse three-dimensional reconstruction of scenes as an output using multiple images as input. At the same time, image segmentation was solved using variations of graph time. With the increased reciprocal action between the fields of computer graphics and computer vision, there was a notable change witnessed in image recognition, toward the end of the 1990s. This included image-based rendering, early light-field rendering, image morphing, panoramic image stitching, and view interpolation.

Recent advancements have seen the revival of feature-based techniques, used in connection with complex optimization frameworks and machine learning techniques. The field of computer vision was further improved with the progression of deep learning techniques. The methods that were presented before have been replaced in tasks ranging from classification, optical flow, and segmentation by the precision of deep learning algorithms on several tests' computer vision data sets.

8.2.3 Robotic Navigation

Before we explore object detection and recognition, we will learn the concept of robotic movement as it is integral to implement it with the computer vision system that we develop to result in a complete bot.

Robotic navigation is defined as the process of path planning used to navigate through an environment through the input provided by the computer vision system it is integrated with. Usually, these systems follow the block diagram as shown in Figure 8.2.

There are distinct types of robotic navigation systems based on their inputs such as vision-based navigation, inertial navigation, acoustic navigation, radio navigation, etc.

Focusing on vision-based navigation, they use two types of systems: one to detect the range of the targeted object (usually laser-based) and the other to recognize the features of the targets. Indoor navigation is a branch of vision-based navigation where the goal is simple – to get the robot to move to the target location using the inputs from the sensor array of the bot.

In this project, we will try and implement robotic navigation with computer vision to automate the process of picking up waste and disposing tit.

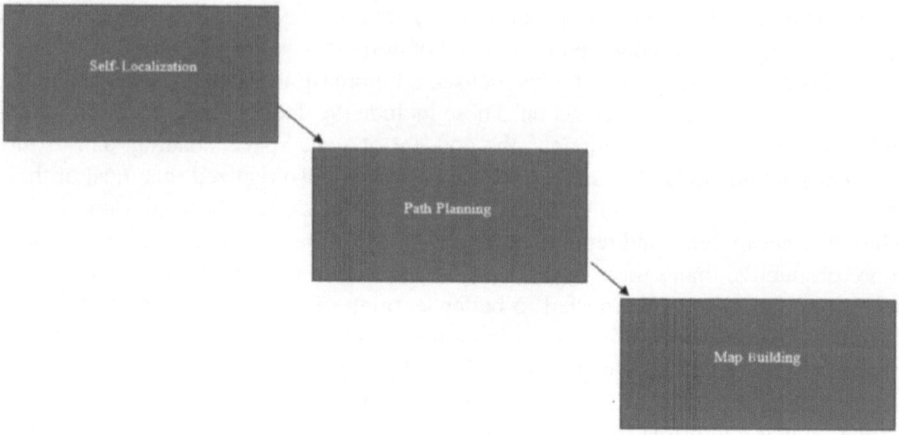

FIGURE 8.2 Block diagram that system follows.

8.2.4 Object Detection

Object detection is a technology used to detect and demarcate specified objects in digital images or sequences of digital images. It is used in various fields like face detection, video segmentation, and tracking objects.

Object detection uses various distinctive features of an object class to achieve detection of the specified object class. This unique feature is crucial while detecting an object as it defines the classification of that object into its said class, for example, while looking for boxes one must use features like the color of the box, length of the sides of the box, and features which define that it is a box (like sides in corners are mutually perpendicular).

There are various methods to implement object detection, but they generally fall into one of the two categories which are neural network and non-neural network based.

In a non-neural network-based approach, it is necessary to define the features of the class that we are trying to classify the objects into and use various techniques to classify them. On the other hand, neural networks are not required to define the features and provide direct classification and are usually derived from convolutional neural networks (CNNs) [11].

8.2.5 Object Detection without Neural Networks

Object detection without neural networks is the predecessor to their counterparts with neural networks. While the methods implemented with neural networks provide better accuracy, methods implemented without neural networks provide efficiency and compact size which enables it to run on computers with low computational power.

One such system is the Viola–Jones framework, initially proposed to counter the problem of face detection in 2001. The Viola–Jones framework is still in use due to its ability to run in even small computers with low computational power. The Viola–Jones framework uses three major components. They are as under:

1. Feature classifiers
2. Learning algorithms
3. Cascade architecture

The Viola–Jones framework can be effectively used to even detect moving objects with features like rotation implemented in the classifiers and using tracking algorithms for learning.

8.2.6 Object Detection with Neural Networks

Object detection with neural networks uses CNNs to generate the output. To understand the use of CNNs in object detection, we must first understand what are CNNs and how they work.

8.3 CONVOLUTED NEURAL NETWORK

CNN is the model of deep learning that implements deep learning using neural networks which are convolutional in nature. In CNN, each layer is known as a feature map and each feature map is associated with a number of neurons. Neurons from each layer are connected to their adjacent layers which result in various transforms of the input. The transforms applied on the input (in this case pixel intensities) include filtering, pooling, contrast normalization, etc. These transforms enable the neural network to extract and detect the features from the provided rudimentary input [12–14].

8.3.1 Region-Based Convolutional Neural Network

Region-based convolutional neural network (R-CNN) is a regional proposal-based framework introduced by Ross Girshick to achieve object detection using CNN. Selective search is used in R-CNN to produce about 2000 regional proposals for an input image. Selective search applies appropriately sized boxes to significantly reduce the searching time period.

The R-CNN uses feature extraction to derive high-dimensional data from the rudimentary input using its hierarchical structure and learning capacity due to its feature maps.

The pre-trained networks which follow will score each of the regional proposals based on constraints which will result in the decision-level output.

CNNs like SPP-net, faster R-CNN, Retina Net, YOLO (you only look once), single shot multibox detectors are also used for object detection [15–17].

8.4 INTRODUCTION TO ROBOTICS

Robotics is a combination of computer science like programming and electronics engineering. It usually involves designing, developing, operating, and using of robots.

Its objective is to design machines that help everyone in most of the fields. It integrates various fields such as mechanical, electrical, information, mechatronics, electronics, computer science, and mathematics [18].

Robotics helps us to design and create robots that can be an alternative to humans and simulate human actions. Robots can be used in various fields, but they are usually used in dangerous environments (radioactive areas), manufacturing, or where humans might die (underwater, in space, in critical temperatures). Robots take many forms but are usually made to look like humans.

8.4.1 Environmental Interaction and Navigation

A lot of robots that are in use today are either human-manipulated or function in a non-moving environment; there is growing interest in robots that operate independently in a dynamic (changes each and every second) environment. These kinds of robots use a blend of hardware and software that help in navigating to move in the demanded environment. Some highly advanced robots have especially good navigation hardware and software. Most of these robots use global positioning system navigating devices with waypoints simultaneously while using radar, which is sometimes used in tandem with other sensors such as light detection and ranging (LiDAR), inertial guidance systems, and cameras for better navigation [19, 20].

8.4.2 Control

The mechanical parts of the robots need to be controlled to perform the required tasks. Managing a robot involves three different and important phases: understanding, processing, and operation. Sensors are used to obtain information about the environment or the location of its joints and the end-effector of the robot itself. This information is then processed to compute the proper signals to the respective actuators, which change the structure to elicit the required motion.

The processing phase can range a lot from extremely easy to overly complex depending on how many effectors are present, degrees of freedom (DOF), etc. It translates raw information from the sensors directly into actuator commands to move them [21].

Commercial robotic control systems right now are overly complex, integrate multiple sensors and effectors, have many DOFs, and need operator interfaces, programming tools, and real-time capabilities. These types of robots are, most of the time, interconnected to larger communication networks, Internet of Things (IoT)-enabled and mobile.

8.4.3 Locomotion

For the most simple case, most of the robots which move have four wheels or a few continuous tracks. This makes it simple as we do not need any sensors or code for balancing the robot and moving it properly [22–24]. These types of robots have advantages such as having greater efficiency and reducing the number of parts required, as well as helping a robot to navigate in narrow ways.

Robots usually use a gyroscope to know how much it is dropping and then rotate the wheels considerably in the direction that it is dropping so that inertia makes the falling part counteract the drop at more than once every 10 milliseconds.

8.4.4 Manipulation

Machines mostly need to influence objects such as pick up, modify, destroy, or have an effect on it. This is done by a useful end of a robot arm aimed to make changes in the environment and are usually referred to as end-effectors, whereas the arm which moves the end-effector to the required position is referred to as a manipulator [25]. Most robots have a plug-and-play approach (replaceable) toward the end-effector, which allows users to replay them as needed, each providing them to perform many tasks which reduces the cost of buying multiple robotic arms for each kind of task.

Grippers are the most common type of end-effectors which can be found nearly everywhere around you. It is the simplest form of end-effector which can perform various tasks from picking up an object to crushing it. It just comprises two fingers or objects that can close and open when needed to pick up small objects. They can be created using a chain with a metal wire running via it. Grippers come in several types and sizes such as friction and encompassing jaws. Friction jaws use all the power of the gripper to use as the name says, friction to keep the object in place. The other type, encompassing jaws, makes the object stay in place, using less friction. Robot hands such as the Shadow Hand and the Robonaut Hand resemble and act more like a human hand [26]. These types of robot hands are the most complex end-effectors to date. The Delft hand is of mid-level complexity.

8.5 RESEARCH

Nowadays there is a lot of research in this field due to many technological advancements. They are not only done for particular industrial assignments, but also on investigation into new types of machines, designing them and manufacturing them.

The term "Generation Robots" can be used to specify the level of progression of a robot. Professor Hans Moravec coined this term. According to him, first-generation robots should have a mental capacity similar to perhaps a reptile and should be accessible by 2010. Second-generation robots would have an intelligence comparable to that of a rodent and should be accessible by 2020. Third-generation robots should have an intelligence that is similar to the intelligence of a primate and fourth-generation robots will have an intelligence that is similar to humans and should be accessible around 2040–2050 [27, 28].

8.6 PROPOSED SOLUTION

In our solution, there are three main components: video capture, central unit, and waste collecting bot. The video capture component contains a camera that would capture the video of the surrounding assigned to it and then send the video feed to the central unit for analysis. This component is set up everywhere around you in the form of security cameras, they can be used as a video feeder and there will not be any need for more cameras.

The central unit uses an image recognition model generated by machine learning and matches it with an already existing model of images of waste. When it detects waste it calculates the point at which the waste is present and sends the information to a nearby bot.

This bot then goes to the given point, picks up the waste present there, and stores waste in it. As no image recognition model can be perfect, there is also a Bluetooth module that is present in the bot so that a user can access it and move it to the point and help it pick up the waste. Accessing it can be done through an app that sends the information to a bot, which then analyses the instructions and responds accordingly.

8.7 WORKING

Briefly, it can be said that the cameras in the environment are the video-capturing device; the local server is the data processor and instruction passer, and the robot is the instruction follower and waste storage device of this system.

First, the video captured by the cameras in the environment is sent to a local server through data cables for quality transfer. The captured video stream undergoes analysis using our custom-built image recognition model, which has been trained on the data collected by us. By feeding the model with this data and evaluating its performance, we have ensured its effectiveness in accurately detecting a wide range of waste types present in the current environment. If the server detects any waste in the environment, it calculates the required details like the position, the nearest robot for picking up the waste, and the instructions needed. These instructions are then sent to the robot via Bluetooth.

The robot then receives these instructions, analyzes them, and sends signals to the wheel motors for a moment and to MG996 Servomotors and SG90 Micro Servomotors for the movement of the arm and the end-effector for it to go near the waste that needs to be picked up and then picking it up and storing waste within it. The instructions are processed by the Arduino Uno R3 microcontroller, then the processed data is sent to the L293D Shield. All the motors are connected to this motor shield and get the data from it.

The second way of sending instructions to the robot is by using an app for the movement and changing the position of the end-effector. This way is used when a certain type of waste is not detected by the image recognition model and the user can connect to the robot and make it pick up the waste so that they might be required to touch it.

In this case (as shown in Figure 8.3), it can be said that the HC-05 Bluetooth module is the data receiver, the Arduino Uno R3 microcontroller is the data decoder and assigner, and the L293D shield is the data passer.

First, the Android and iOS apps are connected to the HC-05 Bluetooth module which makes the Bluetooth module LED blink only once every 2 seconds, which means that the connection is established. When a button is pressed forward, backward, left, right, or stop, the app sends an instruction to the robot using a Bluetooth connection which has been established before. The Bluetooth module present in the

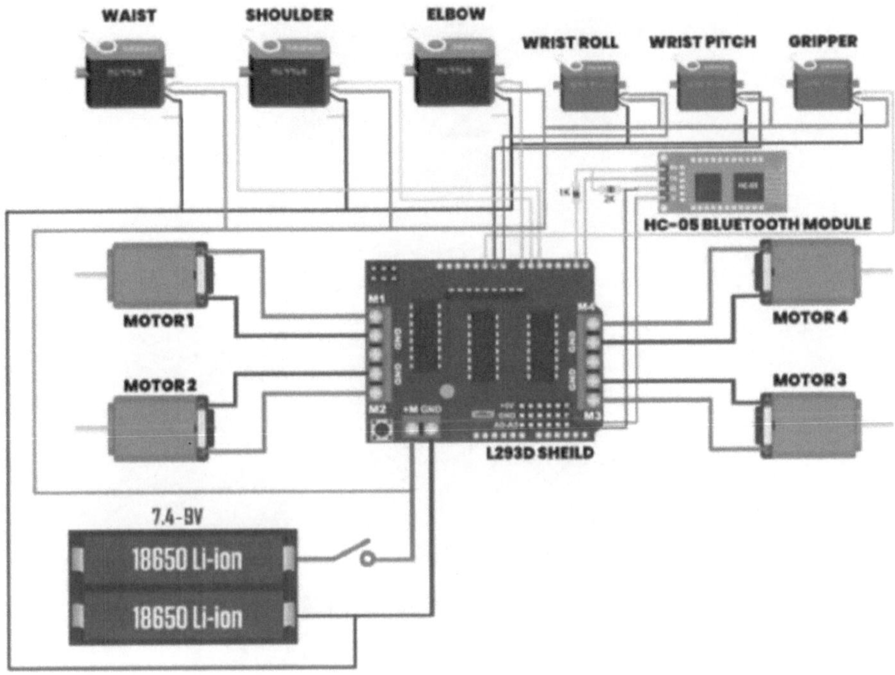

FIGURE 8.3 Circuit diagram of both rover and robotic arm.

robot receives these instructions and sends them to the Arduino Uno R3 microcontroller through the L293D Shield. The Arduino takes the information and checks the specific type of movement. Then the signal is sent to the L293D Shield which then passes the signal to the M1, M2, M3, and M4 motors.

For the movement of the robotic arm, we can change the sliders whose values are in degrees. When we change the value of a slider to say "Servo1" and we change it to 130 degrees, the app takes the slider values, adds two characters in front of that value, and sends it to the HC-05 Bluetooth module. These two characters are different for each slider. For example, for Servo1 the two characters will be "s1," for Servo2 it will be "s2," for Servo3 it will be "s3," for Servo4 it will be "s4," for Servo5 it will be "s5" and for Servo6 it will be "s6." Each slider can take values between 0 and 180 degrees. So, the data sent to the HC-05 Bluetooth module is "s1130," which is then sent to Arduino through L293D Shield. When data is received by the servo it will turn to that position and maintain the same position until it gets another signal to turn from the Arduino.

Using this app, we can pick up light objects which are near the rover and the arm by just manipulating the sliders in the app. If the object is far away, then we can move the rover to the required position by pressing the buttons to move in the needed direction.

These instructions sent by the central unit and the Android and iOS app (Figure 8.4) are of the same type. These instructions include something quite simple such as F for

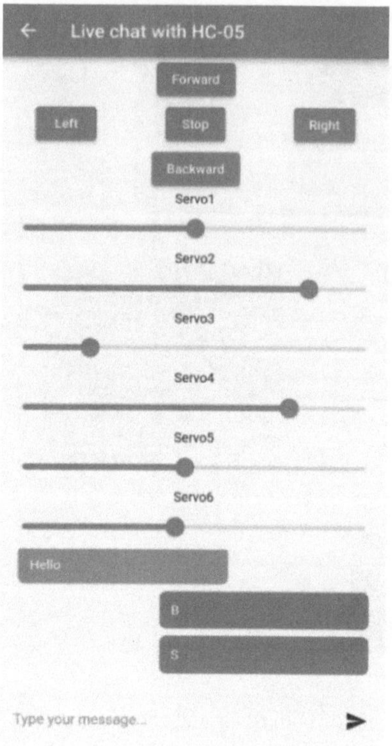

FIGURE 8.4 Bluetooth app to control the rover and robotic arm.

Forward, B for Backward, L for Left, R for Right, and S for Stop, and a little complex type of instructions to move the arm of the robot. The app lets you turn a particular motor about 180 degrees. These instructions are of type "s1149," "s1" means this instruction is for Servo 1, and "149" is the amount of rotation that is to be made by the particular servo motor in degrees. Arduino takes the data and breaks it into two parts: the first part contains two characters (in this case it is "s1") and the second part contains the value (in this case it is "130" degrees). The first part tells the Arduino to which servo to send the data contained in the second part of the data received.

8.8 RESULT AND DISCUSSION

Currently, it works well enough to be deployed in smaller areas and to pick up some types of waste. This can be fixed by improving communication between the robot and the central unit using a long-range communication system such as LoRa. The battery life of the robot can be improved, and a solar panel can be added so that it can sustain itself in summer [29] (see Figures 8.5 and 8.6).

We can increase the database of the images upon which the image recognition model is trained upon to improve the accuracy of the robot to detect waste. Additional improvements include making the exterior skeleton of the robot sturdier using strong

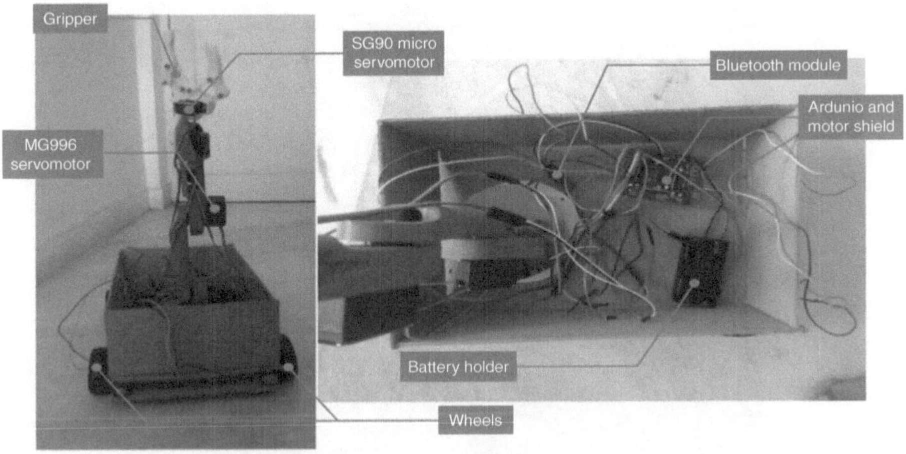

FIGURE 8.5 Components of robot.

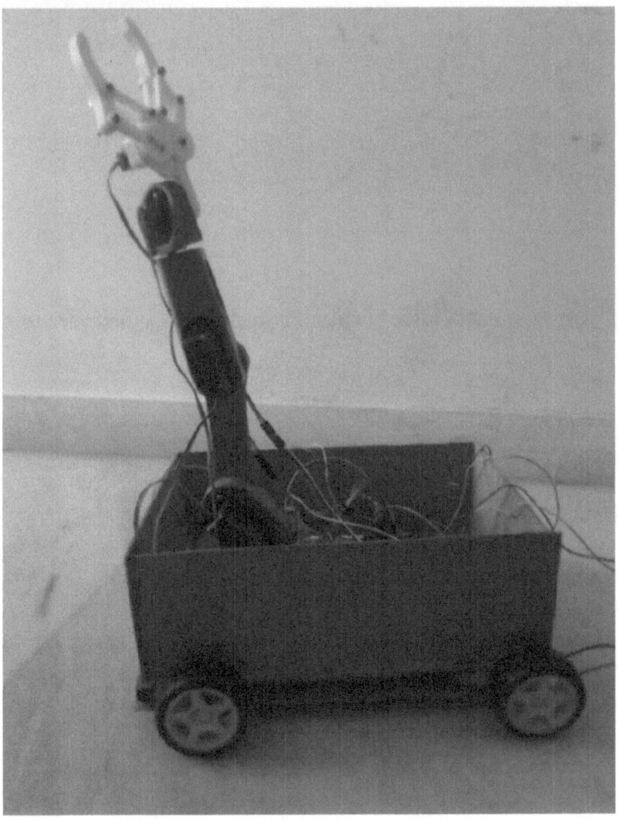

FIGURE 8.6 Side view of both rover and robotic arm.

FIGURE 8.7 Rover and robotic arm picking up trash and throwing it into a box.

aluminum and decreasing the weight of the components which are already used [30] (see Figure 8.7).

8.9 CONCLUSION

This waste-picking robot was developed using Arduino Uno, HC05 Bluetooth module, and servo motors, along with the integration of security cameras, and represents a significant step toward reducing the waste we encounter in our everyday lives. This system proves especially beneficial in closed environments such as office spaces, universities, and gated communities, where vehicle movement is limited, allowing robots to operate efficiently.

However, it is important to recognize that a robot alone cannot address the entirety of the waste management challenge. Once the robot reaches its capacity, human intervention becomes necessary. Therefore, it is crucial for us to take responsibility and actively reduce the waste we generate. By practicing waste reduction, reuse, and recycling, we can make a significant impact on waste management.

It is imperative that we dispose of trash in designated bins and maintain clean surroundings to ensure our own well-being and contribute to a healthier environment. By doing so, we not only benefit ourselves in the present but also pave the way for future generations to enjoy a clean and healthy life.

In summary, while the waste-picking robot system represents a remarkable technological solution, its effectiveness is magnified when coupled with individual efforts to minimize waste generation. Together, we can work toward a cleaner and greener future, improving the quality of life for ourselves and generations to come.

REFERENCES

1. Doron, A. (2018). *Waste of a Nation: Garbage and Growth in India*. Harvard University Press.
2. Ferronato, N., and Torretta, V. (2019). "Waste Mismanagement in Developing Countries: A Review of Global Issues." *International Journal of Environmental Research and Public Health*, 16(6), 1060. https://doi.org/10.3390/ijerph16061060.
3. Wilson, D. C., Velis, C., and Cheeseman, C. (2006). "Role of Informal Sector Recycling in Waste Management in Developing Countries." *Habitat International*, 30, 797–808.
4. Dhana, R. (2021). "Waste Management in India – An Overview." *United International Journal for Research and Technology (UIJRT)*, 2(7), 175–196.
5. Iacovidou, E., Millward-Hopkins, J., Busch, J., Purnell, P., Velis, C. A., Hahladakis, J. N., Zwirner, O., and Brown, A. (2017, December 1). "A Pathway to Circular Economy: Developing a Conceptual Framework for Complex Value Assessment of Resources Recovered from Waste." *Journal of Cleaner Production*, 168, 1279–1288. https://doi.org/10.1016/j.jclepro.2017.09.002
6. Ballard, D. H., and Brown, C. M. (1982). *Computer Vision*. Prentice Hall.
7. Huang, T. (1996, November 19). "Computer Vision: Evolution and Promise." In: Vandoni, C. E. (ed.), *Proceedings of the 19th CERN School of Computing* (pp. 21–25). Geneva: CERN. https://doi.org/10.5170/CERN-1996-008.21
8. Jähne, B., and Haußecker, H. (2000). *Computer Vision and Applications, A Guide for Students and Practitioners*. Academic Press.
9. Morris, T. (2004). *Computer Vision and Image Processing*. Palgrave Macmillan.
10. Soltani, A. A., Huang, H., Wu, J., Kulkarni, T. D., and Tenenbaum, J. B. (2017). "Synthesizing 3D Shapes via Modeling Multi-View Depth Maps and Silhouettes with Deep Generative Networks." In: *Proceedings of the IEEE Conference on Computer Vision and Pattern Recognition (CVPR)* (pp. 1511–1519). https://doi.org/10.1109/CVPR.2017.163
11. Chervyakov, N. I., Lyakhov, P. A., Deryabin, M. A., Nagornov, N. N., Valueva, M. V., and Valuev, G. V. (2020). "Residue Number System-Based Solution for Reducing the Hardware Cost of a Convolutional Neural Network." *Neurocomputing*, 407, 439–453. https://doi.org/10.1016/j.neucom.2020.04.018
12. Garg, H. (2020, February 29). "Drowsiness Detection of a Driver Using Conventional Computer Vision Application." In: *Proceedings of the 2020 International Conference on PARC (Power Electronics and IoT Applications in Renewable Energy and Its Control)* (p. 556). Mathura, India: IEEE. https://doi.org/10.1109/PARC49193.2020.236556
13. Balasundaram, A., Ashokkumar, S., Kothandaraman, D., Kora, S. N., Sudarshan, E., and Harshaverdhan, A. (2020, December 1). "Computer Vision-Based Fatigue Detection Using Facial Parameters." *IOP Conference Series: Materials Science and Engineering*, 981(2), 022005. https://doi.org/10.1088/1757-899X/981/2/022005
14. Cheng Lin, K., Huang, T.-C, Hung, J. C., Yen, N. Y., and Ju Chen, S. (2013, June 7). "Facial Emotion Recognition towards Affective Computing-Based Learning." *Library Hi Tech*, 31(2), 294–307. https://doi.org/10.1108/07378831311329068

15. Russakovsky, O., Deng, J., Su, H., Krause, J., Satheesh, S., Ma, S., Huang, Z., Karpathy, A., Khosla, A., Bernstein, M., and Berg, A. C. (2015, December). "ImageNet Large Scale Visual Recognition Challenge." *International Journal of Computer Vision*, 115(3), 211–252. https://doi.org/10.1007/s11263-015-0816-y
16. Kagami, S. (2010). "High-Speed Vision Systems and Projectors for Real-Time Perception of the World." In: *2010 IEEE Computer Society Conference on Computer Vision and Pattern Recognition – Workshops. IEEE Computer Society Conference on Computer Vision and Pattern Recognition – Workshops (CVPRW 2010)* (pp. 100–107). San Francisco, CA, USA: IEEE. https://doi.org/10.1109/CVPRW.2010.5543776
17. Li, L. H., and Du, J. F. (2012, December). "Visual Based Hand Gesture Recognition Systems." *Applied Mechanics and Materials*, 263, 2422–2426. https://doi.org/10.4028/www.scientific.net/AMM.263.2422
18. Randell, B. (1982, October). "From Analytical Engine to Electronic Digital Computer: The Contributions of Ludgate, Torres, and Bush." *IEEE Annals of the History of Computing*, 4(4), 327–341. https://doi.org/10.1109/MAHC.1982.10003.
19. Tosun, F. E., and Patoglu, V. (2020, June). "Necessary and Sufficient Conditions for the Passivity of Impedance Rendering with Velocity-Sourced Series Elastic Actuation." *IEEE Transactions on Robotics*, 36(3), 937–951. https://doi.org/10.1109/TRO.2020.2977255
20. Burn, K., Short, M., and Bicker, R. (2003, July). "Adaptive and Nonlinear Fuzzy Force Control Techniques Applied to Robots Operating in Uncertain Environments." *Journal of Robotic Systems*, 20(7), 391–400. https://doi.org/10.1002/rob.10093
21. Roozing, W., Li, Z., Tsagarakis, N., and Caldwell, D. (2016). "Design Optimisation and Control of Compliant Actuation Arrangements in Articulated Robots for Improved Energy Efficiency." *IEEE Robotics and Automation Letters*, 1(2), 1047–1054. https://doi.org/10.1109/LRA.2016.2534137.
22. Collins, S., Ruina, A., Tedrake, R., and Wisse, M. (2005, February 18). "Efficient Bipedal Robots Based on Passive-Dynamic Walkers." *Science*, 307(5712), 1082–1085. https://doi.org/10.1126/science.1107799
23. Collins, S. H., and Ruina, A. (2005). "A Bipedal Walking Robot with Efficient and Human-Like Gait." In: *Proceedings of the 2005 IEEE International Conference on Robotics and Automation*. (pp. 1983–1988). Barcelona, Spain: IEEE. https://doi.org/10.1109/ROBOT.2005.1570355
24. Jaulin, L., and Le Bars, F. (2013, February). "An Interval Approach for Stability Analysis: Application to Sailboat Robotics." *IEEE Transactions on Robotics*, 27(5), 1005–1011. https://doi.org/10.1109/TRO.2012.2217794
25. Wullenkord, R., Fraune, M. R., Eyssel, F., and Sabanovic, S. (2016). "Getting in Touch: How Imagined, Actual, and Physical Contact Affect Evaluations of Robots." In *Proceedings of the 25th IEEE International Symposium on Robot and Human Interactive Communication (RO-MAN 2016)* (pp. 980–985). New York City, USA: IEEE. https://doi.org/10.1109/ROMAN.2016.7745228
26. Nishibori, et al. (2003). "Robot Hand with Fingers Using Vibration-Type Ultrasonic Motors (Driving Characteristics)." *Journal of Robotics and Mechatronics*, 15(6), 588–595. https://doi.org/10.20965/jrm.2003.p0588
27. Walters, M. L., Syrdal, D. S., Koay, K. L., Dautenhahn, K., and Te Boekhorst, R. (2008). "Human Approach Distances to a Mechanical-Looking Robot with Different Robot Voice Styles." In: *RO-MAN 2008 – the 17th IEEE International Symposium on Robot and Human Interactive Communication. (RO-MAN 2008)* (pp. 707–712). Munich, Germany: IEEE. https://doi.org/10.1109/ROMAN.2008.4600750

28. Waldherr, S., Romero, R., and Thrun, S. (2000, September 1). "A Gesture Based Interface for Human–Robot Interaction." *Autonomous Robots*, 9(2), 151–173. https://doi.org/10.1023/A:1019115100515
29. Dowling, K. *Power Sources for Small Robots*. Carnegie Mellon University.
30. Wang, T.-M., Tao, Y., and Liu, H. (2018, April 17). "Current Research and Future Development Trend of Intelligent Robot: A Review." *International Journal of Automation and Computing*, 15(5), 525–546. https://doi.org/10.1007/s11633-018-1115-1

9 Industrial Automation and AI in Mechatronics Industry

Vinoth S., Rajasekar C., Gowdhamkumar S., and Muthukumar K.

9.1 INTRODUCTION TO INDUSTRIAL AUTOMATION AND ARTIFICIAL INTELLIGENCE IN MECHATRONICS INDUSTRY

Automation and mechatronics engineers are involved in every stage of product development, from design and testing to manufacturing. As electronics continue to advance, modern mechanical systems progressively incorporate electronics and controls. This distorting of boundaries between self-control has led to a rising need for mechatronics professionals, who possess robust expertise in all these fields.

Mechatronics engineers put on fundamental principles from mechanical, electronics and electrical, computer, and control engineering to design and develop a wide range of products, from purchaser goods like microwaves, automotive systems and mobile devices to industrial systems like Computer Numerical Control (CNC) machines, automated machines, and Magnetic Resonance Imaging (MRI) and X-ray machines. Their goal is to create simpler, more economical, and reliable solutions for these applications.

Automation and mechatronics engineers design machines composed of various components: mechanical assemblies, sensing and actuation elements, control framework, and software. A prime example of a mechatronics system is an industrialized robot, which integrates electronics, mechanics, and computing to perform its day-to-day tasks efficiently [1].

Industrial automation machines often involve the unified collaboration of mechanical and electrical components, which is where the discrepancy between mechatronics and industrial automation can become unclear. However, unlike industrial automation, certain mechatronic devices may not include automated elements. Some academic programs offer courses that bring together students in both mechatronics and industrial automation, providing a well-versed education in both fields.

Industrial automation is a fourth key element of the Industrial Revolution, focusing on incorporating computers with physical machinery to improve performance. Companies that invest in industrial automation typically seek benefits like increased production and improved scalability. By minimalizing or eradicating human involvement, automation reduces issues related to fatigue or human error. Some systems can independently adapt to changes, while others notify operators of maintenance needs, helping to avoid downtime [2].

DOI: 10.1201/9781032615790-9

As manufacturing demands grow, industrial automation will become increasingly relevant and evolve alongside technological advancements to better suit production needs. While there is not always a clear distinction between mechatronics and industrial automation, mechatronics specialists frequently contribute to automation projects, alongside other endeavors. Mechatronics covers a broad range of disciplines, whereas industrial automation focuses on automating tasks once performed manually.

Despite their differences, both fields significantly enhance manufacturing. As the industry becomes more advanced and reliant on specialized machinery, the roles of mechatronics and industrial automation experts will remain crucial for the foreseeable future.

9.2 FUNDAMENTALS OF INDUSTRIAL AUTOMATION AND AI IN THE MECHATRONICS INDUSTRY

For many, artificial intelligence (AI) conjures images of robots performing complex tasks in sci-fi movies, and while this is partially true, AI's real-world impact extends far beyond that. AI enables industrial machines to perform highly intelligent tasks, addressing the new challenges faced by global industries and decision-makers. As industries grapple with the rapid pace of change, there is a pressing need to modernize control systems and industrial assets, both software and hardware. AI has the potential to meet these demands.

AI is already somewhat widely used in industries including finance, gaming, retail, and entertainment. Driving the Fourth Industrial Revolution is advanced automation using better sensors, controllers, Input/Output (IO) modules, Programmable Logic Controllers (PLCs), gateways, and enterprise systems. This revolution is also transforming human and machine collaboration, so strengthening and merging the digital ecosystem [3].

Cost control is a top priority for any manufacturing company, especially in the competitive market of today when companies are constantly looking for ways to cut costs without sacrificing output or quality. One effective strategy for getting steam is industrial automation. Modern machinery and technology will enable businesses to reduce costs, streamline operations, and increase output by means of simplification of tasks.

In industrial automation, AI is increasingly playing a transformational role. Startups in AI are automating repeated labor-intensive activities and reducing reliance on human effort, therefore transforming sectors. This not only reduces human error risk but also lets employees focus on more difficult yet value-adding tasks. Staff members might be taught to manage and maintain automated equipment instead of spending time creating goods on an assembly line so as to ensure their best performance and lower downtime. This change increases output as well and supports long-term operational efficiency.

Automation allows companies to reach higher production rates and maintain quality. Machines free from breaks or tiredness operated continuously produce more and reduce the dangers of defects or inconsistent end products. This great degree of quality control finally helps save time and money by enabling manufacturers to avoid the costs connected to rework or customer returns.

In the fast-changing industrial automation scenario, AI has evolved as a transformational agent. From quality control to production line optimization, AI is transforming developed processes. This section looks at the numerous ways AI is revolutionizing the manufacturing sector and offers several points of view on its transformational potential.

9.2.1 Predictive Maintenance

Maintenance of conventional techniques has absolutely no benefit. Combining numerous sensor data, historical performance, and enough parameters enables AI-driven systems to forecast equipment breakdown before it starts. For instance, General Electric (GE) tracks gas turbines using AI and bases maintenance on real-time data. By early resolution of probable problems, industrialists can reduce unplanned downtime and increase production output.

9.2.2 Quality Control and Defect Detection

Maintaining product quality is vitally crucial in manufacturing, and AI-driven machine vision systems are rather crucial in this sense as well. These systems can quite precisely spot errors, departures from technical criteria, and standards violations. Being strong on large-scale data, deep learning methods could find defects in more complex items including electronic components or automobile parts. For example, a very well-known electronics company, Foxconn, uses AI-driven cameras to observe printed circuit boards (PCBs). Automation of quality control allows businesses to reduce waste rates and raise customer satisfaction [4]. The role of AI in industrial automation is depicted in Figure 9.1.

9.2.3 Process Optimization

AI systems improve produced processes by maximizing several factors for the highest efficiency. Data analysis by AI helps to find the best clarifications for changing machine settings, improving supply chain logistics, or lowering energy consumption. Globally running industrial company Siemens uses AI, for example, to maximize production lines by dynamic parameter modifications in response to real-time demand and resource availability. This method guarantees better use of resources, less waste, and more quantity generated.

9.2.4 Supply Chain Management

By studying historical data, market strategies, and environmental events such as weather and geopolitical events, AI increases supply chain visibility and resilience. These AI systems maximize swings in inventory and demand. For example, Amazon handles its large logistics network using AI to guarantee timely delivery and lower stockouts. Furthermore, blockchain-based AI systems improve traceability and openness—qualities absolutely important for ethical procurement and compliance.

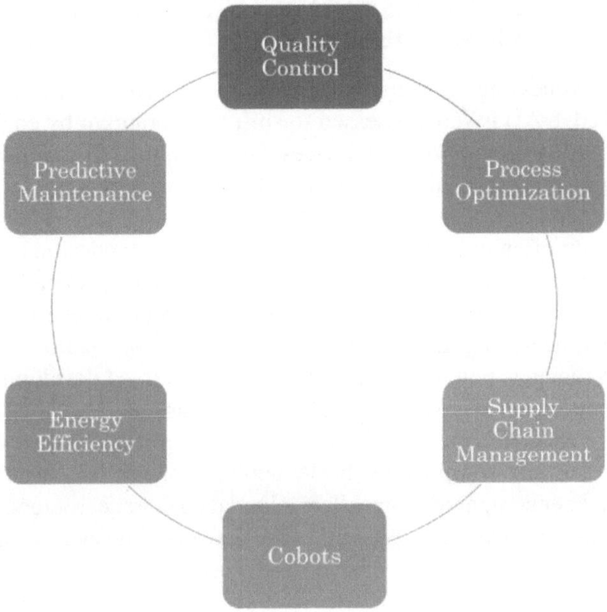

FIGURE 9.1 Role of AIs in industrial automation.

9.2.5 Collaborative Robots

Working with human operators, cobots or collaborative robots expose the synergy between manufacturing and AI. From human examples, these robots learn and grow used to rich surroundings. Cobots lower worker stress and increase output by helping with boring jobs on manufacturing lines. BMW promises safety, for instance by using cobots to help create car doors, therefore attaining both speed and precision.

9.2.6 Energy Efficiency

Supervision systems driven by AI are transforming energy consumption in industrial units. By analyzing real-time data from sensors, AI recognizes energy-intensive practices and recommends approaches for reducing consumption. For instance, Google's DeepMind worked with the National Grid to cut energy usage in data centers using AI, and similar techniques can be applied to industrialized abilities to lower costs and reduce environmental consequences.

In summary, AI's influence on industrial automation makes their efficiency improvements. It allows industrialists to quickly adapt, improve product quality, and manage multifaceted supply chains. As AI technologies advance, the industrialized sector is set for a transformative excursion toward increased productivity and long-term viability.

9.3 INDUSTRIAL SENSORS AND ACTUATORS USING AI IN MECHATRONICS INDUSTRY

Currently, data processing is primarily handled by enormous peripheral servers, but embedded AI (edge AI) is self-possessed to shift this paradigm by enabling data and algorithm processing directly on end devices. However, the effectiveness of AI on the edge, particularly in small-scale devices, is still quite inadequate.

Sensors and actuators play a pivotal role in the development of AI across various levels, yet they are often overlooked when discussing the expansion and future of AI. This oversight could disrupt the interface and impact the widespread adoption of AI. Addressing this gap might be key to achieving the broad uptake that many anticipate. Conversely, neglecting the integration of sensors and actuators could introduce unnecessary challenges, potentially leading to setbacks in AI development.

The hardware layer of AI systems includes various actuators, sensors, and robotics that can tenuously be able to monitor and control equipment such as robots, automobiles, electronic components, home appliances, and industrial sensor nodes. The network layer encompasses controllers, cloud data storage, routers, and numerous wireless protocols, connecting nodes within the network. Entryways and base stations often link to the cloud for more advanced data processing.

Seven AI tools have proven effective with sensor systems, each offering unique capabilities:

- *Knowledge-Based Systems:* Utilize expert knowledge to make decisions and solve problems.
- *Fuzzy Logic:* Handles uncertainty and imprecision by using "fuzzy" rather than binary logic.
- *Automatic Knowledge Acquisition:* Involves methods for automatically gathering and updating knowledge.
- *Neural Networks:* Copycat the human brain's structure to recognize patterns and make predictions.
- *Genetic Algorithms:* Use evolutionary techniques to optimize solutions.
- *Ambient Intelligence:* Creates intelligent environments that respond to users in a natural way.

The increasing power and affordability of modern computers have expanded the applications of these AI tools in sensor systems, making them more prevalent and effective in various domains.

AI systems have been advancing rapidly, leading to increasing unified communication between people and digital systems. While the integration of AI into the industry has been gradual, it holds the promise of enhancing adaptability, re-configurability, and dependability. Novel machine systems are now outperforming humans in a growing number of tasks. As we merge our cognitive abilities with computational power for deliberation, analysis, deduction, communication, and invention, we might be on the brink of a new era in intelligent machines.

AI, or an intelligent machine, integrates a range of sophisticated technologies to enable machines to learn, acclimatize, make decisions, and exhibit novel behaviors.

This is accomplished through technologies such as self-organizing maps, fuzzy logic, neural networks, expert systems, and genetic algorithms. The development of intelligent machine technology has been significantly driven by its application to the extent wherever sensor information needs to be interpreted and processed.

9.4 INDUSTRIAL CONTROL SYSTEMS USING AI IN MECHATRONICS INDUSTRY

The incorporation of AI into industrial control systems is transforming manufacturing processes within the mechatronics sector. AI boosts the accuracy, efficiency, and flexibility of mechatronic systems, driving progress in smart manufacturing and automation. Its impact is becoming increasingly obvious in modern society and is found in consumer electronics, business, medical, and other fields.

Mechatronic systems' motors are controlled utilizing AI methods like neural networks, therefore enhancing positioning precision and lowering mechanical flaws-related inaccuracy. The quick growth of industrial production enables mechatronic systems to be embraced and spread, hence improving production quality and efficiency. Many times, conventional industrial control technologies are not sufficient for modern mechatronics systems. Reducing human mistakes and enhancing the mechatronic system's performance thus depend on more study and application of intelligent control technologies. This paper explores the role of intelligent control in mechatronic systems and examines design methodologies based on intelligent control principles.

The integration of AI into industrial control systems within the mechatronics sector is advancing precision, efficiency, and adaptability significantly. AI techniques enhance production processes by optimizing real-time decision-making and increasing the autonomy of mechatronic systems. However, continued research and development are essential to address current encounters and fully unlock AI's potential in smart manufacturing.

AI technology supports comprehensive recognition and systematic resolution of issues in industrial automation control, making it increasingly intelligent, driverless, and autonomous. Key AI approaches applied in the development of industrial automatic control systems take in neural network control platforms, fuzzy control, and adept decision systems, as shown in Figure 9.2.

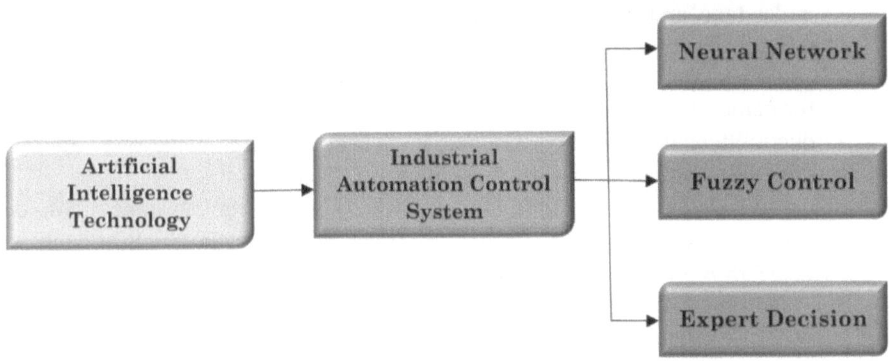

FIGURE 9.2 The AI applied in the construction of industrial control system.

AI technology is emerging as a central component in the next phase of industrial automation systems. However, current technological limitations and application methods present challenges. To enhance the performance of industrial automation control systems, experts and researchers need to strategically apply AI technology in line with the evolving characteristics and specific requirements of industrial automation. This approach will help overcome existing limitations and advance the overall effectiveness of these systems.

9.5 INDUSTRIAL ROBOTICS USING AI IN MECHATRONICS INDUSTRY

In the 21st century, AI has arisen as a pivotal force in the development of robotics. AI systems, which are capable of learning and adapting to their surroundings, excel in tasks that involve multifaceted decision-making and problem-solving. Notable examples of AI-powered robots include autonomous cars, cargo drones, and healthcare assistants.

The next phase of robotics is expected to be significantly influenced by advances in AI, along with progress in materials science and other critical fields. Some specialists foresee robots evolving to become progressively human-like, interacting with people in a more usual and instinctive manner. Others predict a trend toward more dedicated robots, with each machine tailored for specific tasks or industries. From simple mechanical tools to sophisticated, AI-driven systems, the evolution of robotics has been distinguished generally by a slow change.

AI combined with industrial robots is revolutionizing mechatronics, thus improving system efficiency, adaptability, and intelligence. Within mechatronics, AI is enhancing industrial robots as follows:

- *Enhanced Safety and Efficiency:* Working with firms in areas including electronics and automotive, AI-driven cooperative robots (cobots) improves safety and efficient manufacturing. These AI-driven robots provide maximum labor sharing and adapt to human behavior by means of sensors and AI, therefore offering safe and successful interactions.
- *Natural Language Processing (NLP):* It enables AI to support the robots to understand and react to human commands, hence it is improving human–robot interaction and communication.
- *Topology Optimization and AI:* AI-powered machines create a perfect standard for manufacturing of precision operation part designs. Through dynamic production techniques, robots can guarantee best part alignment, lower material utilization, and enhance structural integrity.
- *Advanced Sensor Fusion:* To provide a complete awareness of the surroundings and hence enable real-time robotic action adjustment, AI employs sophisticated sensor fusion, that is, data from numerous sensors such as pressure, temperature, and force.
- *Edge Computing for Real-Time Processing:* Instant data analysis and decision-making made possible by AI integrated into robotic systems help to

provide edge computing for real-time processing and so reduces reliance on centralized processing.

Generally, AI adds intelligence, flexibility, and real-time adaptability to demanding systems, therefore improving industrial robots. This convergence is what drives sectors aiming at higher degrees of automation, reduced costs, and more output to decrease costs.

9.6 INTELLIGENT MANUFACTURING SYSTEMS USING AI IN MECHATRONICS INDUSTRY

Most industrial processes are defined as their principal objectives by high throughput, quality maintenance, cost control, and a friendly operational environment. However, meeting these objectives has become increasingly challenging due to the growing intricacy of products and processes, higher inconsistency in customer demands and predilections, and intense competitive heaviness.

From a positive perspective, this challenging environment presents a prospect for AI to offer matchless advantages over traditional tools and methods. Specifically, AI excels in problem-solving tasks that involve identifying root causes by analyzing and classifying complex, multivariate, and nonlinear patterns in operative and recital data patterns that may be hidden from plant engineers.

Today, vast amounts of data are unremittingly generated by machines, sensors, controllers, and labor records. This data can be categorized as follows:

1. Ecological data acquired from ambient sensors, for example, room temperature and humidity.
2. Process data acquired from sensors on process machines or stations, for example, machining and grinding and heat treat temperature/energy,
3. Production operation data acquired in controller systems, for example, timestamps or elapsed time of each part in each starvation or blockage, idle time, operation station, machine downtime, and shift scheduling.
4. Measurement or inspection data acquired from product quality checks, such as product diameter, shape, and balance, contain valuable prospects for pattern discovery. This data can reveal important intimations for solving complex problems and offer deeper insights into the relationship between various physical parameters and system characteristics. By effectively handling high-dimensional data, AI can transform the extensive and intricate manufacturing data common in modern factories into actionable and insightful information.

An intelligent industrialized process has the capability to self-regulate and control to ensure that products are produced within design specifications. The concept of integrating such systems into future factories, everywhere products are created in an artificial life situation, adds significant value. Researchers have explored this idea through the employment of expert systems, initially tested on computer-aided drafting systems at the University of Texas. Recent advancements in feature-based solid modelling techniques and AI applications suggest that expert systems could potentially replace some decisions currently made based on an engineer's knowledge of design procedures and practices [5].

Early efforts with knowledge-based systems and expert systems fell short of achieving the desired level of intellect due to their reliance on representative knowledge representation and the need for human proficiency to encode this information effectively. However, the advent of artificial neural networks (ANNs) in the 1980s, which act as self-organizing energetic systems and model-free estimators, along with the development of fuzzy logic systems for handling uncertainty and human-like reasoning, and the adoption of genetic algorithms (GA) for global optimization, has made the apprehension of intelligent industrialized systems a feasible goal in the current decade.

Intelligent industrialized systems can be achieved in three uncomplicated ways.

- Prevailing manufacturing procedures can become intellectual by monitoring and controlling the state of the industrialized machine.
- Prevailing processes can be made intellectual by adding sensors to monitor and control the state of the product being processed.
- Novel processes can be perceptively premeditated to produce parts of preferred quality without the necessity of a sensing process.

Intelligent manufacturing system

1. Usages of technology that can minimalize the routine of the human brain
2. Regulation for product blend and precedence fabrication, self-regulated.
3. Self-controlled maneuvers with programmed feedback mechanisms.
4. Monitoring and control of the industrialized machine.
5. Novel processes with intellect can be made to produce parts of desired quality without the need for a sensing process.

9.7 PREDICTIVE MAINTENANCE USING AI IN MECHATRONICS INDUSTRY

Predictive maintenance powered by AI aims to prevent equipment failures by forecasting when maintenance will be necessary. While rule-based systems rely on predefined thresholds for maintenance, machine learning (ML)-based systems use advanced analytics to predict failures with greater accuracy. Anomaly detection in rotating machinery is a critical component of ML-driven predictive maintenance.

As the demand for the Internet of Things (IoT) rises in the digital age, optimizing maintenance through data-driven approaches has become a key focus for maintenance professionals. Initially, data collected from various sensors installed on or within machines is pre-processed. This data is then utilized in decision-support systems enhanced by AI tools. This article will explore how AI applications can be effectively integrated into the predictive maintenance field. AI-powered predictive maintenance offers numerous advantages that revolutionize practices across industries.

First, AI knowingly reduces maintenance costs by facilitating pre-emptive measures to discourse potential equipment failures beforehand they happen. By analyzing data from input devices and monitoring systems, AI algorithms can identify initial warning signs of malfunctions, allowing for timely interferences that avoid equipment failures and decrease the need for expensive emergency maintenance.

AI-based predictive maintenance also extends the lifetime of the equipment. Early problem solutions and minimization of unwanted machine strain aid in lowering replacement frequency, machine downtime, and capital expenses. This approach ensures more efficient and long-term usage of their assets, therefore optimizing the return on investment for businesses.

AI-based predictive maintenance improves operational performance by way of better maintenance plans and resource allocation. Predictive analytics and real-time data analysis let AI systems match jobs and more evenly distribute resources. For service providers, this lowers personnel costs and boosts output, therefore enabling maintenance teams to maximize results with little resources.

AI-driven predictive maintenance technologies are altering business equipment maintenance practices. Fundamental in this transformation is the scalability and adaptability of these systems, which significantly increase their efficacy in various industrial situations. The degree of freedom and efficacy AI-based predictive maintenance provides above conventional rule-based systems results in more efficient and successful maintenance procedures [6].

9.8 QUALITY CONTROL USING AI IN THE MECHATRONICS INDUSTRY

As industrial processes become more complex, companies aiming to maximize their operations are increasingly turning to AI-powered solutions. One area where AI can be especially valuable is in quality control (QC). AI-driven systems can identify defects and reduce waste more precisely than traditional methods, helping to raise standards [7].

Including AI and ML in manufacturing processes presents a special set of difficulties, however. These include the need for great training to ensure effective use, the difficulty of merging with present systems, and the need for major expenditure.

9.8.1 Benefits of Using AI for QC in manufacturing

AI in quality control offers several significant advantages:

- *Reduces Human Error:* A well-trained AI system can routinely detect little product defects that the human eye would overlook, unlike human workers who might ignore problems due to exhaustion or haste. This helps human employees focus on carrying out AI-based remedial activities.
- *Improves Brand Reputation:* As consumer satisfaction depends on high-quality items, AI could detect and correct problems before items even reach consumers, therefore avoiding disappointment, negative reviews, and brand reputation damage.

- *Helps Meet Regulations:* Many items follow strict guidelines meant to comply with laws. By pointing up any issues and ensuring things meet the necessary standards, AI allows companies to guarantee compliance.
- *Reduces Waste:* Using defective components and drugs, ML may stop manufacturing and thus lower waste. Early AI-based filtering of defective inputs serves to reduce waste by minimizing the number of rejected goods originating from inferior components.
- *Keeps Goods Viable:* Monitoring storage and transportation conditions—including temperature—allows AI to keep items viable, therefore guaranteeing their optimal condition for their lifetime. AI can assure, for instance, safe food storage to maintain its quality.

Apart from basic rule-based automation, AI improves analytical capacity, thereby improving automated inspection tools. This development motivates more intelligent and sophisticated decision-making, especially in the handling of complex data and scenarios like spotting little defects that conventional automated systems can ignore.

Moreover, providing great scalability in response to changes in manufacturing volume are AI-driven quality control systems. Their real-time fast processing of massive amounts of data generates faster feedback loops for dynamic production modifications and faster inspection processes.

The fundamental advantage of AI above traditional automated systems is its ability to always learn from new data and experience. This continuous learning process throughout time improves the performance and accuracy of AI algorithms, unlike fixed automated systems that could need manual changes or retraining to meet new sorts of issues.

9.9 CYBERSECURITY IN INDUSTRIAL AUTOMATION USING AI IN MECHATRONICS INDUSTRY

Cybersecurity has become a main issue in industrial automation, particularly in the mechatronics sector where networked systems and automation are basic [8]. AI and ML are gradually added into these systems to ensure reliability, confidentiality, and obtainability of required activities; hence cybersecurity issues have to be addressed.

Here in industrial automation in the mechatronics field, cybersecurity and AI cross:

1. AI-driven threat detection and response
2. Vulnerability identification in automation systems
3. AI in securing industrial protocols
4. Role of AI in protecting mechatronic systems
5. AI-enhanced resilience and recovery
6. Security for AI-based mechatronic applications
7. Edge computing and secure data handling

Smart Industry 4.0 has greatly improved mechanicalization, output, and efficiency of industrial environments. Meanwhile, these developments also bring fresh cybersecurity problems. Higher system and equipment connectivity of Smart Industry 4.0 attitudes increases their exposure to security breaches and cyberattacks. AI has to be brought into cybersecurity plans absolutely if we are to overcome these obstacles. Among them is an anomaly detection and predictive analytics. This all-encompassing approach enables Smart Industry 4.0 apps to have improved security posture, therefore lowering the data leak and cyberattack risk. Still, cautious application and design of these technologies are quite important to prevent inadvertent deference. Governments can protect their industrial settings and guarantee operational dependability and safety by means of well-considered deployment of AI-driven cybersecurity solutions. Fourth generation of smart companies are merging cybersecurity with AI.

AI is highly important in the mechatronics sector to enhance the cybersecurity of industrial automation. Industries could better protect important systems, maintain operational continuity, and lower the concerns about cyber threats in ever-automated environments by including AI into the cybersecurity architecture. If one is to fully enjoy these advantages, one also has to pay considerable attention to protecting the AI systems themselves and follow privacy and safety rules.

9.10 HUMAN–MACHINE INTERACTION USING AI IN MECHATRONICS INDUSTRY

Industrial human–machine interfaces (HMIs) greatly benefit workers by way of their simplicity of production system control and monitoring. Delivering real-time alarms, machine inputs, status data, and reports via tactile controls and touchscreens has long been part of this. Industrial HMIs are changing as AI advances to offer more flexible interactions, thereby enhancing user experience, productivity, and accessibility [9]. Use of generative artificial intelligence (GI) and NLP helps to facilitate this development.

HMI technology decisions in industrial settings must take into account the specific responsibilities operators carry combined with the actual surroundings. These elements determine the degree of operator performance and safety maximizing ability.

Kenn states,

> Recent developments have greatly improved the capacity of AI-based systems to distinguish human voices even in noisy industrial environments because of the vast training data and examples now accessible. This development is generating multimodal systems that can combine motions and voice, allowing operators to adjust operations to promote more efficient job completion.

Hummel points out,

> AI often exemplifies the saying, "when you have a hammer, everything looks like a nail." It excels at tackling open-ended problems, such as enhancing the accuracy of parts inspection on a production line. However, it's less effective for fixed, repeatable tasks like pick-and-place operations for a pad printer. If existing machines or automation are already effective and user-friendly, manufacturers should consider applying AI to other areas where it can provide more significant benefits.

The main challenge for HMI providers and users is determining where AI can most effectively improve worker productivity, efficiency, and accessibility. Identifying these optimal use cases is crucial for tackling broader industry issues, such as worker shortages, skill gaps, and employee satisfaction.

9.11 BIG DATA ANALYTICS IN INDUSTRIAL AUTOMATION USING AI IN MECHATRONICS INDUSTRY

The incorporation of data from several sources such as production units, enterprise systems, consumer feedback, and product request systems facilitates real-time decision-making in smart manufacturing. Modern manufacturers are increasingly interested in gathering customer feedback to refine their product designs and cater to diverse customer needs. By analyzing big data, manufacturers can gain insights into product performance and failure causes, understand consumer buying habits, and leverage data-driven marketing to enhance predictive manufacturing strategies.

Big data analytics in industrial automation harnesses vast amounts of data generated by industrial systems to enhance operations. AI processes this data to boost efficiency, predict machine failures (predictive maintenance), and detect anomalies in real time. This continuous learning from historical and real-time data aids in optimizing processes, minimizing downtime, and improving quality control. AI also supports smarter decision-making across supply chains, energy management, and human–machine collaboration, driving industries toward more intelligent, efficient, and adaptive systems in line with Industry 4.0.

Big data analytics in industrial automation driven by AI is revolutionizing sectors by raising productivity, reducing downtime, and optimizing overall operational effectiveness. This integration is what modernizing manufacturers, making them smarter, more flexible, data-driven, and so supporting sustainable and competitive development depends on.

9.12 INTELLIGENT SUPPLY CHAIN USING AI IN MECHATRONICS INDUSTRY

AI is the simulation of human intelligence directed toward the completion of difficult tasks. ML is a primary focus of AI as systems learn from vast quantities instead of dependent on explicit programming. Using this learning ability, AI systems outperform conventional software in tasks including processing video feeds, understanding spoken and written material, forecasting market trends, making judgments in challenging environments, and revealing insights from big data collections.

In the frantic business world, success depends just on production; consequently, supply chain management is mostly dependent on this. Standard approaches of supply chain management are showing insufficient as businesses grow more complicated and worldwide. AI helps to transform solutions and revolutionize this industry. From applying predictive analytics to increase inventory control to increasing agility with real-time tracking, AI may greatly improve almost all aspects of supply chain operations and help find out how companies using AI are improving and modernizing their supply chains.

Driven by AI, a new approach to improving and automating the whole supply chain—including procurement, inventory control, transportation, and customer delivery—is an intelligent supply chain. Supply chains have to be flexible, effective, and sensitive to changing customer expectations in the competitive and always-changing corporate environment. By processing vast amounts of data in real time, demand trend projections drive decision-making and automate repetitive operations, with AI enabling this shift. From this flows cheaper costs, higher operational efficiency, more customer satisfaction, and better risk control. AI's applications within supply chain and logistics are extensive. Key areas include:

- *Operational Procurement:* Simplifying procurement processes with chatbots and cognitive data.
- *Supply Chain Planning:* Employing AI to predict demand and optimize supply strategies.
- *Warehouse Management:* Leveraging AI to enhance stock management and operational efficiency.
- *Shipping and Transportation:* By means of accelerating and raising accuracy in delivery, lead times will be shortened and transportation expenses will be saved.
- *Supplier Selection:* Real-time data allow one to make better purchasing decisions and select vendors with the best match.

AI is being applied by companies increasingly to enhance specific aspects of their supply chain operations. Among the important uses are:

- *Optimizing Shipping and Delivery:* AI improves route planning and delivery scheduling, therefore faster and more reasonably priced logistics.
- *Managing Warehouse Capacity:* Two ways AI help are by optimizing warehouse space and inventory storage.
- *Tracking Inventory:* AI-powered real-time tracking systems ensure accurate inventory control and help to reduce stock discrepancies.
- *Forecasting Demand:* AI better plans and uses resources by estimating some component or part needs.
- *Improving Worker Safety:* AI-based systems track and enhance safety processes, therefore reducing occupational accidents.
- *Ensuring Transaction Integrity:* AI helps to reduce fraud and error risk by allowing transaction integrity over global supply chains.

Notwithstanding these advantages, using AI in supply chains can be difficult and expensive, especially when constructing and training original machine learning models using owned data. Manufacturers and logistics companies must be equipped with their supply chains if they want to completely apply AI into their systems of operations such as:

1. *Assessing Readiness:* By means of present infrastructure and systems analysis, one can determine their readiness for AI integration.

2. *Data Preparation:* Provide high-quality, relevant data to AI models in training.
3. *Skill Development:* Staff training will help them to correctly handle and implement AI technologies.
4. *Choosing the Right Tools:* Choosing AI technologies and platforms that fit certain operational needs and goals will assist you in matching them.
5. *Continuous Monitoring and Improvement:* Regular adjustments in AI systems enable maximum outcomes by means of frequent monitoring and development of them.

9.13 INDUSTRIAL INTERNET OF THINGS USING AI IN THE MECHATRONICS INDUSTRY

The AI deployment with Industrial Internet of Things (IIoT) technology by offering advanced maintenance and operational efficiency is transforming the industrial sector. This is a major change from thinking of AI as a futuristic idea to realizing it as a useful tool for handling actual industrial problems.

AI's fast and accurate assessment of massive amounts of data offers an insightful analysis that guides better-informed operational decisions on cost reduction and new money-generating prospects.

Using AI inside the IIoT helps manufacturers greatly with smart equipment, improved analytics, and more user-friendly interfaces. Predictive maintenance mostly helps AI identify possible equipment defects before they occur, thereby enabling quick interventions and minimum downtime. Another advantage of smart manufacturing—in which improved automation and real-time data processing simplify production lines—is higher efficiency. Real-time monitoring guarantees ongoing system maintenance and facilitates quick identification and fixing of problems, therefore improving the quality of products. AI-driven insights also assist more informed decisions, streamline processes, and help to find new prospects, thus streamlining operations.

Manufacturers will have to handle some difficulties, though, if they are to completely exploit AI. As correct and complete data is absolutely necessary for effective application of AI, data quality and availability are quite important. Including AI systems in current infrastructure causes several technological challenges. Another issue is security as the operational integrity of AI systems is defined by the defense against cyberattacks. Other ethical issues addressed are algorithmic preconception and data secrecy.

Manufacturers may use AI in the IIoT to help innovation, obtain operational efficiency, and keep a competitive advantage in the market by tackling these challenges and customizing solutions to their particular requirements.

9.14 DIGITAL TWIN TECHNOLOGY USING AI IN MECHATRONICS INDUSTRY

Keeping relevance and competitiveness for companies operating in the data-driven business environment of today by means of digital twin technologies and AI is a challenging topic. The fast spread of these technologies raises this question and demands swift response on behalf of businesses. In fields including phonological conception, pattern recognition, decision-making, and investigative learning as well as in others, AI develops in powers. Moreover, evermore complex are digital twins, computer replicas of real items. For companies, this junction of digital twin technologies and AI offers a challenging route.

"Digital Twin Technology and AI Employments in Future-Focused Businesses" addresses the notable difference by means of how AI and digital twin technology are influencing corporate processes. The book examines the main ways modern technology affects efficiency, creativity, and decision-making in many various fields. By staying up to date with these technological advancements, the book offers readers vital knowledge and impressions to survive in the fast-changing corporate environment.

9.14.1 Benefits of Combining Generative AI and Digital Twins

Modern approaches to problem-solving contribute to raising sustainability and efficiency simultaneously. Modern approaches to problem-solving help enhance both sustainability and efficiency. The benefits of these technologies include improved resource allocation through AI models, which assist in asset tracking and repair, thereby modernizing operations and boosting operational effectiveness.

Thousands of bits of data can be converted into useful plans providing vision and foresight for guiding decisions. This is a cultural shift toward development based on smart efficiency.

Digital twins and AI have a tight link that lets changing systems be both mirroring their physical counterparts and foretelling and adaptable as well. Combining these technologies allows one to create a dynamic model that learns, grows, and provides useful intelligence highlighting actual decisions.

Supply chains have been increasingly complex, integrated, and large in recent years, which presents major challenges for companies in appropriate management. AI offers an effective solution by looking at the enormous volumes of data generated inside modern supply chains. With this information, AI might produce quite accurate predictions, identify operational insights, and increase the efficiency of storage and transportation networks. This capacity improves general supply chain management by spanning vast-scale logistics networks with many independent partners.

9.15 CHALLENGES AND FUTURE DIRECTIONS IN INDUSTRIAL AUTOMATION AND AI IN MECHATRONICS INDUSTRY

Industrial automation is fast advancing to the front stage in the digital age as the foundation of the global economy [10]. Ground-breaking technical developments such as robotics, AI, and ML are driving this revolution and significantly changing our manufacturing, running, and application of industrial processes. Although they provide fascinating new prospects, these technological advancements also bring novel challenges the industry has to address.

Fundamental to this transformation is the development of human–robot cooperation. AI is being increasingly used in modern manufacturing systems to lower human presence in hazardous environments, thereby improving safety. AI-driven robots can undertake occupations in risk zones and complete activities traditionally unsafe or impractical for humans when teamed with human operators more effectively. This kind of cooperation not only increases general operating efficiency but also reduces the risk to human workers.

Furthermore, AI transforms maintenance systems in industrial environments. AI included in these technologies helps manufacturers more effectively monitor and spot machinery and product flaws. As they are competent to make self-decisions and optimizations dependent on real-time data, AI systems can adapt immediately to changes in the production environment. Changing production plans, running machining equipment, automatically replacing machine tools, and warning systems in case of possible problems before they become major sources of concern are among these activities.

Mechatronics engineering in all its applications has a very bright future. As technology advances to encourage innovation in many spheres, mechatronics experts will become ever more important. Increasing robots and automation, enhancing IoT connectivity, and so allowing the development of creative manufacturing techniques all depend on their understanding [12–15]. Mechatronics engineering will also be very beneficial for autonomous cars to grow and for healthcare technology. Using these advances will help industries reach improved degrees of efficiency, productivity, and sustainability, thus improving their quality of life and directing the creation of smarter, more integrated systems.

9.16 CONCLUSION

Despite having a base in mechanical engineering, industrial automation and mechatronics is a multidisciplinary science that incorporates principles from mechanical engineering, telecommunications engineering, electronic engineering, computer systems engineering, and control engineering into a single coherent discipline using the principles of all of these disciplines.

REFERENCES

1. H. R. Chi, A. Radwan, N.-F. Huang and K. F. Tsang, "Guest Editorial: Next-Generation Network Automation for Industrial Internet-of-Things in Industry 5.0," *IEEE Transactions on Industrial Informatics*, vol. 19, no. 2, pp. 2062–2064, February 2023, doi: 10.1109/TII.2022.3216903
2. M. Zheng, Y. Zhao, H. Shi and W. Liang, "A Flexible Retransmission Scheme for Reliable and Real-Time Transmissions in Industrial Wireless Networks for Factory Automation," *IEEE Transactions on Vehicular Technology*, vol. 72, no. 8, pp. 10867–10878, August 2023, doi: 10.1109/TVT.2023.3259945
3. Y. Ma, Y. Wang, S. Di Cairano, T. Koike-Akino, J. Guo, P. V. Orlik, X. Guan and C. Lu, "Smart Actuation for End-Edge Industrial Control Systems," *IEEE Transactions on Automation Science and Engineering*, vol. 21, no. 1, pp. 269–283, January 2024, doi: 10.1109/TASE.2022.3216217
4. Z. Yang and Z. Ge, "On Paradigm of Industrial Big Data Analytics: From Evolution to Revolution," *IEEE Transactions on Industrial Informatics*, vol. 18, no. 12, pp. 8373–8388, December 2022, doi: 10.1109/TII.2022.3190394

5. J. Li, Y. Zhou, X. Zhang, S. Liu and Q. Li, "Assessment of Industrial Internet Platform Application in Manufacturing Enterprises: System Construction and Industrial Practice," *IEEE Access*, vol. 9, pp. 103709–103727, 2021, doi: 10.1109/ACCESS.2021.3098940
6. B. Babayigit and M. Abubaker, "Industrial Internet of Things: A Review of Improvements over Traditional SCADA Systems for Industrial Automation," *IEEE Systems Journal*, vol. 18, no. 1, pp. 120–133, March 2024, doi: 10.1109/JSYST.2023.3270620
7. S. R. Siraparapu and S. M. A. K. Azad, "A Framework for Integrating Diverse Data Types for Live Streaming in Industrial Automation," *IEEE Access*, vol. 12, pp. 111694–111708, 2024, doi: 10.1109/ACCESS.2024.3441114
8. V. Lesi, Z. Jakovljevic and M. Pajic, "Security Analysis for Distributed IoT-Based Industrial Automation," *IEEE Transactions on Automation Science and Engineering*, vol. 19, no. 4, pp. 3093–3108, October 2022, doi: 10.1109/TASE.2021.3106335
9. J. Li, J.-J. Qiu, Y. Zhou, S. Wen, K.-Q. Dou and Q. Li, "Study on the Reference Architecture and Assessment Framework of Industrial Internet Platform," *IEEE Access*, vol. 8, pp. 164950–164971, 2020, doi: 10.1109/ACCESS.2020.3021719
10. R. Tallat, et al., "Navigating Industry 5.0: A Survey of Key Enabling Technologies, Trends, Challenges, and Opportunities," *IEEE Communications Surveys and Tutorials*, vol. 26, no. 2, pp. 1080–1126, Second Quarter 2024, doi: 10.1109/COMST.2023.3329472
11. H. Wang, C. Li, Y.-F. Li and F. Tsung, "An Intelligent Industrial Visual Monitoring and Maintenance Framework Empowered by Large-Scale Visual and Language Models," *IEEE Transactions on Industrial Cyber-Physical Systems*, vol. 2, pp. 166–175, 2024, doi: 10.1109/TICPS.2024.3414292
12. Sharma, V., Balusamy, B., Sabharwal, M., and Ouaissa, M. (Eds.). (2023). *Sustainable Digital Technologies: Trends, Impacts, and Assessments* (1st ed.). CRC Press. https://doi.org/10.1201/9781003348313
13. M. d. M. Fernandes, J. A. Bigheti, R. P. Pontarolli and E. P. Godoy, "Industrial Automation as a Service: A New Application to Industry 4.0," *IEEE Latin America Transactions*, vol. 19, no. 12, pp. 2046–2053, December 2021, doi: 10.1109/TLA.2021.9480146
14. R. Liu, et al., "Industrial Big Data Analytical System in Industrial Cyber-Physical Systems Based on Coarse-to-Fine Deep Network," *IEEE Transactions on Industrial Cyber-Physical Systems*, vol. 1, pp. 359–370, 2023, doi: 10.1109/TICPS.2023.3331331
15. D. Gunapriya, N. Pusphalatha, S. Sudharsan, S. Pandi, R. SivaKumar and P. V. Kumar, "A General Investigation of Battery Handling Techniques Driven by IoT," *2023 Fourth International Conference on Smart Technologies in Computing, Electrical and Electronics (ICSTCEE), Bengaluru, India*, 2023, pp. 1–6. IEEE. doi: 10.1109/ICSTCEE60504.2023.10584895

10 AI-Based Sustainable Solution for Energy Management in the Mechatronics Industry

Mohan Raj C., Rajanbabu W., Hemananth B., and Sheikameer Batcha S.

10.1 INTRODUCTION

Let's take an automobile manufacturing plant for example. Various machines are used in it. These machines are usually electrically powered. In these machines, Alternating Current (AC) motor, Direct Current (DC) motor, stepper motor, and permanent magnet DC (PMDC) motor, which can be called special machines, are used. We use many types of machines in the mechatronics industry. So, we use artificial intelligence (AI) technology to reduce this time rush. Before manufacturing a vehicle, we decide in advance what it should look like, what color it should have, and how it should be designed. As per our decisions, we determine the size of the vehicle, the color, and the engines that can be installed in the vehicle. AI in mechatronics industry is used to quickly identify the cause of the malfunctions in the vehicles manufactured in this way, to inform us precisely what caused the malfunction, when it occurred, and how long it will take to fix it. The general block diagram of AI integrated factory is shown in Figure 10.1.

10.2 ROLE OF AI SENSORS IN MECHATRONICS INDUSTRY

There are many types of sensors. Sensors are used according to the needs of the application. How to choose a sensor? What is the sensor used for? What is the sensor used to measure? How far does this sensor work? At what input voltage does this sensor work? All these depend on the use OF the sensor in the factory. Apart from that, the output of the sensor is very important. This output can also be digital. Sometimes it can be hot. Sometimes called a communication port, it can run on bus, ethernet, and some other form of communication. The accuracy of the sensor we use to digitize machines is very important. To measure the light, to measure the frequency, and all possible variations in the sensor must be properly recorded. The signals must be communicated to the controller in an appropriate manner. Inside this controller is a digital twin. Based on the data received from the sensor, this digital twin works. Sensors typically produce 0 to 5 volts. Sometimes it works in millivolts and milliamperes. A few sensors work at lower voltages than this. Such sensors are

DOI: 10.1201/9781032615790-10

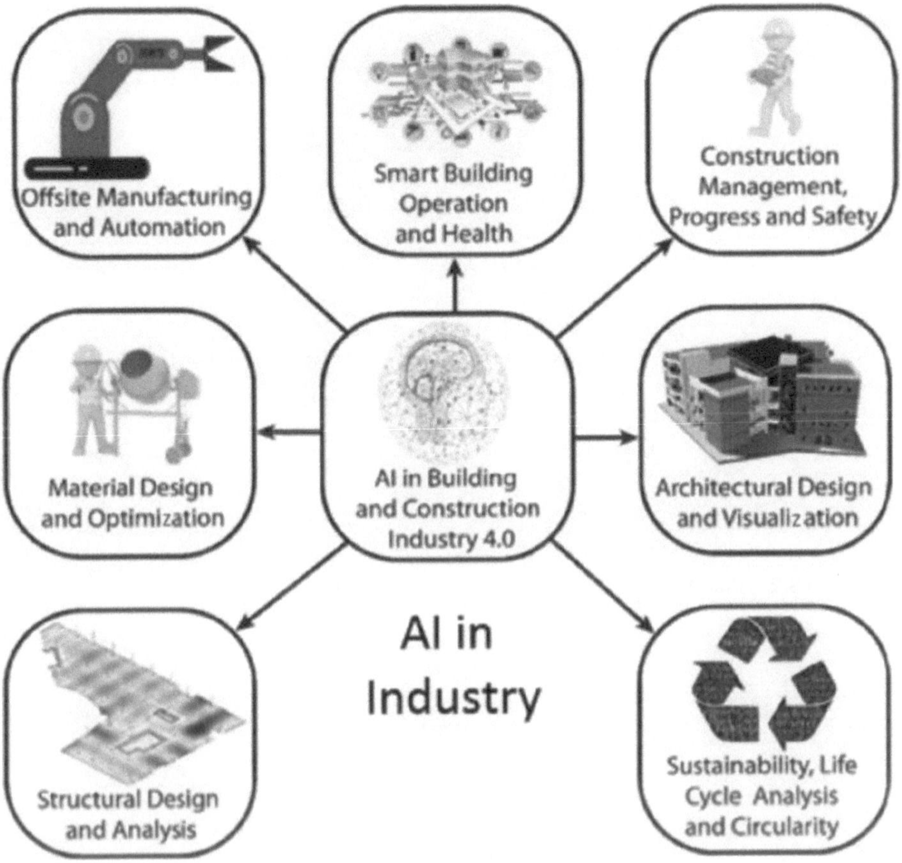

FIGURE 10.1 AI in industry generalized block diagram.

used in the field of mechatronics [1, 2]. The different types of AI sensors used in the industries are shown in Figures 10.2–10.12 [3].

10.3 ROLE OF AI CONTROLLERS IN THE MECHATRONICS INDUSTRY

These controllers are completely different from the existing controllers. The power supply to these varies with the way they operate and their size. Mechatronics industries often lack AI controllers because existing robots do not support AI controllers. The controllers are more advanced. These controllers can think and make decisions on their own. The power supply voltage to this control is much lower than 5 volts. These controllers can integrate sensors that support AI. Programs written in this controller are written in Java or Python language. These controllers will first execute based on the program given to them. It will then automatically learn to measure the machine's parameters as appropriate for the machine to operate. The data thus learned

FIGURE 10.2 Photoelectric sensors through beam.

FIGURE 10.3 Diffuse proximity sensor for sensing metals.

are passed into the controller as a lesson. These controllers can easily detect any variation in measurements or measuring instruments in the future. The results thus detected are immediately reported to the operator of the machine. By doing this, we can predict engine failure in advance. Also, we can avoid material losses and time losses that may occur if the machine breaks down. These AI controllers typically have 40 or more inputs and outputs.

FIGURE 10.4 Light grid sensor with 70 cm working range.

FIGURE 10.5 Infrared-based radar sensor.

The power that this controller can draw is much less than millivolts. All processes invented by and after the Yes processor are currently being designed to support AI. We can easily program these AI controllers. We need a smaller amount of hardware. If our mobile phone has Java or Python language, then we can program it using our mobile phone [4].

10.4 COMMUNICATION BETWEEN THE AI CONTROLLERS AND ACTUATORS

Actuators receive the signal from the controller and transmit it to the outside world. These actuators can be actuated manually, hydraulically, or electromechanically. These actuators are widely used in the field of mechatronics. Electromechanical actuators are most commonly used followed by hydraulic or pneumatic. These

FIGURE 10.6 Laser sensor.

FIGURE 10.7 Safety light curtain for avoiding accidents near the machine's working area.

electromechanical actuators act like a contact eye. These electromechanical actuators typically use electrical signals to magnetize the coil. Thus, magnetized coils or relays turn on whatever load we want to turn on. These actuators generally require 5volts, 12 volts, or 24 volts to operate. DC voltages are more commonly used than AC voltages. Some contactors or relays are designed to operate on AC voltage. In such cases, we can use AC voltage instead of DC voltage. These actuators usually receive low voltage signals from the control and turn on an electromagnet through it and then supply the required power to the load through a contactor. Any damage to these coils or contracts should be repaired immediately [5]. If it is not fixed, then the whole system will run into problems. Thus, the various types of actuators used are male and female using AI. When must an actuator be male? Controllers decide when to switch according to the program they are designed to follow.

FIGURE 10.8 Pulse counter sensor with IoT.

FIGURE 10.9 Thermocouple-type sensor with IoT.

FIGURE 10.10 Vibration-type sensor with IoT.

10.5 PROGRAMMABLE LOGIC CONTROLLER IN MECHATRONICS INDUSTRY

PLC or Programmable Logic Controller contains an input power supply and memory. The memory is of two types: one is temporary memory, which stores data temporarily, and the other is permanent memory, which stores programs and retains them even

FIGURE 10.11 Ampere or current sensor with IoT.

FIGURE 10.12 IoT gateway for sensor interfacing.

when the power is turned off. It is widely used in the field of mechatronics. All types of machines currently running are controlled by this PLC. Most machines are driven by an AC motor, DC motor, or servo motor. The speed and torque of these machines are precisely controlled with the help of PLC. We can write our own programs in PLC. These written programs can be monitored and further data transferred with the help of our computer. This PLC system is connected to other PLC systems through communication. A set of PLCs is thus connected and signals exchanged among themselves. A machine interacts with many machines. Once a process is completed the signal is sent to the next machine and the next process is started. Processes are done in chain mode. Once the process is complete, eventually all PLCs emit signals. The signals thus emitted can be viewed as HML data that a human–machine base can interpret. This data can be stored by operators or transmitted from one location to another [6–8]. This is done with the help of a technology called Internet of Things (IoT). The Mitsubishi PLC is shown in Figures 10.13–10.15 using an AI-integrated facility.

10.6 COBOTS IN MECHATRONICS INDUSTRY

Collaborative robots, or cobots, are completely different from robots. We need more time to automate robots. The reason this amount of time is required is that we have to

FIGURE 10.13 Sample Mitsubishi PLC with AI support.

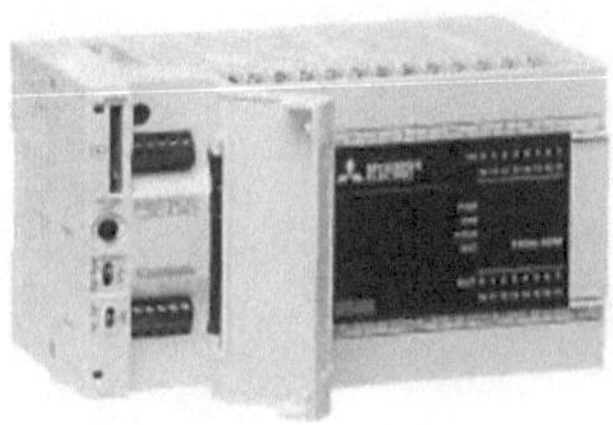

FIGURE 10.14 Sample Mitsubishi FX5U PLC with AI support.

FIGURE 10.15 Sample Mitsubishi Q Series PLC with AI support.

specify in the programs the exact point at which the robots should move and where they should move. The method mentioned in such programs is called teaching. This method of teaching is done by the programmer using teaching pendants. Therefore, we need a teaching method to send a robot's path from one place to another. With this teaching we can teach almost 30,000 points to a machine or a robot. Usually, robots operate with two, three, or more than three axes. However, teaching cobots does not take much time. Mechatronics industrial robots are now widely used. Teaching these cobots does not require a lot of equipment, known as Teaching Pendant. If we hold and pull this robot to the place where we want to move our robot, it will easily reach that point. We can

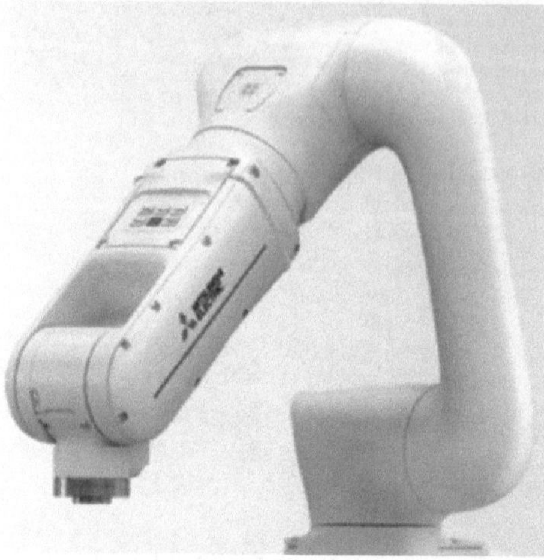

FIGURE 10.16 Sample Mitsubishi Cobot for the mechatronics industry.

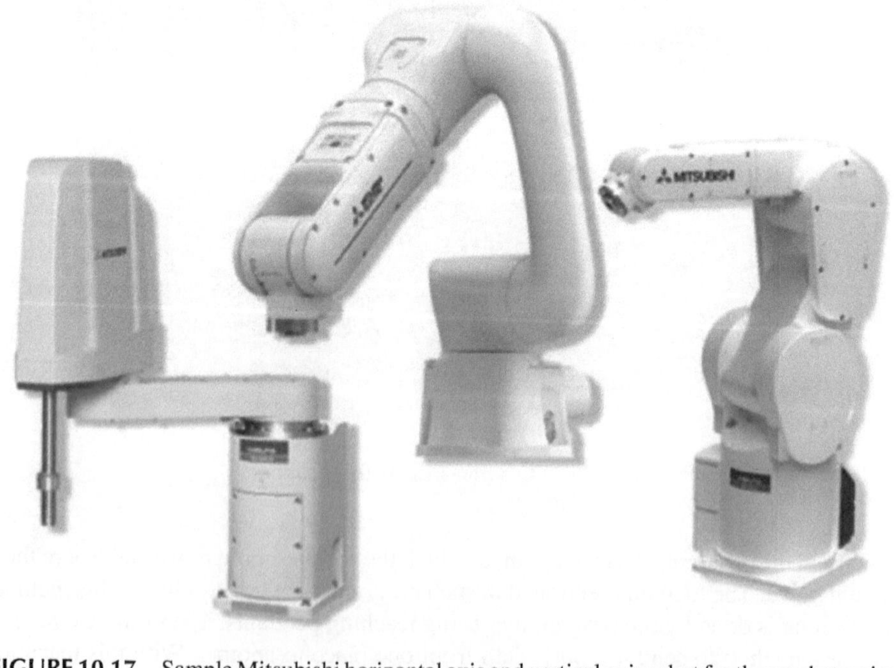

FIGURE 10.17 Sample Mitsubishi horizontal axis and vertical axis robot for the mechatronics industry.

easily store that point in the computer. Thus, wherever we want to stop the robot we can easily save those points by keeping the tool at the respective points. Then we need to download the saved points. Finally, when the cobot is activated, the robot starts its operation at the points we have taught it. A cobot is a machine that can work together with a human. A cobot is a machine that can work together with a human. Figures 10.16 and 10.17 illustrate the use of Mitsubishi Electric India cobots in the mechatronics industry today [9–11].

10.7 HUMANOID IN MECHATRONICS INDUSTRY

Humanoid robots, which are known as humanoids, are completely different from industrial robots and cobots. These humanoids can look like humans. These humanoids can perform the tasks that humans do with greater efficiency and speed. These humanoids are currently being used extensively in factories in countries like China, Japan, and North Korea. These humanoids have a brain system similar to humans, and their brains are made of AI. Arms and legs act as actuators for human organisms. The controller fitted to these humanoids is more than one. Humanoids normally transmit signals electrically to actuators. The signal from the humanoid's brain or spinal cord is amplified before it goes to the actuators.

Once the amplified signals reach the actuator, the actuator does the job perfectly. Actuators are often operated using an electromagnetic system. These actuators are often made up of a stepper or servo motor. These humanoids are designed to communicate with multiple humanoids simultaneously. These humanoids are designed to run on a battery, usually called DC power. Once a humanoid completes its task, it knows what the next task is from the commands given to it. In mechatronics car factories, there are various types of jobs such as assembly, welding, and painting [12]. A humanoid robot is designed to perform many of these tasks. These human IDs are usually operated using a microcontroller or a microcomputer and their memory system is connected to cloud computing. We do not need much electricity when we work with such artificially intelligent humanoids. AI only expends precisely as much energy as it needs to do a job. A humanoid was created in a system that allows it to move from one place to another. While moving in this way, they calculate the minimum distance and start their movement as per the program. Hence, the energy required to move from one place to another is largely conserved [11–15]. Humanoids make a major contribution to energy conservation [13]. The humanoid robot model is shown in Figure 10.18.

10.8 ROLE OF RENEWABLE ENERGY IN THE MECHATRONICS INDUSTRY

The field of mechatronics is generally based on mechanical engineering, electrical engineering, and electronics. Energy conservation is one of the most essential aspects in these mechatronics industries. Power required by the machines is usually supplied through the electricity board itself, but the amount of electricity used results in high

FIGURE 10.18 Sample humanoid robot for mechatronics industry.

electricity bills. The biggest challenge is to reduce the electricity bill burden. We need electricity to run the motors used in machines, PLC, and other automatic systems that can control such motors. This uninterrupted electricity can be achieved by connecting electricity boards with renewable energy such as solar panels, wind turbines, and other renewable energy sources. By installing solar panels on the rooftop of factories we can generate large amounts of power. Also, through this, we can charge the batteries that can be used for autonomous lines. All industrial power is not 100 percent energy efficient and must be powered by at least 40 percent renewable energy [16–19]. The off-grid and on-grid solar energy system is shown in Figures 10.19 and 10.20.

Although the initial installation costs of solar panels are high, we can recover these costs within a given year. Similarly, wind power plays a major role in supplying electricity to the industry during certain months. Power electronic controllers required to combine wind and solar energy must be installed in more and more factories. When power from one source is not available, it can be obtained through the other. Furthermore, 24-hour electrical power is required for the critical loads that run factories. In such times when we use renewable energy instead of relying entirely on the electricity board, we can reduce our utility bills and always get uninterrupted electricity. An IoT-based solar energy system is shown in Figure 10.21.

In the same way as machines are controlled by AI, the amount of electricity obtained from solar panels and wind turbines, the rate at which they are stored in the battery, and the rate at which they are spent are monitored and controlled by AI, enabling us to achieve the goal of optimal energy efficiency. AI plays a major role in determining the maximum electricity that can be obtained from solar panels and the maximum electricity that can be obtained from wind turbines. Also, the amount of electricity stored in the battery, how long they can draw power from the battery, the temperature of the battery, the temperature of the solar panel, and the speed of the wind turbine are all monitored by AI [20–22]. The IoT-based wind energy system is shown in Figure 10.22.

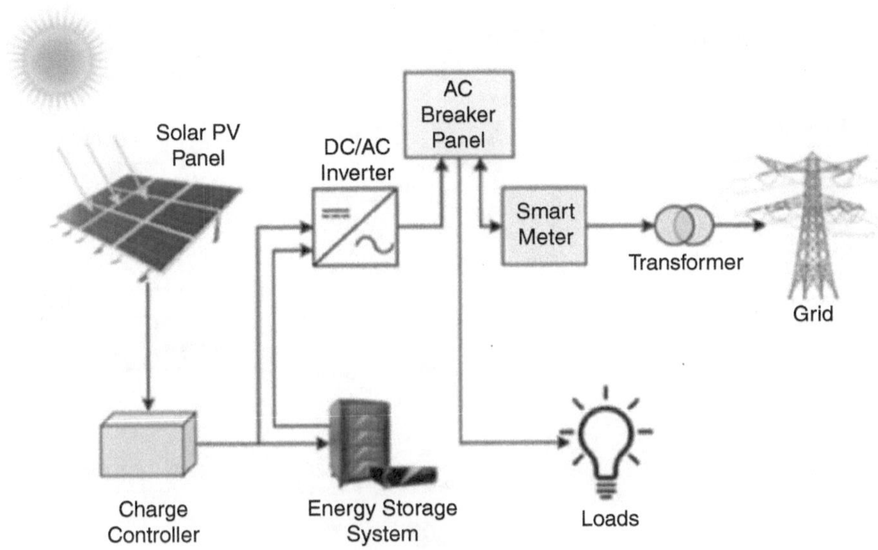

FIGURE 10.19 Off-grid solar system for mechatronics industry.

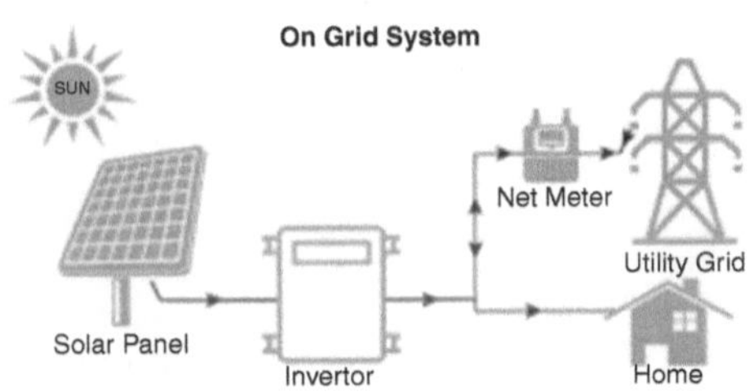

FIGURE 10.20 On-grid solar system for mechatronics industry.

10.9 THE ROLE OF AUTOMATION AND AI IN THE INDUSTRIES OF TODAY

AI-powered automation is not an optional tool anymore but a requisite in the existing industrial competitive world (see Figure 10.23). AI's ability to automate mundane tasks, process and analyze big data, and make in-process decisions has significantly decreased the need for human input while increasing accuracy and productivity. This change not only improves the efficiency of processes but also standardizes quality, a requirement for the industry.

10.10 REDUCTION OF ENERGY USAGE AS A PRIORITY FOCUS

As energy consumption increases across the world, it is imperative for industries to focus on conserving energy. Adaptive control systems, real-time monitoring, and other AI applications help factories pinpoint problems and take precise actions to

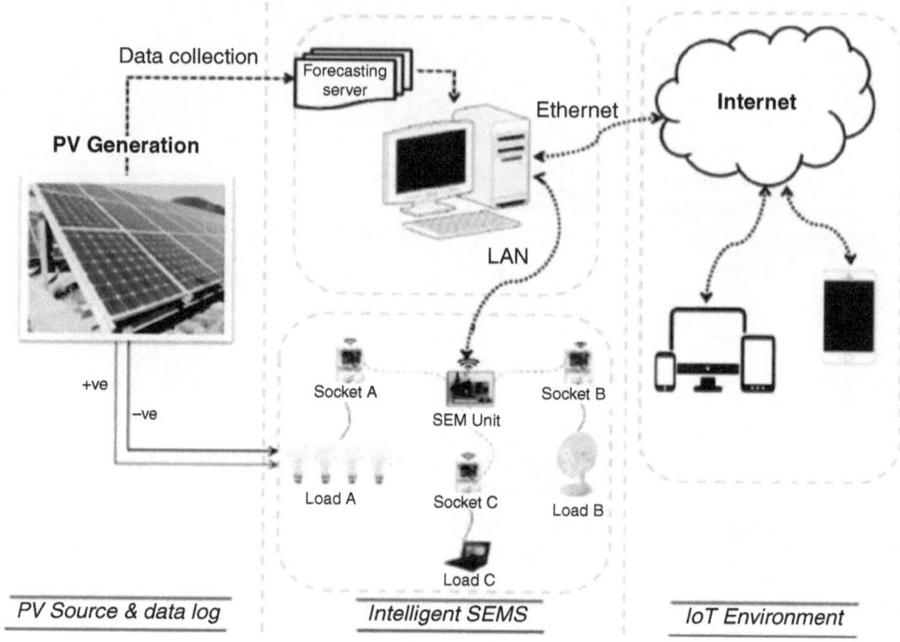

FIGURE 10.21 IoT-based solar renewable energy system.

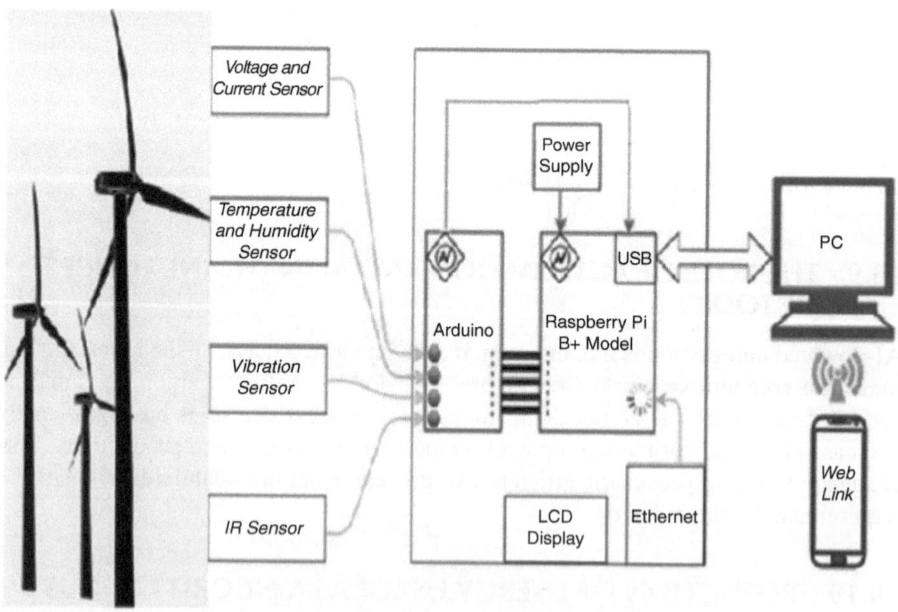

FIGURE 10.22 IoT-based wind energy system.

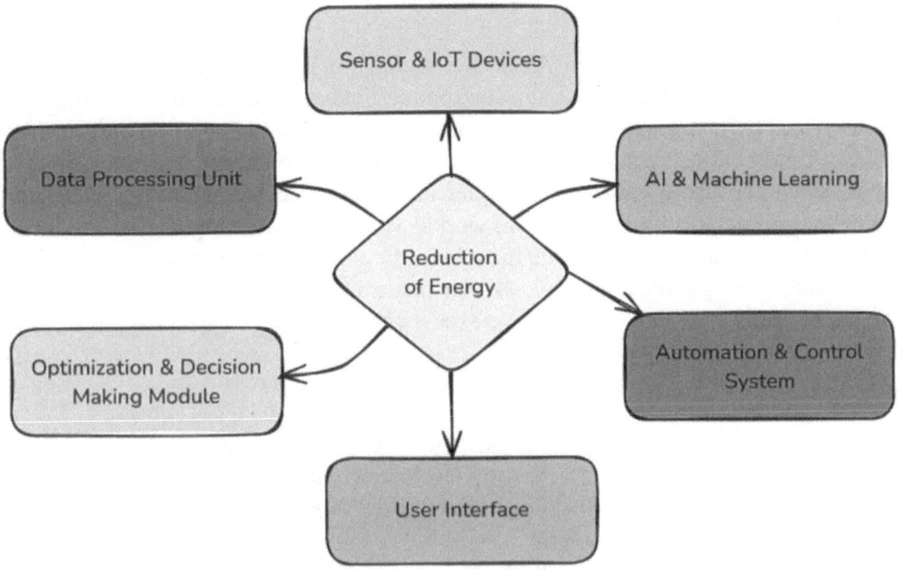

FIGURE 10.23 Reduction of energy.

prevent the unnecessary use of energy. This emphasis and many others fit into global sustainability objectives and enable industries to aid in protecting Mother Nature (see Figure 10.23).

10.11 COLLABORATIVE ROBOTS IN THE MANUFACTURING PROCESS

Cobots have enhanced the dual role of humans and machines in workplaces. AI-enhanced cobots can work without the risk of injury to human beings in an environment of risk by taking over difficult and dangerous chores. This partnership increases productivity, creativity, and efficiency (see Figure 10.24).

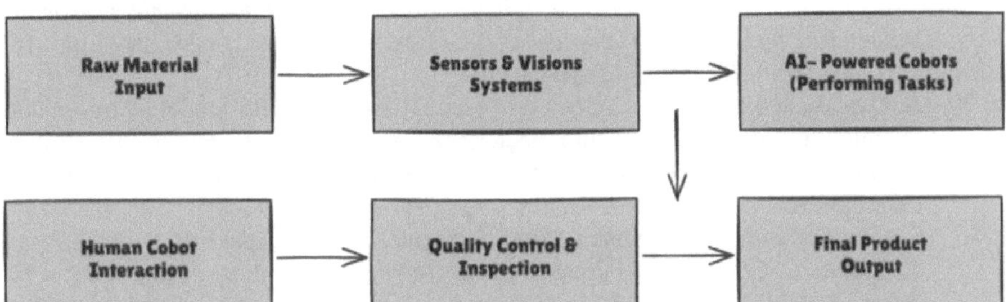

FIGURE 10.24 Workflow of Cobots in Manufacturing.

10.12 CONCLUSION

The advantage gained through AI is increasingly being used in the mechatronics industry. All machines and all controllers in the field of mechatronics are controlled by AI. Thus, by using AI technology, time, manpower and errors in manufacturing a product can be reduced. Also, AI is playing a huge role in energy conservation. This AI is developed using various techniques known as machine learning. What algorithm we use dictates how our system works. So, choosing the right machine for the mechatronics industry is one of the most important things before opting for AI. Thus, by choosing the right machine tool, we can create a high level of efficiency in mechatronics machines. By this, we can save excess energy.

REFERENCES

1. A. T. Tsirkunenko, S. S. Bukhanov and V. V. Platonov, "Ways to Increase Energy Efficiency of Traction Mechanisms by Mechatronic Systems," *2018 International Multi-Conference on Industrial Engineering and Modern Technologies (FarEastCon), Vladivostok, Russia*, 2018, pp. 1–5. Institute of Electrical and Electronics Engineers (IEEE). doi: 10.1109/FarEastCon.2018.8602922
2. J. Liu and Y. He, "Research on Energy Management and Optimization Algorithm Based on Artificial Intelligence in Energy-Saving Buildings," *2024 International Conference on Telecommunications and Power Electronics (TELEPE), Frankfurt, Germany*, 2024, pp. 985–990, doi: 10.1109/TELEPE64216.2024.00183
3. Sensor Partners, "IoT Sensors" [Online]. Available: https://sensorpartners.com/en/products-cat/iot-sensors/. [Accessed: Apr. 18, 2025].
4. N. M. S, B. Veronica, H. Shaheen, S. Azahad and L. Jayanthi, "Role and Interval Changing Algorithm (RICA) for Energy Consumption Node in Wireless Sensor Network," *2023 International Conference on Intelligent and Innovative Technologies in Computing, Electrical and Electronics (IITCEE), Bengaluru, India*, 2023, pp. 348–350. Institute of Electrical and Electronics Engineers (IEEE). doi: 10.1109/IITCEE57236.2023.10091040
5. I. Georgievski, T. A. Nguyen and M. Aiello, "Combining Activity Recognition and AI Planning for Energy-Saving Offices," *2013 IEEE 10th International Conference on Ubiquitous Intelligence and Computing* and *2013 IEEE 10th International Conference on Autonomic and Trusted Computing, Vietri sul Mare, Italy*, 2013, pp. 238–245, doi: 10.1109/UIC-ATC.2013.106
6. Z. Zhang, J. He, A. Kumar and S. Rahman, "AI-Based Space Occupancy Estimation Using Environmental Sensor Data," *2024 IEEE Power and Energy Society Innovative Smart Grid Technologies Conference (ISGT), Washington, DC, USA*, 2024, pp. 1–5, doi: 10.1109/ISGT59692.2024.10454176
7. A. Gautam and S. Sharma, "Open AI based Differential Gain Model of Information Sharing in Industry 5.0 Health Sector," *2023 4th IEEE Global Conference for Advancement in Technology (GCAT), Bangalore, India*, 2023, pp. 1–7, doi: 10.1109/GCAT59970.2023.10353462
8. S. Teli, T. Kanase, K. Ravindra Desai, S. Ingale, M. Sohani and W. R. Babu, "Auto Temperature Control of Furnace Using Intelligent Modular Controller," *2024 International Conference on Expert Clouds and Applications (ICOECA), Bengaluru, India*, 2024, pp. 375–379, doi: 10.1109/ICOECA62351.2024.00073
9. R. P. Kumar, C. Mohanraj, R. Karthikeyan and J. Prameeth, "Demand Side Management System Using Programmable Logic Controllers," *2023 3rd International Conference*

on Innovative Mechanisms for Industry Applications (ICIMIA), Bengaluru, India, 2023, pp. 1374–1377, doi: 10.1109/ICIMIA60377.2023.10426365

10. C.-Y. Chen, S.-L. Liao and Q. Chen, "Mechanistic Study of Anthropomorphism of Collaborative Robots and Their Acceptance in the Era of Industry 5.0 – A Model of Chain Mediation," *2023 6th International Conference on Mechatronics, Robotics and Automation (ICMRA), Xiamen, China*, 2023, pp. 81–85, doi: 10.1109/ICMRA59796.2023.10708126
11. N. Pusphalatha, B. P. Devi, V. Sharma and A. Alkhayyat, "A Comprehensive Study of AI-Based Optimal Potential Point Tracking for Solar PV Frameworks," *2023 IEEE IAS Global Conference on Emerging Technologies (GlobConET), London, United Kingdom*, 2023, pp. 1–5, doi: 10.1109/GlobConET56651.2023.10149990
12. R. Raffik, R. R. Sathya, V. Sushmita, N. Jyothi Lakshmi and S. Gautham, "Enhancing Human Cobot Interaction with Mixed Reality: A Futuristic Review," *2023 2nd International Conference on Advancements in Electrical, Electronics, Communication, Computing and Automation (ICAECA), Coimbatore, India*, 2023, pp. 1–6, doi: 10.1109/ICAECA56562.2023.10199911
13. A. S. M. Sahan, S. Kathiravan, M. Lokesh and R. Raffik, "Role of Cobots over Industrial Robots in Industry 5.0: A Review," *2023 2nd International Conference on Advancements in Electrical, Electronics, Communication, Computing and Automation (ICAECA), Coimbatore, India*, 2023, pp. 1–5, doi: 10.1109/ICAECA56562.2023.10201199
14. J. Oh, O. Sim, B. Cho, K. Lee and J.-H. Oh, Online delayed reference generation for a humanoid imitating human walking motion, *IEEE/ASME Transactions on Mechatronics*, February 2021, 26(1), 102–112, doi: 10.1109/TMECH.2020.3002396
15. P. Aruchamy, S. Gnanaselvi, D. Sowndarya, and P. Naveenkumar, An artificial intelligence approach for energy-aware intrusion detection and secure routing in internet of things-enabled wireless sensor networks, *Concurrency and Computation: Practice and Experience*. 2023, 35(23), e7818, doi: 10.1002/cpe.7818
16. S. Yahnian, M. Sleiman, N. A. Oufroukh, H. Su and S. AlFayad, "Mechanical Design of a Humanoid Robotic Head for Human Robot Interaction," *2024 WRC Symposium on Advanced Robotics and Automation (WRC SARA), Beijing, China*, 2024, pp. 153–157, doi: 10.1109/WRCSARA64167.2024.10685778
17. M. A. Lasemi, F. Kholardi, A. Arabkoohsar and A. Rajizadeh, "Point Estimation-Based Optimization for Operation of Multi-Generation Energy System," *2022 International Conference on Electrical, Computer, Communications and Mechatronics Engineering (ICECCME), Maldives, Maldives*, 2022, pp. 1–5, doi: 10.1109/ICECCME55909.2022.9987923
18. A. Sayal, A. Gupta, J. Jha, N. Chaithra, O. Gupta and V. Gupta, "Renewable Energy and Sustainable Development: A Green Technology," *2024 1st International Conference on Innovative Sustainable Technologies for Energy, Mechatronics, and Smart Systems (ISTEMS), Dehradun, India*, 2024, pp. 1–6, doi: 10.1109/ISTEMS60181.2024.10560344
19. D. Gunapriya, N. Pusphalatha, S. Sudharsan, S. Pandi, R. SivaKumar and P. V. Kumar, "A General Investigation of Battery Handling Techniques Driven by IoT," *2023 Fourth International Conference on Smart Technologies in Computing, Electrical and Electronics (ICSTCEE), Bengaluru, India*, 2023, pp. 1–6, doi: 10.1109/ICSTCEE60504.2023.10584895
20. P. Pawar, M. TarunKumar and K. Panduranga Vittal, An IoT based Intelligent Smart Energy Management System with accurate forecasting and load strategy for renewable generation, *Measurement*, 2020, vol. 152, 107187, ISSN 0263-2241, https://doi.org/10.1016/j.measurement.2019.107187; www.sciencedirect.com/science/article/pii/S026322411931053X

21. S. Ahmad, S. Mishra and V. Sharma (2023). "Green Computing for Sustainable Future Technologies and Its Applications," Grima, S., Sood, K. and Özen, E. (Eds.) *Contemporary Studies of Risks in Emerging Technology, Part A (Emerald Studies in Finance, Insurance, and Risk Management)*, Emerald Publishing Limited, Leeds, pp. 241–256. https://doi.org/10.1108/978-1-80455-562-020231016
22. D. Kalyanraj, S. L. Prakash and S. Sabareswar, "Wind Turbine Monitoring and Control Systems Using Internet of Things," *2016 21st Century Energy Needs – Materials, Systems and Applications (ICTFCEN), Kharagpur, India*, 2016, pp. 1–4. Institute of Electrical and Electronics Engineers (IEEE). doi: 10.1109/ICTFCEN.2016.8052714

11 Advancements in Path Tracking

Integrating Q-Learning, Reinforcement Learning, and Artificial Neural Networks for Autonomous Robotics

Vidhaya Datta M., Sumit Kumar Jindal, Shubham Mangesh Lad, and Hemprasad Yashwant Patil

11.1 INTRODUCTION

Road safety is a serious issue in many countries, and its respective administrations are trying their best to reduce the number of accidents every year [1]. Automated vehicles are one of the best solutions to reduce accidents and make things easier for humans. There is tremendous progress when automakers, academics, and governments have worked on autonomous driving. The deployment of an automated driving environment involves complicated automotive technology as well as human behavior, ethics, and traffic management techniques. As a result, there are still many uncertainties and difficulties to be resolved [2].

Automated vehicles face challenges while interacting with the environment around them. Obstacles might be stationary or moving. The self-driving car has to be very precise with distance calculation and should have control over its speed to avoid a crash [3]. As road traffic crashes are increasing rapidly all over the world (13.5 million every year and resulting in up to 50 million fatalities), the car must also have the crowd-sensing ability and must be able to slow down or stop accordingly [4].

The car uses reinforcement learning (RL) algorithms to get both quantity and quality-based understanding to make a decision in the face of punishments and rewards [5]. In active learning systems, exploration plays a principal role. The car has to spend time moving around to understand the environment. To reduce the costs (in terms of punishment or a negative reward) of learning action, selection must also take into account the experience gained during learning [6].

Understanding several behaviors of the agent to determine the one that performs well in a given environment is a very important part of RL. We use statistical techniques to understand data and estimate the data through dynamic programming methods. The agent is related to the outer environment through sense and action in standard RL

DOI: 10.1201/9781032615790-11

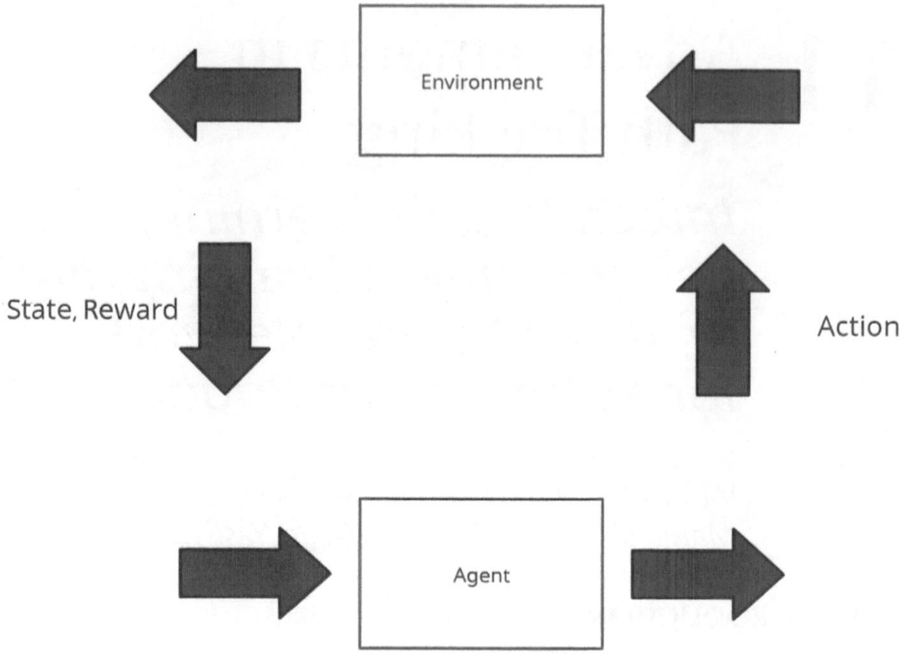

FIGURE 11.1 Reinforcement learning model.

models. Present state and action are provided as input to the current state to predict the next state (output). Rewards and punishments are given based on the action [7].

Many algorithms in RL can be optimized to produce the best results by using tuned hyperparameters. However, the selection of the ideal hyperparameter configuration and the range of values taken into consideration are frequently not documented in the linked literature. Furthermore, choosing the wrong hyperparameters makes it difficult to compare algorithms fairly. In this report, we look into a number of performance-related effects of RL [8].

Figure 11.1 shows us a working model of RL. Correct action taken in correct states gives a correct reward. During the process, our agent is going to learn the environment and help the learner understand if the impact of the model is negative or positive in the environment. We are further going to understand Q-learning, deep learning, and neural networks in this report.

11.2 Q-LEARNING INTUITION

11.2.1 The Bellman Equation

The issue of exploration/exploitation in RL is solved by the Bellman equation. The Bellman equation interrelates the current state of the agent at any time instance to the anticipated state at succeeding time instances [9]:

$$V(s) = maxa(R(s, a) + \gamma V(s')) \tag{11.1}$$

Here, we are presenting output as a function of state, action, and reward. We also present the output as a function of the following state, the value of the current state, and the discount rate. The discount rate is essentially a hyperparameter that aids in fine-tuning and fitting our decision-making process. It indicates whether rewards will be desirable in the environment,thus directing the agent's attention toward long- or short-term benefits. Many actions are implemented by the agent and we only take the action that is repeated maximum number of times.

Agent explores the environment and traces the path to the destination. As soon as it reaches the destination, agent comes back to the starting line, marking values in a way that when the object retraces the path, value of destination is "1" (V = 1).

This current version of the Bellman equation fails in non-deterministic process (random processes). Before discussing what can be done to the equation to make it good for random processes, let us see the difference between deterministic and non-deterministic processes.

11.2.2 Deterministic versus Non-Deterministic Processes: An Understanding

In a deterministic algorithm, the result is easily predictable if the state of object is known. However, detecting output is not easy in a non-deterministic process as events occur in a very random fashion. In real life, we see more random processes than deterministic processes.

In theory, we generally start understanding a process as deterministic process for the ease of calculation; however, in real life, almost all the processes are non-deterministic in nature – the reaction is only for the actual outcomes of actions. We look into how to get the agent to any destination by altering the Bellman equation. This is a sufficient requirement for many real-world issues because the agent's path need not be ideal or repeatable [10]. Table 11.1 shows the difference between deterministic and non-deterministic processes.

TABLE 11.1
The Differences between Deterministic and Non-Deterministic Processes

Deterministic Process	Non-Deterministic Process
• Each activity's progress is determinable. • Longest path in the network is the Critical Path • Time period is fixed. (Time taken for a particular action to be implemented is fixed.) • We can easily find out total time and cost, which is going to be the sum of all time periods of the actions and sum of all the costs for each and every action. • Risks are easy to handle as they are statistic entities in the case of deterministic process.	• Each activity's progress is hard to determine as the processes are random. Random process is represented by random variables. • Generally, we use three-point estimates to determine the time taken by a random process. • Time period is going to be a random number in a random process. • Total cost and time taken are not determinable. We can only write certain algorithms to limit the process and get profit out of it. • Risks are hard to handle as they are random entities in the case of non-deterministic processes.

11.2.3 Markov Decision Process

A Markov decision process or Markov chain is a stochastic model stating all the possible outputs of an event based on previous results. It indirectly states, "What occurs next depends exclusively on the situation right now."

Markov property is only applicable when the Cumulative Distribution Function (CDF) of a stochastic process depends on the present state only. This condition is in both past and present states.

Markov decision processes offer a mathematical approach for making decisions in situations where outcomes are determined by both chance and the decision maker.

$$V(s) = \text{maxa}(R(s, a) + \gamma\ \Sigma P(s, a, s')V(s')) \tag{11.2}$$

Here, P(s, a, s′) is the probability and the equation above is the Bellman equation for random processes. Being statistical representations of processes in the real world, Markov chains have various uses. Examples include researching airport customer lines, currency exchange rates, animal population dynamics, and cruise control systems in moving cars [11].

The exactdefinition of Markov process changes as it has either a discrete state space or discrete index set. To define a Markov process, we have to understand that the process happens in discrete time with a set of states irrespective of nature of time or it happens in discrete time in either countable or continuous set of states regardless of the state space [12].

The states E and F in Figure 11.2 represent the probability of change from one state to its subsequent state. Arrows indicate the direction. The probability that state stays the same at F is greater than the probability that state changes from F to E.

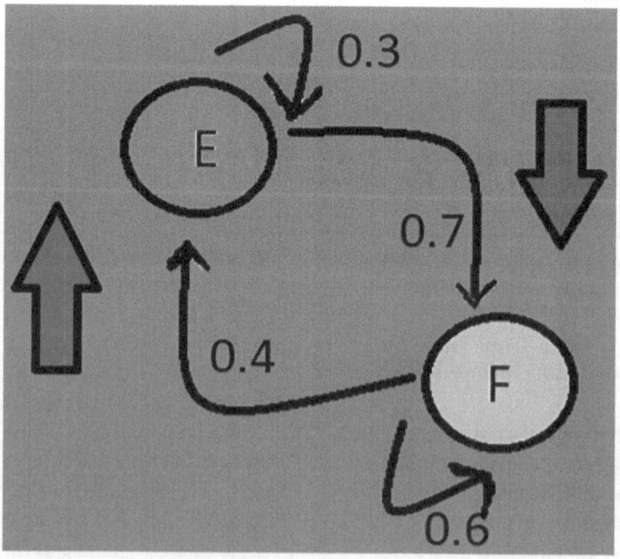

FIGURE 11.2 Two-state Markov process.

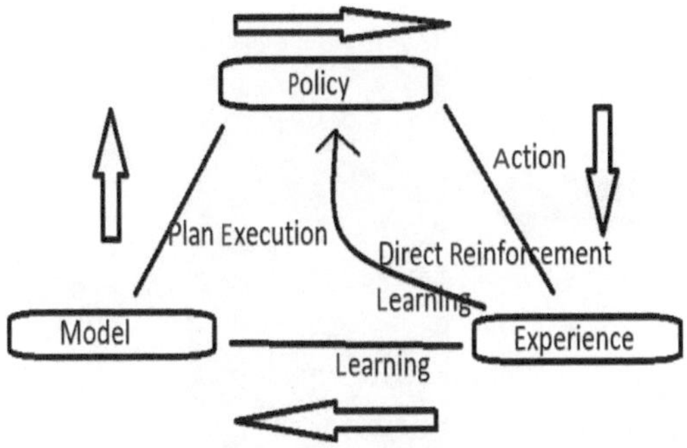

FIGURE 11.3 Execution of policy, plan, and direct reinforcement learning.

11.2.4 Policy versus Plan

A series of actions leading from the initial state to the destination can be defined as a plan. The presence of concurrency can make it more complex but this is the fundamental idea. Policy is a set that maps from state to action, which allows from any surrounding state. Planning involves the execution of policy through time and we must alter the policy based on the resultant trajectory.

In RL, the policy explains how an agent acts in an environment. Representation is done by P(s, a, s′) (probability of performance for each action at a given state). RL's main objective is to increase the reward through practice and interaction with the surroundings.

This can be accomplished in a variety of ways. One of them entails creating a model of the world first, and then using that model to find the optimal course of action. The planning phase is the next step.

Planning is the process of moving from a blueprint to a policy, whereas learning is the process of moving from expertise to a policy.

Figure 11.3 shows how a policy and its execution work. It is a better process than direct RL as it helps us understand our environment better.

11.2.5 Q-Learning

Q-learning is an algorithm that helps us figure out the quality of action in RL. It determines the value of a given action (quality of action) in a given state. Q-learning is also called as model-free RL model as it is independent of the model or type of environment the agent is in. It only cares about the path that our agent has to follow in a given environment, but not the environment.

The Q-learning process supersedes the Markov decision process in terms of accuracy and is a much more effective algorithm when it comes to path tracing. There is no time limit for exploration, and this is a random learning process. In Q-learning, we maximize the value of total reward throughout the process across all steps [13]. "Q" is the algorithm that computes the total reward for a specific action and state:

$$V(s) = \text{maxa}(R(s, a) + \gamma \, \Sigma P(s, a, s')V(s')) \tag{11.3}$$

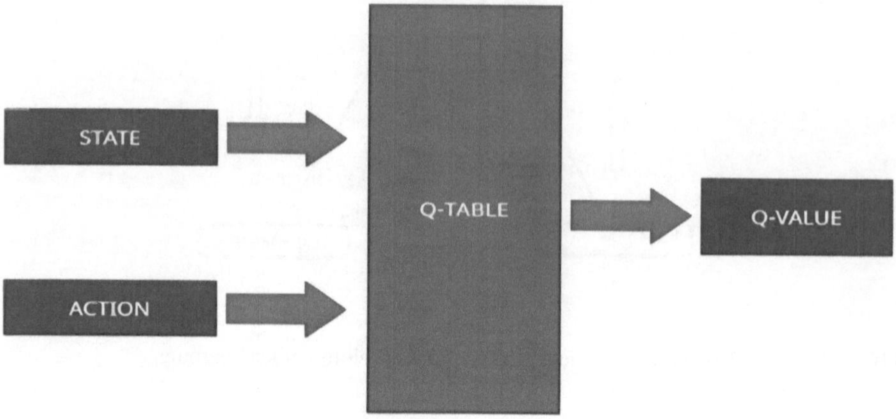

FIGURE 11.4 Basic Q-learning model.

$$Q(s, a) = (R(s, a) + \gamma\ \Sigma P(s, a, s')V(s')) \tag{11.4}$$

where Q(s, a) represents the quality of state and action, and from the above equation, we can also realize V(s) is max of all Q(s, a) values. We can also change Q(s, a) equation by representing V(s′) as maxa′(Q(s′, a′)).

Q-Values or Action-Values: Both states and actions define Q-values. Q(S, A) is an estimation of the likelihood that taking an action A at state S would give expected results (see Figure 11.4).

Q-tables have the values that are held by a particular state and action.

11.2.6 Temporal Difference

Temporal difference (TD) learning is a type of model-free RL that starts learning from the more newer estimates of the value function.

Similar to dynamic programming and Monte Carlo techniques, the models are fined-tuned using the newer estimations and sample data assembled from the surroundings [14].

TD learning updates the predictions based on its recent learnings to determine precise forecasts before the final event. We are going to introduce a new term here called learning rate [15].

The amount of newly acquired knowledge superseding prior learned data depends on the learning rate or step size. If the step size is 0, it indicates that the agent learns nothing new and uses only its previous knowledge, whereas a step size of 1 indicates that the agent only considered the most recent knowledge. Alpha_t = 1 is the learning rate (ideal) in situations that can be predicted entirely. In stochastic problems, the algorithm tends to change under certain technical constraints. Alpha_t = 0 says that no learning is done. In reality, a fixed learning rate is frequently used, such as alpha_t = 0.4 for all t:

$$Q(s, a) = Q(s, a) + \alpha\ TD(a, s) \tag{11.5}$$

α = Learning rate

TD(s, a) = Temporal difference of state and action.

Let us understand TD(a, s). Let us say current Q-value of the agent (car) is Q(s, a) and next Q-value according to Q-learning formula will be R(s, a) + γ (maxa′(Q(s′, a′)). The difference between these two equations is TD(s, a).

$$TD(s, a) = R(s, a) + \gamma\ (maxa'(Q(s', a')) - Q(s, a) \tag{11.6}$$

Q(s, a) is something that we have through imperial calculations and R(s,a) + γ (maxa′(Q(s′, a′)) is something that we have already calculated. So TD takes both present value and next value into consideration. The unsupervised learning technique helps us predict total reward expected as time progresses. So, it is used a lot in RL.

11.2.7 Artificial Neural Networks

The four main components of an artificial neural network (ANN) are as follows:

Learning inside an ANN is actually happening on three basic levels:

Input Layer: It is where the network takes the input data not processed and is mainly called "features." Inputs here are essentially the basic matter from which the network generates findings and predicts.

Hidden Layers: These units perform computation by making use of weights and biases. The hidden layers process the input data to find patterns that are crucial for a proper decision. The number of hidden layers and neurons in each layer can be varied according to the complexity of the task.

Output Layer: This layer produces the final output or prediction, which is commonly known as "labels." These processed data are made understandable to humans, which can include classifications, regression values, or even a choice between two options that are binary.

Label and Features: These are input data values the network computes for meaningful insights. In the case of image recognition, pixel intensities constitute features. The expected values, that is, labels, or ground truth values, are used in training the network; that is what makes supervised learning possible.

Loss Function: It represents a measure of error, as indicated in the difference between predictions and actual values, through which a direction in which corrections might have been provided could be deduced. These would involve common applications like mean squared error in regression or cross-entropy loss in cases with classifications.

The loss function drives the network to include parameters that "push toward" reducing error and improving performance across successive iterations on a basic level.

Optimizer: The optimizer supports the learning process by changing weights and biases that minimize the error determined by the loss function. These algorithms include stochastic gradient descent: Updates the weights on the go for every data point. It combines the momentum advantage with adaptive learning rates, thus accelerating convergence. Optimizers enable the network to effectively explore the loss landscape to settle on the most favorable configuration of the parameters.

The input data is provided to the neural network's bunch of layers. A loss function must be used to evaluate the network's performance. The loss function notifies the network of the path to follow prior to gaining complete knowledge. The network increases its dataset with the help of an optimizer [16].

11.2.8 Deep Q-Learning

The influence of receptive fields is emulated by the DeepMind system in an architecture of deep CNN, which consists of multiple layers of tiled convolutional filters. Such designs help capture spatial hierarchies and patterns, which exist in input data for the efficient processing of a high-dimensional input of a system. Such an approach is well-suited for tasks that involve visual and sensory data, where it is critical to understand localized features and their relationships in a broader spatial context. Deep CNN architecture is based on the human visual system's ability to detect edges, shapes, and complex structures by progressively extracting higher-level features from raw input data [17].

RL faces severe challenges when Q, the quality function, is represented using a non-linear function estimator like a neural network. The primary challenge is in the basic instability and incompleteness of these representations in dynamic environments. There are many factors that exacerbate this problem. Sequential correlations in the observational data can cause bias in the entire training procedure as the experiences of the agent are neither independent nor identically distributed. Also, small perturbations in the Q values can cause drastic changes in the behavior of the agent, making its decisions erratic or suboptimal. This sensitivity is undesirable in situations where consistent and reliable actions are important to succeed. Moreover, the tight correlation between the computed values of Q together with their related target values is highly unstable at update time especially when the environment has limiting or otherwise undiversified training examples. Problems from these aspects make it generally difficult for an agent to learn effectively, also generalize actions of the same across various instances.

Addressing the foregoing challenges, DeepMind formulated experience replay. The procedure is actually borrowed from some biological phenomena. This way, experience replay enhances learning because it breaks the sequential correlation that is inherent to the agent's observations. Instead of relying on recent actions and states, this approach picks samples randomly from a buffer that is filled with experiences that were previously encountered. This method, therefore, ensures the increased diversity of the training set – the model will learn through a wide variety of circumstances that it has experienced in its lifetime. Through decorrelation of training examples, experience replay reduces the effects of observation order and hence bias due to sequential dependencies.

This method also increases the uniformity of the data distribution as the sampling procedure incorporates experiences from diverse conditions and temporal contexts. Such diversity is important for the development of resilient models that can generalize well to new situations. Experience replay enables iterative modifications to the Q-function which is directed toward convergence using periodically updated target values, which stabilizes the learning process so as not to lead to overfitting or divergence [18].

The addition of experience replay has significantly enhanced the reliability of deep Q-learning algorithms. This is due to its capability to eliminate the instability caused by correlations and allow the model to learn from a greater variety of environmental

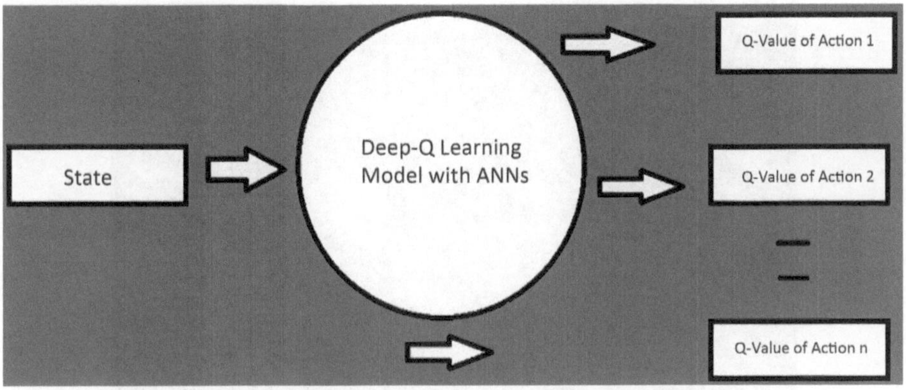

FIGURE 11.5 Basic deep Q-learning models.

situations, which makes it an important component of modern RL architectures. Its addition has led to the creation of more efficient and scalable artificial intelligence (AI) applications within complex and dynamic environments, thus enhancing the capabilities of autonomous systems (see Figure 11.5).

11.3 UNDERSTANDING REINFORCEMENT LEARNING THROUGH A SIMULATION

An autonomously controlled vehicle simulation has been developed that makes round trips between a given origin and a destination. The main goal of the project is to ensure that the vehicle is able to understand, navigate, and track any route effectively connecting the starting and ending locations. This is a representation of the learning capacity of an autonomous agent in a controlled environment.

It utilizes a reward-and-punishment system to train the car. Whenever the car traces the path with precision, it reinforces positively; however, in cases of deviations from the path and obstacles, negative reinforcements in the forms of punishments are administered. The dynamic loop of feedback allows the car to iterate upon its behavior and to hone its navigation skills in the environment itself.

The simulation produces visual representations that yield insights into the process of learning. These visualizations depict metrics including cumulative rewards, rates of punishment, and the advancement of learning over time. In the beginning, the car's pattern can appear unpredictable as it explores the environment and acquires experience. Upon progress in learning, the visual representations reveal a trend toward stabilization, indicating that the car has learned the most optimal route.

To better elucidate the learning process of the agent, we use a set of assessment metrics that involve the mean score curve and the time evolution of the rewards. The mean score curve, for example, will describe the evolution of cumulative rewards in time; hence, giving an intuitive visualization of the learning trace. A flatter curve indicates that the vehicle has reduced errors and developed an efficient route to the goal.

It indicates how RL helps in the adaptive behaviors of autonomous systems across varied conditions by iterative improvement. Simulating real-world challenges within

FIGURE 11.6 A visual of our simulation.

a controlled virtual environment will allow for the assessment of development and optimization of algorithms that can be applied to a greater extent in autonomous navigation, robotics, and other real-time decision-making systems.

Figure 11.6 shows the test scenario for the demonstration of practical applicability of the RL. The graph model has an agent who is an autonomous vehicle charged with traveling from a particular source located at the bottom-right point to a destination at the top left-hand side. The environment has types of terrains and likely obstacles, so the automobile had to explore, change the strategy, and learn by trial and error what it really wanted to do.

The simulation captures the cyclic nature inherent in the learning process where the vehicle first explores the environment with no notion of the optimum path. Over time, with rewards for staying on the path and penalties for leaving the path, the vehicle progressively improves its understanding of the path. This figure shows one frame taken during the simulation, which shows the position of the vehicle, the path, and the challenges the environment poses.

Figure 11.7 shows the agent learning trajectory in terms of the average score curve over timesteps. The curve plots the cumulative rewards earned by the car as it goes about its navigation exploration. At an early stage, the curve is jagged because it represents exploratory time during which the car often goes off track, and gets penalized at such points.

As the simulation goes on, the vehicle begins to detect patterns within its environment and modifies its behavior in such a way that maximizes rewards while minimizing penalties. Once the vehicle has stabilized, identifying its most effective strategy to traverse the environment, it will be at the point where the vehicle has learned enough knowledge regarding the best route and will then continue to follow it with minimal errors.

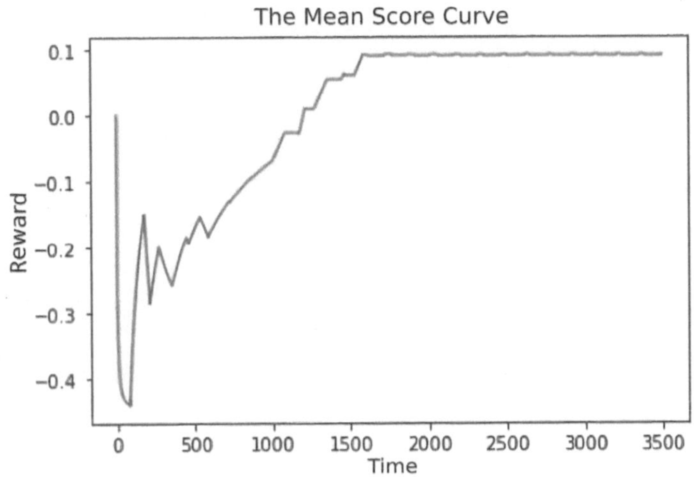

FIGURE 11.7 The mean score curve of the car.

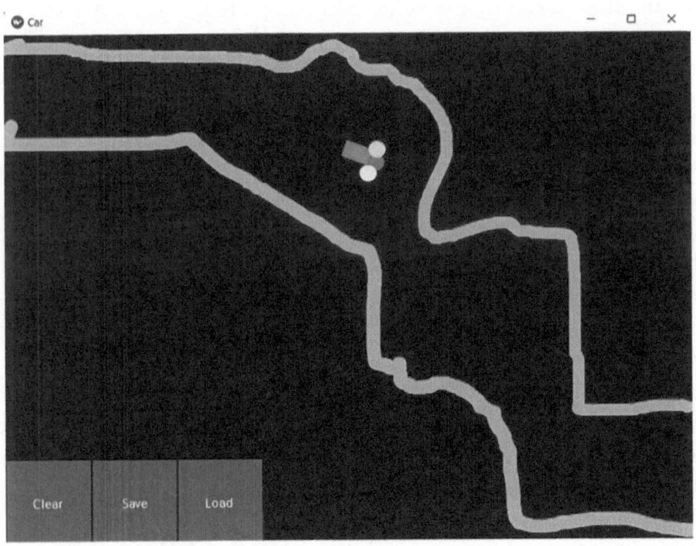

FIGURE 11.8 Time graph.

A curve describing the mean score behaves as a quantitative measure to describe the learning process as it reflects the changes and adjustments of the agent along time. This curve also can be used to recognize the effectiveness of the integration of the RL algorithm with the reward–punishment framework in determining the action of the agent.

Figure 11.8 shows a temporal analysis of the navigation efficiency of the vehicle showing the challenges experienced as the complexity of the route increases. The

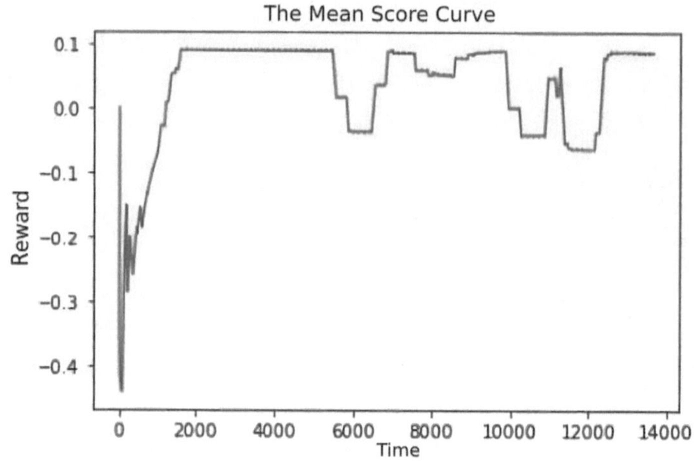

FIGURE 11.9 Mean graph.

graph shows the time taken by the vehicle to complete the journey from the origin to the endpoint in successive experiments.

The time graph in the initial periods of learning is filled with fluctuations because the vehicle does not fully learn the art of properly passing through the environment, keeps deviating from the determined course, and needs rectifications each time. Poor results and increased traveling time result from each deviation. At later periods of learning, the vehicle starts developing shortcuts and thereby reduces travel time for the task.

In the later stages, it converges to a constant that reflects that the car has learned the optimal path. At any point when the environment complexity increases – obstacles pop up or routes change – the peaks within the graph demonstrate additional learning the car does in adapting to these alterations.

The temporal graph further enriches the mean score trajectory by providing a time perspective of the learning process. Altogether, these examples demonstrate the interplay between exploration, adaptation, and optimization in the RL framework (see Figure 11.9).

Now we can see that the car took a lot of punishment as the path got complex but it was figured out at the end (graph got flat; see Figure 11.10).

We can see that as the graph gets complex, the car's learning gets harder. The car takes more punishment before figuring out the way.

11.3.1 Scope

RL has demonstrated significant potential in various sectors by optimizing processes, enhancing productivity, and reducing operational costs. By leveraging its capability to learn from dynamic interactions within an environment and improve decision-making through iterative refinement, RL has emerged as a transformative technology across multiple domains [19].

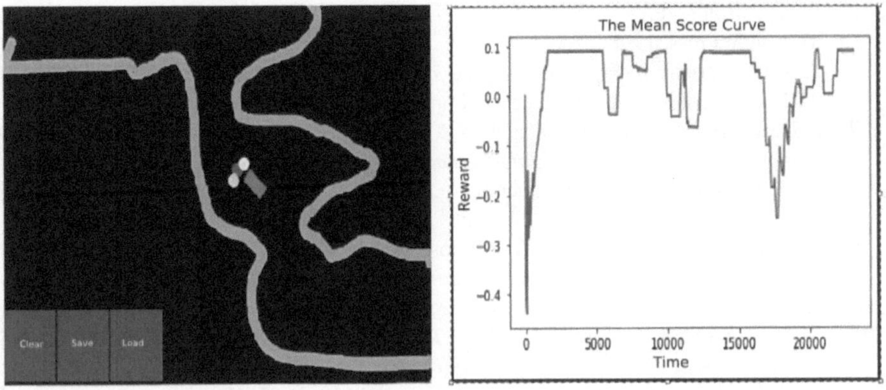

FIGURE 11.10 Visual of our simulation and the corresponding graph.

We can deal with increasing production and lowering costs by using RL. Uber and other virtual ride-providing services have radically altered transportation and have the ability to increase productivity in transportation. One of the fundamental tasks, like delivery requests, interest forecasting, and planning a course are examples of such ridesharing services. It chooses which options to distribute available drivers for nearby unassigned travelers [20].

There is also a lot of scope for AI in the medical field. The use of AI in the medical field is growing at an exponential rate. Healthcare AI initiatives were the most funded sector in 2016.

However, some people are still doubtful and recommend precautions because of unrealistic benchmarks. However, with proper precautions, we can use AI to positively transform healthcare and medicine [21, 22].

REFERENCES

1. S. Chatterjee, P. S. Bandyopadhyaya, and S. Mitra, "Identifying Critical Safety Issues on Two-Lane National Highways in India – A Case Study from NH 117 and NH 60," *Transportation Research Procedia*, 2020, vol. 48, pp. 3908–3923. doi: 10.1016/j.trpro.2020.08.029
2. M. Martínez-Díaz and F. Soriguera, "Autonomous Vehicles: Theoretical and Practical Challenges," *Transportation Research Procedia*, 2018, vol. 33, pp. 275–282. doi: 10.1016/j.trpro.2018.10.103
3. S. C. Calvert, I. R. Wilmink, A. M. G. Soekroella, and B. van Arem, "Considering Knowledge Gaps for Automated Driving in Conventional Traffic," in *Proceedings of the Fourth International Conference on Advances in Civil, Structural and Environmental Engineering – ACSEE 2016*, December 2016, pp. 102–111. doi: 10.15224/978-1-63248-114-6-33
4. M. el Barachi, F. Kamoun, J. Ferdaos, M. Makni, and I. Amri, "An Artificial Intelligence Based Crowdsensing Solution for On-Demand Accident Scene Monitoring," *Procedia Computer Science*, 2020, vol. 170, pp. 303–310. doi: 10.1016/j.procs.2020.03.044

5. R. S. Sutton and A. G. Barto, *Reinforcement Learning: An Introduction*. Cambridge, MA: MIT Press, 1998.
6. S. B. Thrun, *Efficient Exploration in Reinforcement Learning*, Technical Report CMU-CS-92-102, School of Computer Science, Carnegie Mellon University, Pittsburgh, PA, 1992.
7. L. Pack Kaelbling, M. L. Littman, A. W. Moore, and S. Hall, "Reinforcement Learning: A Survey," *Journal of Artificial Intelligence Research*, 1996, vol. 4, pp. 237–285.
8. P. Henderson, R. Islam, P. Bachman, J. Pineau, D. Precup, and D. Meger, "Deep Reinforcement Learning That Matters." [Online]. Available: www.aaai.org
9. B. O'donoghue, I. Osband, R. Munos, and V. Mnih, "The Uncertainty Bellman Equation and Exploration," *Proceedings of the 35th International Conference on Machine Learning (ICML)*, vol. 80, pp. 3839–3848, 2018.
10. S. Koenig and R. G. Simmons. "Real-Time Search in Non-Deterministic Domains," *Proceedings of the 14th International Joint Conference on Artificial Intelligence (IJCAI)*, Montreal, Quebec, Canada, 1995, pp. 1660–1667, 1995.
11. A. (Xinye and) Yang, "Markov Chain and Its Applications." [Online]. Available: https://ssrn.com/abstract=3562746
12. F. S. Melo, *Convergence of Q-Learning: A Simple Proof*, Technical Report. Institute of Systems and Robotics, 2001, pp. 1–4.
13. B. Gerald Tesauro and T. Keith, "Temporal Difference Learning and TD-Gammon," *Communications of the ACM*, vol. 38, no. 3, pp. 58–68, 1995. doi: 10.1145/203330.203343
14. N. Masuda and H. Ohtsuki, "A Theoretical Analysis of Temporal Difference Learning in the Iterated Prisoner's Dilemma Game," *Bulletin of Mathematical Biology*, 2009, vol. 71, no. 8, pp. 1818–1850. doi: 10.1007/s11538-009-9424-8
15. O. Isaac Abiodun, A. Jantan, A. Esther Omolara, K. Victoria Dada, N. AbdElatif Mohamed, and H. Arshad, "State-of-the-Art in Artificial Neural Network Applications: A Survey," *Heliyon*, 2018, vol. 4, p. 938. doi: 10.1016/j.heliyon.2018
16. S. Ohnishi, E. Uchibe, Y. Yamaguchi, K. Nakanishi, Y. Yasui, and S. Ishii, "Constrained Deep Q-Learning Gradually Approaching Ordinary Q-Learning," *Frontiers in Neurorobotics*, 2009, vol. 13. doi: 10.3389/fnbot.2019.00103
17. W. Fedus, M. E. B. Babb, M. Azar, I. Osband, R. Munos, "Revisiting Fundamentals of Experience Replay," July 2020, [Online]. Available: http://arxiv.org/abs/2007.06700
18. B. Singh, R. Kumar, and V.P. Singh, "Reinforcement Learning in Robotic Applications: A Comprehensive Survey," *Artificial Intelligence Review, 2022*, vol. 55, pp. 945–990. doi: 10.1007/s10462-021-09997-9
19. E. Akanksha, Jyoti, N. Sharma, and K. Gulati, "Review on Reinforcement Learning, Research Evolution and Scope of Application," in *Proceedings of the 5th International Conference on Computing Methodologies and Communication (ICCMC)*, Coimbatore, India, April 2021, pp. 1416–1423. IEEE, 2021. doi: 10.1109/ICCMC51019.2021.9418283
20. V. H. Buch, I. Ahmed, and M. Maruthappu, "Artificial Intelligence in Medicine: Current Trends and Future Possibilities," *British Journal of General Practice*, 2018, vol. 68, no. 668, pp. 143–144. doi: 10.3399/bjgp18X695213
21. A. Kumar, A. Gupta, and U. Raj, "New Era of Intelligent Medicine: Future Scope and Challenges," in *Proceedings of the 11th International Conference on Reliability, Infocom Technologies and Optimization (Trends and Future Directions) (ICRITO)*, Noida, India, Apr. 2024, pp. 1–6. IEEE, 2024. doi: 10.1109/ICRITO61523.2024.10522148
22. Smith, J., & Lee, A. (2021). *Ethical Considerations and Best Practices in the Use of AI for Healthcare. Journal of Medical Innovation*, 15(4), 245–259. https://doi.org/10.1234/jmi.2021.0456

12 Industrial AI for the Smart Manufacturing Sector

Jayaraj M., Praveen Kumar M., Sivakumar R., and Kiruthiga Devi V.

12.1 INTRODUCTION

Among the many issues that industries are facing in the current period are rising prices, maintaining safe operations, improving performance and dependability, handling data security issues, and dealing with employee turnover. To address these challenges, companies are rapidly moving toward smart manufacturing. Smart manufacturing may greatly benefit from the integration of emerging technologies such as cloud computing, sophisticated big data analytics, cyber–physical systems, and the Industrial Internet of Things (IIoT). These technological advancements allow industry processes to be more flexible and efficient [1, 2].

Smart sensors have been thoughtfully positioned at every manufacturing step in an intelligent factory. Artificial intelligence (AI) must be used to derive valuable insights from the massive volumes of data generated by these sensors. AI is essential in enabling machines to not only understand this data but also spot anomalous trends, draw lessons from the past, adjust appropriately, estimate the lifespan of equipment, and assess and predict demand trends in the industry. Consequently, the total efficiency of the workforce is improved [3].

AI technologies are transforming traditional factories into "smart" factories that can make decisions quickly with minimal human intervention. Remember that the first three industrial revolutions were triggered by the introduction of mechanical, electrical, and digital technology, respectively. However, a fourth industrial revolution, commonly known as "Industry 4.0," is expected to develop in the future, and AI is expected to lead this revolutionary journey [4].

Although the idea of incorporating AI into different industries has potential, many sectors are notably reluctant to use it in real world, practical applications. This reluctance is caused by a number of worries.

Limited Use of AI Techniques: One major issue is that many AI applications are primarily focused on image processing, speech recognition, social networks, robotics, and self-driving cars. A thorough assessment by Wuest et al. [5] has illuminated the versatility and variety of possible applications for machine learning algorithms in the manufacturing sector. They have persuasively argued that not all industrial problems can be solved by machine learning approaches. As a result, there have been certain limitations on the success of AI in industrial applications [5].

DOI: 10.1201/9781032615790-12

12.2 ABSENCE OF PROVEN SUCCESS STORIES

Another issue that encourages caution is the lack of credible evidence of AI's quantifiable benefits in the business sector. The subfield, however, shares concepts and theories with other disciplines such as computer science, cognitive science, psychology, neuroscience, and linguistics, among others. The progression of AI is based on sharing ideas between these subjects. Each discipline offers a different approach and idea for developing and defining AI research.

In creating and implementing AI, however, one pivotal concern is the soundness of the systematic discipline that makes them. Reliability of the AI systems is proper if all of them are from the sturdiness and validity of algorithms used and from the competence of researchers, engineers, and scientists working on the particular problem and engaged in the improvements of the AI systems. Such knowledge lets them develop algorithms that will accurately reflect the cognitive, decision-making, and problem-solving capabilities of the person.

However, there is a rather heated discussion about skills that the AI is capable of emulating or even outperforming human brains in. This interest in deepening the knowledge of the human cerebral machine is the common interest of scientists in cognitive science and of AI that want to design artificial systems capable and able to imitate or to enhance the performances of the brain. However, the best results have been achieved in some narrowly defined functions, in particular in tasks related to machine learning and other related cognitive systems, including pattern recognition, natural language processing (NLP), and data mining. Yet, AI specialists are far from agreeing that these systems can potentially continue to grow, becoming a general intelligence to themselves, and, above all, consciousness as we get it from people.

Respect for the multiplicity of AI across these societies accompanies demand from the multiplicity of fields, present and available people, and continuous discussion about efficiency and also the potential of AI systems developed. Despite the remarkable progress demonstrated in the past few years, it could be seen that the integration of AI, as well as the elimination of numerous issues connected with it, depends upon various levels and branches of cooperation and development of numerous branches of science and technology.

12.2.1 Absence of Panoramic Algorithm Toolbox

The need for an organized toolkit of algorithms, each with specific capability suitable for a wide range of industrial applications, is one notable problem. The use of AI in manufacturing has been hampered by the lack of diversity in algorithmic solutions.

12.2.2 Inadequate Benchmarking Systems

First, data acquisition and management are critical issues during the implementation of AI innovation in manufacturing industries. The data in many industries are fragmented across multiple systems, and consolidating the data into one comprehensive system is not easy. However, independent data might contain information deemed confidential or restricted, so it cannot be easily offered and normalized across a wide range of sectors. This approach has made it challenging to create broad benchmark systems, as the datasets used vary in comparison of the AI algorithms. Secondly, in manufacturing, datasets are noisy and contain anomalies, which makes it hard to develop and test AI-based systems and algorithms. Thirdly, the openness and dynamic setting of manufacturing increases the challenges associated with applying AI algorithms to processes. Moreover, in some applications of AI systems, decisions have to be made within finite periods of time or risk adverse impacts. As many AI algorithms need to operate in real time, processing and analyzing real data, this is especially so. Fourthly, the active environment of manufacturing companies may become a problem for AI algorithm implementation. While manufacturing progresses, the input data may transform, so AI models need to learn new conditions from time to time. Sustaining precision and reliability of AI systems can be quite cumbersome, as it may require continual retraining or recalibration. Finally, the embodiment and retrofitting of the AI systems into the existing industrial structures can be a great challenge. AI solutions may be integrated into manufacturing systems through defined processes and technologies, hence calling for massive changes to the existing architectures. This can be a very lengthy and expensive affair, and this, in a way, poses a challenge to the implementation of AI in some industrial contexts. Overall, the use of AI in manufacturing is an effective means of attaining success, but a number of issues related to data acquisition and analysis, real-time decision-making, the use of AI in conditions of significant variability, and proper integration into the manufacturing system must be addressed. It is noteworthy that the given practical issues hold a strategic importance for obtaining optimal performance of AI in industrial environments by fostering elaborate benchmark systems.

12.2.3 Inadequate Standardized Data

A possible solution toward solving this problem is to come up with a common format for inputs and outputs for the purpose of acceptance across most industries. This format can comprise a few specific fields and structures, which can be further incorporated into the data incorporation, exchange, and analysis processes. By using this standard data format, industries are able to share and provide data to others, and those with the AI are able to go through the data more efficiently to find these details and then come up with meaningful outputs.

The second way of dealing with this challenge is through the use of a single platform that is used in the aggregation of data from several industries. This might make it possible to include data parsed from other sources in this platform, hence simplifying the process of analyzing data from the integrated database. Also, it could aid industries with opening up to one another and exchanging data about the most efficient methods for data processing and utilizing the information throughout these industries.

Finally, additional investments should be made to enhance and acquire methodologies of analyzing unstructured data. This would increase the capability of AI systems to better process different data formats and sources and get a better understanding of the information that is being processed. It is through this investment, possibly through support for collaborations between industry, academia, and government, that research and development on this vital area is stimulated.

12.2.4 Lack of Failure Data

Businesses in the real world typically do not leave their machines running until they malfunction. This technique limits the quantity of failure data that may be used to train machine learning models. Using models trained using filtered or suspended data may reduce the precision of machine health forecasts. Furthermore, in manufacturing datasets, distinguishing between normal and deviant behavior is sometimes challenging.

12.2.5 Non-Availability of Good Quality Data

Outliers, missing numbers, and noise are common in real-world industrial data. It can be difficult to confirm that this data is accurate, reliable, and up to date. In some cases, there may be no reliable means to ensure data quality.

12.3 DYNAMIC APPLICATION CONTEXTS

Real-world industrial environments are impacted by a number of strong points of view in dynamic application contexts. The rapid and erratic changes in these factors make the application context dynamic. Frequently, these changing conditions are not understood beforehand. Human intervention becomes crucial to ensure the applicability of AI-driven insights in evolving situations; apply common sense and assess findings.

Resolving these problems is crucial to the effective implementation of AI in industrial settings as they significantly affect the accuracy and reliability of AI-based solutions in manufacturing and other industrial domains. As a result, a systematic framework for the use of AI in industrial settings was developed, leading to the term "Industrial AI" [6]. This is because the use of AI in manufacturing requires the integration of a number of important technologies and cutting-edge methods.

Industrial AI is "a systematic discipline dedicated to the development, validation, and deployment of various machine learning algorithms specifically tailored for industrial applications, while ensuring sustainable performance." This is a concise definition of the concept. At its core, this concept aims to effectively coordinate distributed production resources through the use of AI algorithms to provide manufacturing services to end users on demand. The ultimate goal is to make industrial systems robust against faults, self-organizing, and easily adaptable to changing demands [7].

Identified four key enabling technologies that are necessary for Industry 4.0-based manufacturing systems to successfully integrate industrial AI [8–10]. AI is one of the key drivers in Industry 4.0-based manufacturing systems. Therefore, the initiative *Make: AI* seeks to define and explain these technologies in detail. These technologies underpin smart factories and are inevitably key enablers of increasing efficiency, adaptability, and production output [11–13]. In this chapter, we will explore four essential enabling technologies: they are now more advanced with the application of big data analytics, cloud, Internet of Things (IoT), and robotics and automation [14, 15].

12.4 BIG DATA ANALYTICS

Business intelligence can be defined as an activity whereby a large amount of data – structured and unstructured – is collected, stored, and analyzed for the purpose of generating new insights that can be used by various decision makers. In the context of industrial AI, big data analysis is used to predict when equipment will require maintenance, how product assembly lines should be run efficiently, and quality control. Using complex analysis, including statistical and machine learning methods and other AI tools, manufacturers can turn this particular data into useful information and gain a competitive advantage.

12.5 CLOUD COMPUTING

Cloud computing is the practice of using computers across the internet to store and process data and to run applications. When it comes to industrial AI, users running it in real-life industrial production processes can easily get the data and AI algorithms without installing everything themselves. The flexible deployment means that manufacturers can implement AI as desired, and at the same time, it means cost reduction and simplification. This also aids in data sharing and centralized storage of data, and where data can be used throughout departments, sites or even companies, making it more complex but interlinked AI solutions.

12.6 INTERNET OF THINGS

IoT incorporates physical objects, devices, and machines capable of transmitting or receiving information through an internal or external network. In its application toward industrial AI, IoT allows for the acquisition of large amounts of information from different aspects of the manufacturing world, including production processes and equipment, as well as the supply chain. It can further be utilized for also creating the models for maintaining the machinery, improving the productivity of the processes involved, and promising the quality of the products in -real time. IoT also helps in the building of smart factories because it also supports the integration of AI solutions at various stages of the manufacturing chain.

12.7 ROBOTICS AND AUTOMATION

Robotics and automation put emphasis on the utilization of machines, robots belonging to equivalent classes, and software use in executing functions commonly handled by human beings. Robots and automation in manufacturing processes are applied in the context of industrial AI applications to help in increasing productivity and decreasing mistakes and risk. It is the process of using computer vision and machine learning algorithms on robots to provide them with the capability to work with less errors and more efficiency. Besides, it enhances the productivity and adaptability of production procedures and assists manufacturers in achieving challenging tasks that are quite hard and time-consuming for people.

Therefore, the four enabling technologies analyzed in this chapter, namely big data analytics, cloud computing, IoT, and robotics and automation, are critical for advancing and deploying industrial AI in Industry 4.0-based manufacturing systems. Such technologies help in data acquisition, storage, and analysis, as well as AI incorporation at the various levels of the manufacturing value chain. Together they indicate that manufacturers more readily have the ability to utilize the strengths of industrial AI to improve business processes and productivity to foster a more competitive environment in the modern economy.

12.8 THE TECHNOLOGIES OF INDUSTRY 4.0

12.8.1 Data Technologies

Data technologies (DT) is undeniably an essential component of Industry 4.0 and digitalization as they help organizations with handling, processing as well as utilizing data. Here are some examples of DT and their applications in various industries.

12.8.2 Big Data Analytics

Big data analysis is a process of analyzing large volumes of structured data and unstructured information to reveal useful information. Examples of big DT like Apache Hadoop and Apache Spark are commonly used by organizations in finance, health, and business such as e-commerce. Such technologies help many companies to obtain a comparative advantage over their competitors by analyzing trends in the data and setting out strategies for future development.

Data Warehousing: Data warehouse, as referred to, is an organized and integrated storage and collection system of business data. Some of the data warehousing technologies are Amazon Redshift, Google BigQuery, and Microsoft Azure Synapse Analytics. These technologies allow enterprises to effectively capture, manage, and analyze huge quantities of historical data for the purpose of making useful information as well as decision-making.

Machine Learning (ML) and AI: The fields of ML and AI are revolutionizing industries as organizations can design smart systems that learn from patterns and notions derived from data. Machine learning and AI technologies are TensorFlow, PyTorch, and Amazon SageMaker. These technologies are applicable in different sectors like healthcare, finance, and manufacturing to enhance patient experiences, identify fraud, and enhance manufacturing procedures.

Data Visualization: Data visualization technologies make it easy for the business organizations to come up with a graphical representation and make it easy for the stakeholders to analyze the data and come up with insights. Tableau, Power BI, Qlikview, and other tools are some of the data visualization technologies.

These tools can be applied to different industries, such as business, healthcare, retail, and financial-related fields to enhance business decision-making, organizational communication, and cooperative actions among employees.

IoT: IoT can be described as an extended system of tangible assets, field devices (FDs), vehicles, structures, and other possibly connected things equipped with sensors, software, and/or openness. These devices produce enormous data, which can be analyzed to increase productivity, decrease costs, and increase the satisfaction of customers. Some of the IoT technology companies include Zebra Technologies, Qualcomm, and Google Nest. These technologies are being applied in manufacturing, agriculture, and transportation sectors to enhance processes, save energy, and enhance their performance.

Cloud Computing: Cloud computing helps businesses to store, process, and analyze their data in servers that are located in any part of the world. Amazon Web Services (AWS), Microsoft Azure, and Google Cloud Platform (GCP) are some of the technologies under cloud computing. These technologies enable business organizations to cut costs, be more flexible, and be scalable to the extent that they need to be.

Blockchain: Blockchain is described as a distributed ledger technology that allows parties to share extensive information comprehensively and securely. This technology was being adopted in the financial, logistic, and medical sectors to create more efficiency, a lower expense, and a higher level of protection.

Finally, it is noted that DT act as enablers to Industry 4.0 and the digitalization progress of many industrial sectors. Therefore, embracing these technologies can improve decision-making processes, boost organizational productivity, increase product quality, and consequently satisfy customer needs.

12.8.3 Gathering Data

Then, the data analysis phase, where processed and interpreted of the resultant data from the data collection phase. It is important because at this stage you can visualize things in data that were not visible in the data itself. Hence, the information gathered through data analysis can be processed and modified in a way that will provide decision makers with the required analysis on decision-making. For instance, when analyzing sensor data of the manufacturing devices and assembly lines, which predict the reliability of the machinery in various conditions, show possible causes of failures, and suggest possible preventative actions that could minimize failures.

12.8.4 Storage and Data Management

Data warehousing is a good solution for archiving and processing a massive amount of structured data, including the customers' purchase history. The use of a data warehousing system will enable organizations to transform consumer purchase data into a structured format covering a wide range of formats and securely store it to reduce/eliminate the risks posed by duplicated storage.

The first prerequisite to that, which pushes toward constructing a data warehousing solution, is to identify the nature of the implementation project. You could think of this as defining what kind of data you are storing and where these data have to be pulled from. For the real-time data, the sources could include point-of-sale systems, e-commerce procuring systems, and customer relationship management systems, among others.

Like any important decision in data warehousing, this is a step-by-step process. Once you have decided on the nature of the data you need and the sources and requirements, then the next necessary step is to choose the right data warehousing solution. The following are some of the existing solutions: cloud-based data warehouses, on-premise data warehouses, and a combination of both. Depending on the features, such as the data size, cost, the scale to which the solution should be adaptable, and security, a specific solution will be preferred over others.

Once the appropriate Data Warehouse (DW) solution has been chosen, the actual data must be extracted, transformed, and loaded (ETL) for storage in the data warehouse. ETL is an important process through which data is extracted from source systems, converted into forms that are easily understandable, and loaded into a data warehouse. Often data transformation may be required where data cleaning, data normalization, data formatting just to mention few is done to ensure the actual data needed is of the right format for analysis.

The data, once placed in a data warehouse, can be manipulated for future analysis through tools such as SQL, charts, graphs, and many more reporting tools. Based on the number of customers' previous purchases, companies and businesses can make important conclusions regarding customer behavior, their choices, and tendencies, which can be definitely used by businesses when implementing effective marketing campaigns, creating new products, and serving customers.

Therefore, data that is in a structured format, such as consumer purchase history data, should be stored and analyzed properly using data warehousing. Through a data warehousing system, organizations are well placed to ensure appropriate storage and management of their data and at the same time get useful information that may be vital to their decision-making and hence the success of their business.

12.8.5 Data Analytics and Data Processing

Data analytics can be referred to as the process of analyzing structuring, sorting, purging, parsing, and transforming large data to come up with beneficial insights, make decisions, and draw conclusions. Among others, big data processing involves data mining, statistical approaches, machine learning, predictive modeling, and business intelligence.

Data analytics assists in organizational decision-making, creates awareness of organizational operations, and assesses possible improvement. Business intelligence refers to data analytics that can be used in industries, for example, finance industry, marketing industry, healthcare industry, and e-commerce industry.

On balance, data analytics might be described as a means of defining the values that are attached to the information in the data and using it as a way of supporting decision-making and achieving business objectives. To help with early disease detection and treatment planning, for instance, healthcare organizations use patient data to spot trends.

12.8.6 Data Visualization and Reporting

Data visualization tools transform complex data into visual representations like charts, graphs, and dashboards, making it easier for decision makers to understand and act upon insights.

Example: Logistics companies use real-time dashboards to track the movement of goods and optimize delivery routes based on traffic and weather data.

12.8.7 Data Security and Conformity

Securing data decency and ensuring permission through data isolation regulations are crucial. This technology uses encryption, auditing processes, and access controls to protect data.

For instance, financial organizations protect client financial information and adhere to laws like General Data Protection Regulation (GDPR) and Health Insurance Portability and Accountability Act (HIPAA) by using cipher and secure substantiation.

12.8.8 Interoperability and Data Integration

Data is used in many businesses from several sources and platforms. Data equity technology makes it possible for the smooth flow of data between numerous places and networks.

For example, in the energy sector, smart grids join details from distribution, electricity production, and sources of utilization to reduce waste and increase energy flow.

12.8.9 Predictive and Prescriptive Analytics

Predictive analytics uses historical data to forecast future events, while predictive analytics recommends actions to optimize outcomes based on those predictions.

Example: Agriculture uses predictive analytics to anticipate crop yields and predictive analytics to determine optimal planting and harvesting times.

12.8.10 Data Governance and Data Quality

Data governance frameworks ensure the accuracy, consistency, and reliability of data by establishing rules and principles for data management.

For instance, insurance firms uphold stringent data governance to guarantee precise policy pricing and claim handling.

12.8.11 Real-Time Data Processing

Few sectors need dynamic or real-time data processing to make fast decisions. Data streaming and improved event processing technologies permit this.

For instance, stock trading platforms soon execute buy and sell orders by processing dynamic market data.

12.8.12 Data Monetization

When promoted, data may be a highly lucrative asset. Companies can generate new revenue streams by selling or sharing data with partners.

Example: Social media platforms like Facebook and Twitter profit on user data by offering employment with customized advertising.

To sum up, new-generation industries are built on DT, which make them use data to increase decision-making, improve efficiency, and boost competitiveness. These technologies are versatile and have usage in a range of industries, including manufacturing, healthcare, finance, and agriculture.

12.9 ANALYTICS TECHNOLOGIES

Analytics technologies are a bunch of tools, approaches, and strategies used to check and clarify data to support decision-making and give useful information across a range of sectors. Businesses may escalate productivity, strategic planning, and vying by using these tools to find interrelationships, motifs, and trends in data. Here, we will provide a complete explanation of analytics tools, backed up with illustrations from a variety of industries.

12.9.1 Descriptive Analytics

Descriptive analytics attempts to give a clear understanding of previous performance and trends by drawing conclusions from historical data.

For instance, Anchor stores use descriptive analytics to get sales data from previous years to find the best-selling products and the assiduous times of year to buy.

12.9.2 Predictive Analytics

The use of enumeration algorithms and previous data, predictive analytics foretells trends or events for the future.

For example, pecuniary companies use predictive analytics to obtain credit risk and anticipate the prospect of mortgage defaults based on cadger characteristics.

12.9.3 Prescriptive Analytics

This kind of analysis goes much beyond projections by offering certain strategies or tactics to improve outcomes.

Healthcare institutions, for example, employ prescriptive analytics to provide customized treatment plans based on a patient's medical history and genetic data.

12.9.4 Diagnostic Analytics

The goal of diagnostic analytics is to determine the rudimentary reasons for puzzles or irregularities in observations.

Industrial organizations, for instance, use diagnostic analytics to identify the causes of equipment failures in order to help prevent such issues.

12.9.5 Real-Time Analytics

Real-time analytics permits fast reactions to overturn the situations by checking data as it is made.

For instance, to find the location and condition of freight and make timely changes to give routes, the logistics and transportation sectors use dynamic or real-time analytics.

12.9.6 Text Analytics

Text analytics is a technique that involves drawing conclusions from unstructured text data, including social media posts or customer reviews.

Social media firms, for example, employ NLP to analyze user sentiment and comments to gauge public opinion and improve user experiences.

12.9.7 Spatial Analytics (Geospatial Analytics)

Location-based data is the target of spatial analytics, which checks geographical data for trends and connections used for future purposes to improve the qualitative analysis of the data for the specific positions.

La localización ocupa un lugar crítico en el sector del urbanismo y la calidad de vida de los ciudadanos ("Localization plays a critical role in the urban planning sector and the quality of life of citizens"). Traffic flow optimization, for instance, applies geographic data to enhance understanding of how people, cars, and products get through a town. In other words, understanding the traffic flow will assist the urban planning agency in finding ways of eradicating congestion, minimizing the time taken on the roads, and ensuring road safety. For example, it can be achieved by developing new and efficient roads, by using IT systems for the organization of transport, and by infusion of public transport buses in traffic flow during rush of hours.

Further, spatial analytics is used to place the public amenities, such as hospitals and schools, in such places that anyone in the community can reach them easily. This depended on taking the population density, the socio-economic status of the population, and the infrastructure to estimate the areas that necessitate services very much. Hospitals and schools can be situated in such a manner that every person in a city or town has a chance to get basic public products and services regardless of his or her class.

In general, decision-making in urban planning could be conducted more effectively when supported by a spatial analytics approach, and services to residents of cities would be more beneficial. In using data in relation to the spatial organization of built forms, urban planners can formulate plausible strategies for traffic management and positioning key facilities in the public domain.

12.9.8 Behavioral Analytics

Behavioral analytics is the technique of collecting and analyzing various behavioral data for a specific user to understand his or her choice, activity, and overall satisfaction with the certain digital environment. The objective of behavioral analytics is to enhance the usability of the programs and inform decision-making. In e-commerce, behavioral analytics also play a very vital role in improving the attractiveness of the products, boosting their sales. Based on the data collected on virtually any user, including past purchases, search history, and interactions with the e-commerce platform, a user is provided with relevant product suggestions. This, in turn, enhances the likelihood of a sale and also the credibility of the user to the platform.

Some ways in which behavioral analytics can be applied in e-commerce include:

1. *Recommending Products:* Recommendations can serve as a means to help the customer buy by being able to provide the customer with recommendations based on their browsing and past purchasing behavior which in turn raises the likelihood of a purchase being made, leading to a more positive customer experience in the e-commerce system.
2. *Improving Website Design:* Through the study of the user activity, the e-commerce platform is able to draw the right conclusions on the appearance of its site and other essential characteristics, including the site layout and the arrangement of various products and services in a way that is most suitable to the targeted users.
3. *Segmenting Customers:* Consumer behavior can be analyzed with the help of behavioral analytics so that the e-commerce platform targets specific customers with several marketing strategies and offers, promotions, etc.
4. *Enhancing Customer Support:* It is important for an e-commerce platform's customer to be able to get the needed help; analyzing its customers' behavior on the e-commerce platform would help improve its customer support services.

Altogether, the presented work emphasizes the role of behavioral analytics in customer experience improvement as well as in decision-making processes in e-commerce. Upon studying customers' behavior, the sites make decisions based on data, providing users with personalized environment, enhancing the design, classifying users, and improving the customer support.

12.9.9 Customer Analytics

Customer analytics focuses on understanding customer behavior, preferences, and requirements in order to customize marketing strategies and boost customer satisfaction.

For example, airlines use consumer data to offer individualized services, segment their passenger base, and enhance pricing methods.

12.9.10 Supply Chain Analytics

To further illustrate the importance of Analytics Technologies (AT) in today's business environment, here are a few more examples of how these technologies are being used across various industries:

Finance: In the finance sector, it is used for assessing risk and detecting fraud. Predictive analytics is also applied by the financial institutions to detect money laundering and undesirable transactions. Also, the company applies data visualization to reduce clients' chances of making wrong decisions when using their investment portfolio.

Healthcare: Use of AT in healthcare organizations has numerous benefits, which include increasing patient outcomes, decreasing costs, and increasing patient care. For example, data collected from patients can be used to predict the next disease outbreak and the best prevention approach to take. On the other hand, EHRs and telemedicine have been used to assist healthcare personnel to provide care to patients and to improve patient satisfaction.

Manufacturing: With regard to the manufacturing facilities, AT is used to enhance production operations, control costs, and improve product quality. Also, for instance, some of the data collected is used to analyze equipment usage, standards of how equipment fails, and when to schedule regular maintenance. Besides, through predictive maintenance, failure in a machine can be prevented, and production time is always on track.

Retail: The AT is applied in the retail sector with the primary purposes of analyzing customers' behavior, managing stocks, and adapting to marketing communications. For instance, the cash receipts of the retailers can be utilized to develop an effective advertisement campaign, applying the customer's purchase history as a base, and to provide the valued clients with convenient recommendations on the products that meet their needs. In addition, demand forecasting can be used to make sure the right products are on hand in adequate quantity to avoid stockout or overstock which are expensive exercises. In conclusion, ATs are significant for present-day organizations as they provide means for collecting data, analyzing it, and making rational decisions about the organizations' future growth and competitiveness. Irrespective of the pointed-out technologies, those organizations which are using these technologies can definitely improve their business operations and managerial competence along with producing a positive change in the customer experience. Besides, these technologies can help organizations to overcome the challenges of the competitors.

12.10 PLATFORM TECHNOLOGIES

Platform technologies (PT) refer to the foundational infrastructure, software frameworks, and ecosystems that facilitate the development, deployment, and administration of several digital services, apps, and solutions across diverse sectors. The foundation for developing scalable, interoperable, and flexible systems that may spur creativity and productivity is provided by these technologies. Let us explore PT in detail, along with examples from various industries.

12.10.1 Cloud Computing

Cloud computing provides on-demand access to computing resources, such as servers, storage, and databases, over the internet. It provides cost-effectiveness, flexibility, and scalability.

Example: Cloud systems such as AWS or Azure are utilized by healthcare institutions to store and process patient data securely while scaling resources as needed.

12.10.2 IoT Platforms

IoT platforms enable the connectivity and management of IoT devices, which make it easier to connect and manage IoT devices and provide device data.

For instance, smart cities employ IoT devices to identify environmental, transit, and public service characteristics for better metropolitan planning.

12.10.3 Data Integration Platforms

Platforms for data integration make it possible for data to easily move between data archives, applications, and systems, improving system functionality.

As an illustration, distributors use solutions from the data incorporation to unify inventory data across their online and physical sites for precise stock management to get maximum profit.

12.10.4 Application Development Platforms (PaaS)

Platform as a Service (PaaS) platforms give services and tools for making, testing, and deploying software applications without the need for infrastructure management.

Example: Financial institutions use PaaS platforms to create and deploy mobile banking apps quickly and securely.

12.10.5 Enterprise Resource Planning Systems

Enterprise resource planning (ERP) platforms combine human resources, finance, and supply chain into a single system to increase data visibility and streamline processes.

Example: Manufacturing companies use ERP platforms like SAP or Oracle to manage production, inventory, and financial processes.

12.10.6 IIoT Platforms

IIoT platforms are specialized platforms designed for industrial environments that allow machines and manufacturing processes to be monitored and optimized.

For instance, automakers utilize IIoT platforms to gather information from manufacturing machinery to improve production lines and avoid malfunctions.

12.10.7 Blockchain Platforms

Blockchain platforms offer decentralized and secure ledger technology for transparent and tamper-resistant record-keeping.

Example: Supply chain companies use blockchain platforms to trace the origin and authenticity of products, ensuring product quality and compliance.

12.10.8 Edge Computing Platforms

Edge computing platforms bring computing resources closer to the data source (at the edge) for real-time processing, reducing latency.

Example: Energy companies use edge computing platforms to monitor and control power grids in remote locations where immediate decisions are critical.

12.10.9 AI and Machine Learning Platforms

AI and ML platforms provide tools and frameworks for developing and deploying AI and machine learning models.

Example: E-commerce platforms use AI and ML platforms to personalize product recommendations based on user behavior and preferences.

12.10.10 E-Commerce Platforms

E-commerce platforms enable businesses to set up and manage online stores, process payments, and deliver digital or physical products.

Example: Online retailers use e-commerce platforms like Shopify or WooCommerce to sell products and services to customers worldwide.

PT are pivotal in modernizing industries, enhancing digital transformation efforts, and fostering innovation. These platforms provide the necessary infrastructure and tools for organizations to adapt to evolving market dynamics and deliver seamless digital experiences to their customers.

12.11 OPERATIONS TECHNOLOGY

Operations Technology (OT) refers to the hardware and software systems used to monitor, control, and automate industrial processes and operations across various industries. OT is essential for optimizing production, ensuring safety, and enhancing efficiency in manufacturing, energy, transportation, and other sectors. Here, we provide a comprehensive explanation of OT, along with examples from different industries.

12.11.1 Supervisory Control and Data Acquisition

As they are capable of handling large industries and processes in every sector, supervisory control and data acquisition (SCADA) systems are very important nowadays. They assist the operators and enable them to take the right actions when other problems are likely to occur. A proper design of SCADA systems should allow real-time data and interface but should also be secure and fairly easy to use.

Utility companies use SCADA systems to control power distribution, manage substations and respond to power grid irregularities. This can also prove useful in reducing time wasted due, to let alone, avoiding outages and increasing the efficiency. Some key features of a SCADA system in the utility sector may include the following.

Real-Time Monitoring: It lets the operators see what the status of the power grid, substations, or other components of the equipment are and allows them to do something about it when the problems occur.

Alarm and Event Management: The system is capable of identifying such conditions and informing the operators, so they can take necessary action to contain any ensuing issues.

Data Acquisition and Historian: From the various input sources of the system, data are accumulated and can be utilized for analysis and reporting as well as for overall future planning.

Human–Machine Interface (HMI): It is a display of the status of the system populated with panels and icons that the operators can use to make or control the decisions on the system.

Security: So to ensure security measures that would stanch unauthorized access, cyber threats, and the likes.

Communication: Marshaling multiple protocols that will allow the different devices and systems within the utility network to communicate with one another.

Supervisory Control: The ability to allow operators to independently manage and control the power distribution network for network performance and network stability. Finally, we observe that SCADA systems are key to the control, monitoring, and addressing disturbances of power distribution networks for producing a secure power supply to the users.

12.11.2 Programmable Logic Controllers

Programmable logic controllers (PLCs) are ruggedized computers used to control machinery and automate processes on the factory floor. They execute logic-based operations in real time.

Example: Automotive manufacturers use PLCs to control robotic arms on assembly lines, ensuring precise and consistent welding, painting, and assembly.

12.11.3 Distributed Control Systems

Distributed control systems (DCSs) are used in process industries to manage complex and continuous processes, such as chemical production or refining. They provide centralized control and monitoring capabilities.

Example: Petrochemical refineries use DCSs to regulate temperature, pressure, and flow rates in various stages of the refining process.

12.11.4 Industrial Control Systems

Industrial control systems (ICSs) encompass both SCADA and DCS technologies and are used in critical infrastructure sectors like energy, water, and transportation to manage and protect essential processes.

Example: Water treatment facilities use ICSs to control the purification process, ensuring safe drinking water for communities.

12.11.5 HMI Systems

HMI systems provide operators with graphical interfaces to interact with industrial machines and systems, enabling monitoring and control.

Example: Pharmaceutical companies use HMI systems to monitor and adjust parameters in pharmaceutical manufacturing equipment for precise drug production.

12.11.6 Manufacturing Execution Systems

Manufacturing execution systems (MES) systems bridge the gap between enterprise-level systems (like ERP) and shop floor operations. They track and manage production processes and provide real-time visibility.

Example: Aerospace manufacturers use MES to ensure efficient production scheduling, quality control, and traceability of aircraft components.

12.11.7 Robotics and Automation Systems

Robotic and automation systems have become paramount throughout the food processing industries for increasing production efficiency, product quality, and minimizing human intervention. These systems are used in the food processing plants for automation of some operations, including sorting, packaging, and even inspection operations. For instance, when using vision systems, robots are capable of sorting fruits and vegetables according to size, shape, and color. Moreover, automation systems allow automation in the selection and packaging process to guarantee standardized packaging of the products.

Further, robots in the food processing industry are crucial for maintaining food processing compliance toward health. However, with the use of Autonomous Mobile Robots (AMRs) in the food processing environment, tasks such as cleaning and sanitation are accomplished with minimal human assistance. They can be useful in cleaning up production lines, processing machinery, or other places in the compounds if manufactured with ANDs and cleaning apparatuses. In conclusion, the use of robots and automated systems in food processing facility has positively affected high productivity, standardization of the products, and less or no human intervention. These technologies are useful to enhance the prospects of food safety for the supply chain so that the monetary cost for production can be enhanced.

Example: Food processing plants use robotics for tasks such as sorting and packaging to improve efficiency and product consistency.

12.12 ASSET MANAGEMENT SYSTEMS

An asset management system is simply a system of organization that provides support to any organization in the tracking of its assets, and these are all the items of economic value that the company either owns or exercises control over. Items like office equipment or machinery can be included along with the likes of a vehicle, inventory and material assets, and even intellectual property.

The objective of asset management is to track and sustain equipment in a manner that outlines and provides for the best possible usage of the equipment and the least frequency of breakdown. This is done through maintenance of records of the condition of specific assets, monitoring of maintenance schedules, and the records of all maintenance operations. It assists in avoiding times when equipment will develop faults that are very expensive to solve, and it lowers the cost of maintaining the equipment to an acceptable level while guaranteeing that such equipment is always functional.

Apart from recording and monitoring the use and care of the assets, the asset management systems can also give an indication of the general performance of any organization. Thus, it will open up the doors for business management to evaluate particular actions that demand the cutting down of costs or require enhancement of the level of efficiency to ensure enhanced business performance and profitability. AM workflows and Preventive Maintenance (PM) systems are very effective for reaching organizational goals because most companies need equipment and assets to operate, from manufacturing industries to construction and governmental companies. By guaranteeing that equipment is optimally utilized and properly maintained, these systems act as a facilitator that improves a company's operational efficiency, thereby increasing productivity.

Example: Wind farms use asset management systems to monitor the condition of turbines and plan maintenance to maximize energy production.

12.13 ENERGY MANAGEMENT SYSTEMS

Energy management systems (EMSs) are used in the energy sector to monitor and optimize power generation, distribution, and consumption.

Example: Utility companies use EMS to balance supply and demand, integrate renewable energy sources, and improve grid stability.

12.14 TRANSPORTATION CONTROL SYSTEMS

In other words, transportation control systems are sophisticated technologies applied to maintain and optimize traffic within urban centers. So, they monitor where cars are and help tackle the traffic planning so people can get from one point to another more easily and there is less traffic. These systems are used in smart cities to improve public transportation and transport systems in general.

OT is a critical part of the execution of business and industrial processes that make the success of many companies and industries. It contributes to keeping other processes going on, checking conformity to security measures, reducing losses, and facilitating better responses to alterations in the environment faster. OT can be done in different fields, and these make organizations work well with their productivity in a given area.

12.15 CASE STUDY: SMART MANUFACTURING IOT 4.0 IN COIR PITH MANUFACTURING

12.15.1 Introduction

- *Collection and Sorting:* Raw material for coir pith is obtained from many coconut processing players throughout the world and then graded into superior and inferior types. The screw is washed and screened to get rid of all foreign or unwarranted materials, such as metal pieces, woods, or any other undesirable materials.
- *Moisturizing:* Another process is the moisture content of the sorted coir pith through watering to get a specific MOISTURE CONTENT. This measure is vital in order to guarantee the right constructive densification of the last coir pith pieces.
- *Mixing and Forming:* The moist coir pith is then blended with an additional material that hardens it for better durability of the product; substances used include calcium hydroxide or gypsum. The mixture is then compacted to block forms that develop coir pith blocks in the precise shape and size.
- *Curing and Drying:* Following formation of the coir pith blocks, they have to undergo a process of curing, which simply means setting them to dry and harden. This can be done naturally or artificially by air drying or by putting the instrument in a hot air oven. After a period of 24–48 hours, the blocks are sufficiently cured, depending on the size of blocks, and the thickness of the bed and block.
- *Packaging and Storage:* Therefore, when coir pith blocks are adequately cured and dried to the expected standard, they are packaged in the appropriate packaging material to maintain the blocks' qualities and protect them from possible moisture content and such other factors. Two, the packaged blocks are also stocked at a dry and cool place awaiting transportation to their final destination.

By implementing Smart Manufacturing IoT 4.0 in the coir pith block manufacturing process, several benefits can be achieved:

- *Quality Control:* The IoT sensors can be employed in real-time tracking of factors such as moisture content as well as temperature, among others, during the manufacturing process. It serves as a means of standardizing the quality of the coir pith blocks and hence also prevents the formation of defects in the final product.
- *Resource Efficiency:* Smart systems with IoT influence can entail resource efficiency by measuring overall usage and identifying opportunities for

enhancement in consumption of, for instance, water and energy. Also, it helps reduce the wastage and optimize the process in totality.

- *Predictive Maintenance:* Manufacturers could connect the equipment and machinery using IoT devices to anticipate when this equipment and machinery may be likely to develop faults, and they can come in early enough to carry out maintenance, which will minimize wastage of time.
- *Automation and Remote Monitoring:* This includes the use of IoT 4.0 in automating several processes in the manufacturing process, like material handling, mixing of blocks, and block pressing. This does not only leads to efficiency improvement but also helps to monitor and control the manufacturing process from a distance, which means that changes can be made in real time.
- *Data Analysis and Insights:* IoT sensors can gather important information about the manufacturing process to improve the understanding of the production process, to detect patterns, and to make further decisions that apply to the general performance of the production.

Finally, based on results from the identified literature, the application of Smart Manufacturing IoT 4.0 to the coir pith block manufacturing process, the following conclusions can be made. Thus, by applying these advanced technologies, the coir pith manufacturers can help in enhancing the demands of agriculture and horticulture industries, reducing wastage and improving standards.

12.15.2 Collection of Coir Pith

Coir pith or coir fiber waste is actually the by-product generated by the coconut processing industries. It is the soft, brown, spongy layer just surrounding the coconut husk and generally discarded during coir fiber extraction. Coir pith is obtained from the coconut industry and has been gaining popularity all around the world because of its multipurpose uses in agriculture, horticulture, and environmental management because of certain special characteristics: its highly organic matter and its water holding capacity.

Coir pith, also known as coir fiber waste, is a by-product of the coconut industry. It is composed of small coconut fiber particles produced while producing the coir fibers that are used in the production of ropes, mats, brushes, and many others. To make it suitable for use in various industries, such as agriculture, horticulture, and landscaping, the coir pith goes through a series of processing steps:

- *Collection:* It procures coir pith from different places, that is, coir retting factories where coir fiber is made.
- *Cleaning:* In the processing of coir pith, the first activity is to wash coir pith because it may contain some unwanted additives and dust as well as large fibers. The coir pith is separated via water and air separation processes, whereby particles heavier than water are siphoned out of the coir pith as it is immersed in water. Once this water is drained, there are repeats of this process until the water becomes clear, then the water is presumed purified.

- *Sieving:* As the coir pith contains microorganisms and dust, it has to be washed properly and is sieved later to get a homogenous mesh size. This step is very important as it enables the product to become standardized, which is well suited for different uses. Further, it is useful for sieving, filtering out larger fibers so that any remaining impurities are also sieved.
- *Drying:* The sieved coir pith is then dried to eliminate the high moisture content of the fiber. This is usually done with the sun or mechanical dryers. The drying process is also useful in the preservation procedure, helloing to control the quality and duration of coir pith.
- *Packaging:* In the last step, the dried coir pith is packed and stocked for use in multiple sectors all around the world.

The processed coir pith can be used in mushroom growing, plant husbandry, landscaping, and to prepare out eco-bricks. Some of the characteristics it possesses include good water retention and aeration; hence, it is a good growing medium for plants and mushrooms. Furthermore, it is a renewable resource and a more environmentally friendly product than conventional mediums such as peat and clay.

- *Mixing with Binders:* The coir pith is mixed with natural binders like latex or synthetic binders to improve its cohesion and make it suitable for compression.
- *Compression:* The mixture is compressed into blocks using hydraulic or mechanical presses. These blocks can come in various sizes and densities, depending on the requirements.
- *Drying:* The compressed coir pith blocks are dried in the sun or through mechanical drying processes. This step reduces moisture content and prevents mold growth.
- *Packaging:* Once the blocks are completely dry, they are packaged and prepared for distribution. Some manufacturers also add fertilizers or beneficial microorganisms to enhance the coir's effectiveness as a growing medium.
- *Distribution:* Coir pith blocks are shipped to agricultural and horticultural markets for use as a soil conditioner, substrate for plant growth, or for hydroponic farming.

Coir pith blocks have gained popularity due to their eco-friendliness, water retention capabilities, and suitability for organic farming. They serve as a sustainable alternative to traditional soil and peat moss in various agricultural and gardening applications (Figures 12.1 and 12.2).

- *Sensors (IoT):* IoT sensors are placed at various points in the coir pith manufacturing process to collect real-time data on factors like temperature, humidity, moisture levels, and machinery conditions.
- *Data Storage:* Thus, there are multiple ways if storing data securely is a major concern for an organization. Firstly, there are data encryption techniques to be adopted in such a way that the information enclosed is not easily retrievable

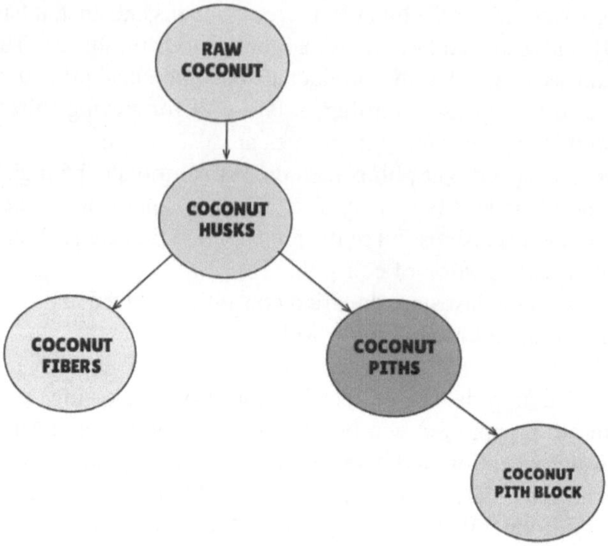

FIGURE 12.1 Coir pith block manufacturing process.

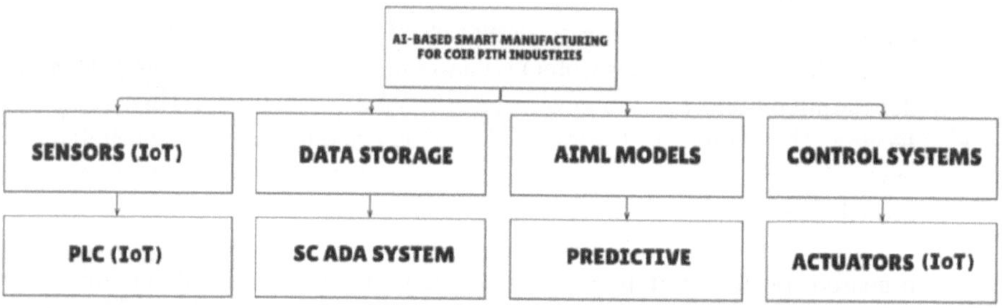

FIGURE 12.2 Block diagram for an AI-based smart manufacturing system for coir pith industries.

to other users with wanton access within the network. This means that even if a third party is able to obtain the data, then he or she cannot really decipher the meaning of such data without having the right decryption key. Another way applied to provide security is the use of a strict access regime, which means that only those persons who have a right to get access to the data. Any unauthorized access to the system can be protected by firewalls, intrusion detection, and secure network communication. Information should also be saved in another place to avoid losing it in a case of system breakdown or hacking. The data safety would, however, be facilitated by the current cloud storage service providers that comply with the security policies and have achieved certain certifications on compliance and sound procedures to protect data.

- *AI/ML Models:* Sensors generate large amounts of data, and AI and machine learning models are used in analyzing these data for industries to make the right decisions. The sophisticated use of mathematics makes it a solution that can handle a huge volume of information and actually improve decisions and processes.

Here are some key applications of AI and machine learning models in analyzing sensor data:

- *Predicting Quality:* It can detect patterns from a large number of sensors and use that information to predict product or service quality. Through the patterns and trends identified in the data, these models can predict future problems and recommend means to keep the quality of these products high.
- *Detecting Anomalies:* Moreover, machine learning models can understand and tag abnormal or unimaginable data points as a sign of a problem or failure. As such anomalies are deeply rooted in real time, businesses are able to address these problems in a real-time manner and therefore minimize breakage of operations.
- *Optimizing Processes:* New ways can be developed and found through live analysis of sensor data, which in turn helps in studying the possibilities for improvement of already used systems. Using actual data and analysis of past records, such models can indicate future possibilities of improving particular processes, cutting costs, and increasing performance.
- *Making Data-Driven Recommendations:* For businesses, the information collected by sensors can be great for AI and machine learning algorithms and can present recommendations. They may comprise suggestions that can touch on everything from the design of the products right from the packaging as well as the services customers get to the manner in which business is carried out.
- AI and machine learning models are highly effective techniques that may be used by any company requiring the interpretation and usage of sensor information to enhance its operations and yield. Thus, these technologies make operations management processes and informed decisions easier for companies to use, which is why they produce positive results.

12.15.3 Control System

- PLC and SCADA are both critical parts of controlling systems in industries today. These systems complement AI models that are employed in order to achieve the best efficiency and performance of techniques and equipment.
- PLC refers to a class of digital computer that is designed to control methods and apparatus manufacturing in industries. That is actually why they can do logic and math, and that will help in controlling and computing actions of the machinery. Equations specialized for these PLCs can be programmed such that they perform in accordance with changeable conditions and inputs from changing environments, and thus these devices can be used to regulate a variety of processes.

- While SCADA systems are the computer systems for industrial process control and monitoring. An example of such a system is the SCADA system, in which it is possible to enable access to the processes the system is controlling and collect data from diverse measurement points, such as sensors, meters, and PLCs. These systems can also calculate and decide upon various parameters of the processes to offer optimal management of the systems.
- The interface between the AI models and PLCs and SCADA systems enables higher forms of control of the machinery and all processes. AI systems can also analyze different data collected from different sensors, PLCs, or SCADA systems and take appropriate decisions based on historical and real-time data. By using these, they can be used to regulate the running of equipment, to set parameters that define processes, to improve performance.
- Therefore, PLCs and SCADA are the key parts of control systems because they help in decision implementation learned by AI models. These systems all control the machinery, processes, and adjustments to achieve and maintain those set standards of performance and usage.

12.15.4 Analytics

Guideline models that form part of the AI/ML elements are essential tools that could be utilized in the coir pith production chain. These models use past data, current operational data and other factors to predict the future problem or changes in quality which might occur. This predictability allows producers to apply timely preventive and improvement actions, which maintain uninterrupted, effective, and quality production.

- The use of predictive models can also be applicable in coir pith production to measure the temperature, relative humidity, moisture content, pH, and efficiencies of equipment that are used in the process. These models use data collected by analytic sensors installed in different areas of the production facility to detect signs of a problem or even a tendency showing that there may be a problem and allow for immediate correction. For instance, failure mode prediction, yield fluctuation potential, or optimal coir pith harvest time for the finest product acquisition.
- Taking the adoption of AI/ML-driven predictive models into account, a lot could be done for the improving industry of coir pith sustainability, efficacy, and financial performance. By reducing the use of time in dealing with unwanted breakdowns and other issues affecting production, predictive models can assist coir pith producers in streamlining their coir pith production and delivery processes, therefore saving their customers costs on low-quality coir pith products.

12.15.5 Actuators (IoT)

- *IoT Sensors:* The first process involved in this system is where data is gathered from various IoT sensors. These sensors collect data on the variables associated with the production process of the manufacturing firm. In the case of a coir pith industry, these sensors may be for temperature, moisture, pH level, humidity, etc.

- *Data Processing and Storage:* Information from the IoT sensors is thereafter forwarded to a data analyzing and handling unit. This unit could be in the form of a cloud hosting an interface or an on-premises server. Here, the specifics of the data are prepared for analysis by operations that include cleaning of data.
- *AI/ML Models:* It is then used for training of AI/ML models. These models are chosen to be able to "train" or discover patterns in the data and relationships between the factors and make future forecasts based on such relationships. For a coir pith industry, these models could identify the right level of moisture, any problem with the machinery, or signs of wastage, among others.
- *Predictive Decisions:* Afterwards, the AI/ML models are able to decide what is best for the manufacturing process based on the analyzed data. For instance, they might recommend changes to a machine's settings that will create a better-quality product.
- *Control System:* Then the decisions made by the AI/ML models go to the control system; this system can be a PLC or a SCALDA system. The control system analyzes the received commands and sends them to the IoT actuators.
- *IoT Actuators:* IoT actuators are used as devices that carry out the instructions from the control system or IoT controller. Depending on the command, some of these actuators will change the settings of the machinery, control the moisture, and perform other operations so as to enhance the manufacturing process.
- *Feedback Loop:* While there is this process being executed, there is an intermediate feedback process that is used to achieve iterative improvements. Finally, as the actions the IoT actuators implemented, observations are made and new information produced. Now, this new data is used to fine-tune the AI/ML models to keep on learning new conditions and tasks or even on constant challenge.
- This block diagram basically shows how IoT, AI, and control systems are integrated in an Industry 4.0 or Smart Manufacturing environment with respect to a coir pith manufacturing sector. Thereby, the industry can apply these technologies to raise efficiency, quality, and sustainability in the production processes.

The adoption of Smart Manufacturing IoT 4.0 enabled the coir pith manufacturer to overcome quality, efficiency, and market demand challenges. This case study illustrates how IoT-driven technologies can revolutionize traditional industries, enhancing quality, sustainability, and competitiveness.

12.16 CHALLENGES IN IMPLEMENTATION OF AI IN THE SMART MANUFACTURING SECTOR IOT 4.0

The challenges of implementing industrial AI in the Smart Manufacturing Sector IoT 4.0 are multifaceted and require careful consideration. Here are some outlines of the key challenges.

12.16.1 Data Integration and Standardization

- *Challenge:* Integrating data from various sensors, machines, and systems across different manufacturing units can be complex. Moreover, ensuring data consistency and standardization is crucial for effective AI implementation.
- *Solution:* Implement data integration frameworks and adopt industry-standard data formats to streamline data collection and processing.

12.16.2 Data Quality and Reliability

- *Challenge:* Industrial AI heavily relies on high-quality data. Inaccurate or incomplete data can lead to incorrect predictions and decisions.
- *Solution:* Establish data quality control measures, including data validation and cleaning, to ensure data accuracy and reliability.

12.16.3 Security and Privacy Concerns

- *Challenge:* IoT devices and AI systems are vulnerable to cyberattacks. Protecting sensitive manufacturing data and intellectual property is a top priority.
- *Solution:* Implement robust cybersecurity measures, including encryption, access controls, and regular security audits, to safeguard data and systems.

12.16.4 Scalability and Infrastructure

- *Challenge:* As manufacturing facilities expand, the scalability of AI systems becomes critical. Additionally, ensuring that the necessary infrastructure can support increased data volumes is essential.
- *Solution:* Plan for scalability from the outset and invest in cloud-based solutions or edge computing where appropriate.

12.16.5 Cost of Implementation

- *Challenge:* Implementing industrial AI can be expensive, from acquiring IoT sensors and AI algorithms to training personnel.
- *Solution:* Develop a clear cost-benefit analysis and Return on Investment (ROI) strategy to justify the investment in industrial AI.

12.16.6 Skill Shortages

- *Challenge:* Finding and retaining talent with expertise in AI, IoT, and industrial automation can be challenging.
- *Solution:* Invest in employee training and development programs and consider partnerships with academic institutions.

12.16.7 Interoperability and Legacy Systems

- *Challenge:* Many manufacturing facilities still rely on legacy systems that may not be compatible with modern IoT and AI technologies.

- *Solution:* Gradually upgrade and modernize existing systems and implement middleware for interoperability.

12.16.8 Change Management

- *Challenge:* Employees may resist adopting AI-driven technologies due to fear of job displacement or lack of familiarity.
- *Solution:* Implement comprehensive change management strategies, including employee training and clear communication of the benefits of AI.

12.16.9 Regulatory Compliance

- *Challenge:* Manufacturing is subject to various regulations and standards, which AI systems must comply with.
- *Solution:* Stay informed about industry-specific regulations and ensure that AI implementations adhere to compliance requirements.

12.16.10 Ethical Considerations

- *Challenge:* As AI becomes more integrated into manufacturing, ethical concerns related to data usage and decision-making may arise.
- *Solution:* Develop ethical guidelines and policies for AI usage, ensuring transparency and fairness in decision-making processes.

Addressing these challenges requires a holistic approach, with careful planning, investment, and ongoing monitoring to ensure that industrial AI enhances smart manufacturing in the IoT 4.0 era while mitigating potential risks.

12.17 CONCLUSION

Industrial AI serves a dual purpose: it not only assists in predicting and preventing potential issues, thus reducing unexpected operational disruptions, but it also ensures that every piece of equipment operates safely and efficiently at its peak capacity. By this metamorphosis, machines acquire intellect, computational, communication, and control (3C) skills, and the ability to negotiate with each other independently. Through self-reconfiguration made possible by this agreement, a variety of items may be produced in a flexible manner while maintaining consistently high quality and lowering costs. The future trajectory of industrial AI involves liberating industrial processes and products from reliance on experiential-based practices and instead shifting toward evidence-based practices and protective systems.

REFERENCES

1. Xu, L.D., He W., and Li, S. (2014). Internet of Things in Industries: A Survey. *IEEE Transactions on Industrial Informatics*, 10(4), 2233–2243. doi: 10.1109/TII.2014.2300753

2. Lee, J., Lapira, E., Bagheri, B., and Kao, H.A. (2013). Recent Advances and Trends in Predictive Manufacturing Systems in Big Data Environment. *Manufacturing Letters*, 1, 38–41. https://doi.org/10.1016/j.mfglet.2013.09.005
3. Zhang, L., Luo, Y.L., Tao, F., Li, B.H., Ren, L., Zhang, X.S., Guo, H., Cheng, Y., Hu, A.R., and Liu, Y.K. (2014). Cloud Manufacturing: A New Manufacturing Paradigm. *Enterprise Information Systems*, 8, 167–187. http://dx.doi.org/10.1080/17517575.2012.683812
4. Lee, J., Bagheri, B., and Kao, H.-A. (2015). A Cyber–Physical Systems Architecture for Industry 4.0-Based Manufacturing Systems. *Manufacturing Letters*, 3, 18–23. https://doi.org/10.1016/j.mfglet.2014.12.001
5. Wuest, T., Weimer, D., Irgens, C., and Thoben, K.-D. (2016). Machine Learning in Manufacturing: Advantages, Challenges, and Applications. *Procedia CIRP*, 52, 105–110. doi: 10.1016/j.procir.2016.07.023
6. Lee, J., Davari, H., Singh, J., and Pandhare, V. (2018). Industrial Artificial Intelligence for Industry 4.0-Based Manufacturing Systems. *Manufacturing Letters*, 18, 20–23. https://doi.org/10.1016/j.mfglet.2018.09.002
7. Ahmad, S., Mishra, S., and Sharma, V. (2023). "Green Computing for Sustainable Future Technologies and Its Applications." In: Grima, S., Sood, K., and Özen, E. (Eds.) *Contemporary Studies of Risks in Emerging Technology, Part A (Emerald Studies in Finance, Insurance, and Risk Management)*. Emerald Publishing Limited, Leeds, pp. 241–256. https://doi.org/10.1108/978-1-80455-562-020231016
8. Zhu, K., Hong, G.S., Wong, Y.S., and Wang, W. (2008). Cutting Force Denoising in Micro-Milling Tool Condition Monitoring. *International Journal of Production Research*, 46, 4391–4408.
9. Naveenkumar, P., Rubikaa, M., Jeeva, M., Kunguma Divya, M., and Jayakumar, S. (2022). "An Intelligence Gathering Cyborg Aided by the Internet of Things (IoT)." In: *2022 8th International Conference on Smart Structures and Systems (ICSSS), Chennai, India*, pp. 1–6. Institute of Electrical and Electronics Engineers (IEEE). doi: 10.1109/ICSSS54381.2022.9782266
10. Sheik Ameer Batcha, S., Pushpalatha, N., Nitinchand, S., Bhavatarini, R.V., Boopathiraja, S., and Sanjai, A. (2022). "Multi-Purpose Security and Protection Padlock with a Tiny, Handheld Device Using IoT." In: *2022 8th International Conference on Smart Structures and Systems (ICSSS), Chennai, India*, pp. 1–4. Institute of Electrical and Electronics Engineers (IEEE). doi: 10.1109/ICSSS54381.2022.9782163
11. Prasanth, A., Lakshmi, D., Dhanaraj, R.K., Balusamy, B., and Sherimon, P.C. (Eds.). (2022). *Cognitive Computing for Internet of Medical Things* (1st ed.). Chapman and Hall/CRC. https://doi.org/10.1201/9781003256243
12. Pushpalatha, N., Sri Dhananjayan, K., Senthooriya, O.S., Jabeera, S., Sudhev, R., and Manojkumar, R. (2023). "IoT based Modern Agriculture Buffer Stock System AAF-Availability Accessibility Feasibility." In: *2023 7th International Conference on Computing Methodologies and Communication (ICCMC), Erode, India*, pp. 1278–1282. Institute of Electrical and Electronics Engineers (IEEE). doi: 10.1109/ICCMC56507.2023.10084174
13. Pushpalatha, N., Mohanram, S., Sivaranjani, S., and Prasanth, A. (2023). "5 Enterprise Data." In: Sharma, V., Balusamy, B., Joshua Thomas, J., and Godlin Atlas, L. (Eds.) *Data Fabric Architectures: Web-Driven Applications*. De Gruyter, Berlin, Boston, pp. 95–114. https://doi.org/10.1515/9783111000886-005

14. Shazib, A., Mishra, S., and Sharma, V. (2023). "Green Computing for Sustainable Future Technologies and Its Applications." In: *Contemporary Studies of Risks in Emerging Technology, Part A*. Emerald Publishing Limited, pp. 241–256.
15. Sri Sangeetha, R., Pushpalatha, N., Gururaj, P., Dharmaselva, R., Hemalatha, N., and Gunapriya, D. (2023). Twofold Base IoT-Based Intelligent Power Quality Monitoring System by Considering Social Stewardship. *Advances in Intelligent Systems and Technologies*, 073–077. doi: 10.53759/aist/978-9914-9946-4-3_12

13 The Contemporary AI-Based Data-Gathering Environment in Smart Manufacturing Enterprise

Santhiya T., Sabareeshwaran K., Gomathi V., and Nandhini B.

13.1 INTRODUCTION TO SMART MANUFACTURING

The manufacturing process represents the way products are produced; however, with advanced technologies like smart industrial, the same can be done with more efficiency, quality, and flexibility. If we simply go for outdated manufacturing practices, it will not be that effective as it depends more on manual labor and conventional machines, which will lead to inefficiencies and limited agility in responding to changing market demands. On the other hand, smart manufacturing with robotics, artificial intelligence (AI), Internet of Things (IoT), and other advanced technologies will build the manufacturing industries with intelligent networked production processes. The main thing in smart manufacturing is the concept of connectivity, with which the machines, systems, and processes are faultlessly interconnected through IoT devices and communication networks. This connectivity option enables simultaneous monitoring, control, and optimization of manufacturing operations.

Using AI with machine learning (ML) algorithms, a huge volume of raw data collected from sensors is analyzed, which allows for useful information extraction and also promotes continual improvement and better decision-making. With predictive analysis, the maintenance can be done very prior to the events, which also optimizes the time and schedules and minimizes the downtime, while prescriptive analytics can be recommended for an optimal course of action to minimize performance and resource utilization.

Digital twins (DTs), nothing but a virtual copy of actual assets, just like simulation models, are integrated into smart manufacturing, which is another big advantage in the industries. These DTs' integration enables the manufacturers to model their systems and improve the production processes virtually before using them in practice in the real world. Moreover, this lowers time to market and improves product quality by enabling smart prototyping, testing, and practicing before facing real-world problems.

DOI: 10.1201/9781032615790-13

13.1.1 Asset Management

In smart manufacturing, asset management is a method-based process by which we can monitor, preserve, and maximize the tangible assets in a production setting by combining digital technologies with other smart algorithms. If the approach used for maintenance is ineffective, the cost associated with the products might increase. The end result is that the particular organization will suffer financial consequences from poor management in the manufacturing sector, where maintenance costs may exceed energy-related costs. It is critical for maintenance managers to maintain high standards to make sure effective and well-organized production asset servicing. By being proactive, this asset management reduces the maintenance needs in advance [1].

13.1.2 Quality Assurance

Another important aspect of the manufacturing process is quality assurance, which is aimed at ensuring the products meet predefined standards and specifications. It involves a series of systematic procedures designed to monitor, assess, and improve the quality of products in their entire lifecycle. Sensor network-based smart manufacturing will do this with a data management system that is open software and securely manages, stores, and analyzes the data streams [2].

13.1.3 National Priority

Undeniably, Industry 4.0 and other advanced technologies like virtual implementation will nurture the smart manufacturing option in a wide perspective. An adeptly crafted network infrastructure stands as a vital remedy for enhancing the flexibility, autonomy, and efficacy of manufacturing and logistics operations. The advent of the 5G communications standard heralds significant prospects for implementing industrial privatization networks [3] and has explored major challenges observed in conventional networks when deploying smart manufacturing such as nonorthogonal multiple accesses, Software-Defined Networking (SDN), Network Function Virtualization (NFV), and cloud computing.

13.2 HUMAN INVOLVEMENT IN INTELLIGENT MANUFACTURING

Many industrialists and experts are optimistic about the potential of Industry 4.0. Utilizing state-of-the-art AI tools, Industry 4.0 is poised to transform manufacturing and distribution processes, with a focus on leveraging the motivational aspects of human involvement [4]. Over time, humans are expected to assume roles primarily centered around overseeing manufacturing and ensuring the quality of goods, while physically demanding tasks will be delegated to robots, either autonomously or in collaboration with human counterparts. Though it is still a difficult task, striking a good balance between people and robots is thought to be essential for success [5]. The area of human and robot collaboration (HRC) has gained the attention of many industries as well as various academic and industry laboratories that are looking for

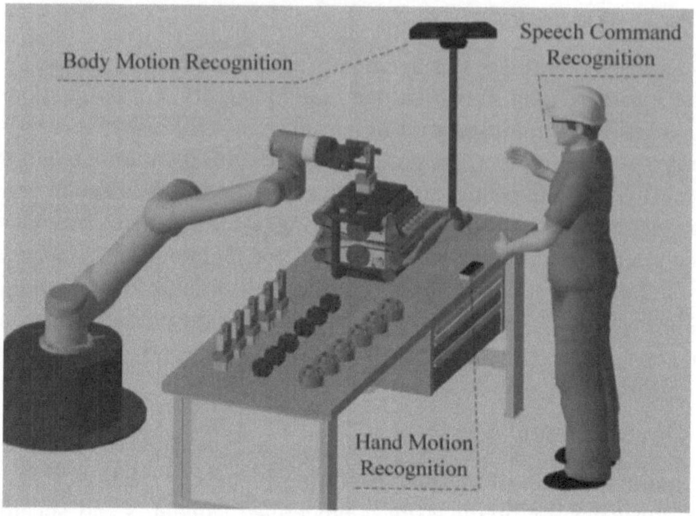

FIGURE 13.1 Human with partial assistance of computer system.

technological advancements around the world and investigating various strategies to make HRC a workable concept.

Many researchers argue that for successful HRC, it is very much needed to carefully balance the human and computer interaction (HCI) interface and appropriately divide tasks between the robot and the human operator, although this area is still in its nascent stages and lacks practical industrial applications [6]. One potential solution currently under investigation is a multimodal fusion system, which integrates various technologies such as body motion identification, hand movement recognition, and voice command recognition, as depicted in Figure 13.1. This system is targeted to achieve the finest balance between human intuition and robot precision.

Despite the apparent decline of human involvement in Industry 4.0, certain human competencies will remain essential. While new employment prospects will arise from these changes, humans will continue to play a vital role, especially in interpersonal interactions that require various psychological and emotional abilities. Manual labor will be largely supplanted by tasks centered on planning and management. Nonetheless, individuals will still be necessary for the upkeep and innovation of these advanced systems, requiring a willingness to acquire fresh skill sets such as data management and technical proficiencies in data-related fields, as well as fostering an open-minded approach to ongoing AI initiatives.

13.3 TECHNOLOGIES USED IN SMART MANUFACTURING

A vast type of cutting-edge technologies is now used in "smart manufacturing," which aims to improve profitability, flexibility, and efficiency in industrial processes. Among the crucial technologies, smart manufacturing includes the following.

13.3.1 Internet of Things

To gather practical data on equipment efficient performance, product quality, record levels, and other operational parameters, IoT devices known as accurate sensing sensors, actuators, and Radio Frequency Identification (RFID) tags are placed throughout the production environment. The term IoT normally refers to a wireless network of interconnected devices, or we can say intelligent nodes, that exchange data with one another through the use of a communication medium. Over the earlier days, IoT has experienced a notable increase in demand, popularity, and commercial availability. Smart cities, the manufacturing sector, healthcare institutions, and governmental organizations are just a few of the many sources from which massive volumes of data are generated by various IoT applications [7].

13.3.2 AI and ML

In an effort to create really intelligent systems, ML is being studied worldwide due to its enormous unrealized potential. Its uses are not limited to a single field; they include automation, recommender systems, anomaly detection, prediction models, and teaching computers how to control helicopters [8]. This study looks at multivariate linear regression, a supervised ML technique, to predict Industry 5.0's effectiveness. The efficacy of the model is contingent upon a multitude of features and constituents, encompassing safeguards, Industrial Internet of Things (IIoT) effectiveness, connectivity, accessibility, reach, and other related aspects. Weight "w" and bias "b" are allocated to each of these elements, which are categorized as algorithm characteristics. These elements can be updated and changed to increase the model's overall performance and efficiency.

13.3.3 Big Data Analytics

The growing big volume of data on intelligent manufacturing creates new difficulties and demands specialized skills in big data from both business and academics. With specific goals like monitoring, scrutinizing, or decision support to minimize potential risks and reduce costs in their manufacturing, several firms have started programs to gather and analyze data inside their own settings. However, given the wide range of application areas that are available, it is necessary to concentrate on the most promising ones [9].

13.3.4 Cloud Computing

A maximum number of devices are connected wirelessly to the internet in a smart manufacturing environment to collect a required amount of data at every stage of the goods lifecycle. A wide variety of applications and services are introduced by the cloud-based intelligent manufacturing strategy to analyze this data and enable widespread manufacturing collaboration. However, there are several obstacles that limit their applicability for real-time applications that need low latency and fast speed, including network outages, bandwidth congestion, and latency constraints.

To overcome these obstacles, fog computing and edge computing bring the cloud's networking, storage, and processing power closer to the edge [10]. This study suggests a hierarchical reference design for edge, cloud, and fog computing in smart system-based manufacturing to improve system responsiveness and efficiency.

13.3.5 Robotics and Automation

The advancement in manufacturing is said to prioritize Industry 4.0, IoT, cyber-physical systems (CPS), and cloud manufacturing (CM), which are the technologies that are needed to achieve Industry 4.0. Despite the fact that Industry 4.0 strives for almost zero faults, its goals do not specifically state how to achieve absolute zero flaws [11]. To accomplish the goals of Industry 4.0 and the lofty goal of no faults, this study proposes a brand-new platform called Advanced Manufacturing Cloud of Things (AMCoT) that will employ automatic virtual metrology (AVM) technologically. And so, Industry 4.1 begins when AVM and Industry 4.0 combine to achieve zero flaws.

13.3.6 Augmented Reality and Virtual Reality

The total cost of items might increase significantly as a result of ineffective maintenance practices used in manufacturing operations. Poor management in this particular area of manufacturing can have a detrimental effect on a company's finances to the point where maintenance expenditures may eventually become energy-related charges. Therefore, to ensure efficient and well-organized servicing of production assets and avoid maintenance chores taking up a sizable percentage of the operating budget, maintenance managers must remain alert. Digitalization technologies have changed the production scene as companies move toward the Industry 4.0 paradigm, boosting many manufacturing particulars to create smart factories. With the expectation that extended reality (XR) technologies, such as virtual reality (VR) and augmented reality (AR), would play a significant part in future maintenance operations, this digital transition has also had an impact on plant maintenance strategy [12]. This study explores the possible effects of AR in a program to improve processes that target reactive maintenance (RM) procedures in automation systems. By means of a process improvement project that resulted in the creation of an AR-based software proof of concept, it has been discovered that this kind of technology can increase automation systems' overall equipment efficiency (OEE) by as much as 11%.

13.3.7 Digital Twins

The traditional assembly line has undergone several changes as a fine-tuning shift from mass manufacturing to mass personalization. One significant modification is the requirement to generate a DT, which is an exact reproduction of the systems, procedures, and physical assets found in an actual manufacturing facility [13]. This project is to give a DT that will help a water bottling factory run more efficiently along its whole production line. The DT uses a Programmable Logic Controller (PLC) to gather sensor data from the production line via an Open Platform Communication server. This data is then processed by the DT to provide solutions, which are subsequently fed to the Open Platform protocols and received by selective PLCs to decrease production delays and relieve bottlenecks.

Configuring a DT with basic three SMART Manufacturing Units devoted to 500 ml water bottle production is part of this process. Pre-made 500 ml bottles are sent into the system in this configuration, where the first SMART Manufacturing Unit fills them, the second seals them, and the third packages them according to requirements. These SMART Manufacturing Units working together in the production line are evaluated using the DT. This performance evaluation provides information on time-to-market estimates and is a predictor of the physical manufacturing line's anticipated efficiency.

13.3.8 Edge Computing

In the context of Industry 4.0 intelligent factories, edge devices can also be used to run complex software programs, often known as manufacturing services, to support production equipment manufacturing. But when sudden high-computation (SHC) jobs arise that require more processing power than these edge devices can provide in a given amount of time, they might not be able to provide the necessary functions as quickly as expected. One critical and challenging problem for smart manufacturing is making sure an edge device can effectively manage SHC duties to deliver manufacturing amenities on schedule. While current job-shop scheduling (JSS) techniques are good at allocating work for production lines in the best possible way, they are not immediately applicable to the management of SHC jobs on edge devices as these activities are irregular and do not have designated processing time slots for JSS techniques. The article [14] presents a unique manufacturing behavior-based map-reduce-style task offloading strategy in response to this challenge. To assist manufacturing operations, a disseminated job processing framework is first developed to allow distributed edge computing devices to work together to handle an SHC task. Then, a software structure in the map-reduce style is developed to make it easier to split an SHC task into several segments so that certain edge devices can execute the segments in parallel. Finally, to attain the best offloading efficiency, a mechanism is suggested for choosing the right edge devices to perform the SHC work, taking into consideration each edge device's past manufacturing practices.

The technologies that are propelling modern manufacturing toward smart factories are edge computing, digital duplicates, VR and AR, robotics and automation, cloud computing, immense data analytics, AI, ML, and the IoT. Figure 13.2 illustrates this.

Figure 13.2 shows the various technologies of smart enterprises. Edge computing enables real-world data processing nearer to the data source, while DTs offer virtual replicas for simulation and optimization. AR and VR enhance visualization and training, while robotics and automation streamline processes and improve safety. Cloud computing offers scalable computing resources, and big data analytics uncovers insights from vast datasets. AI and ML optimize operations, and IoT connects devices for seamless communication and control, collectively empowering manufacturers to achieve efficiency, innovation, and agility in their operations.

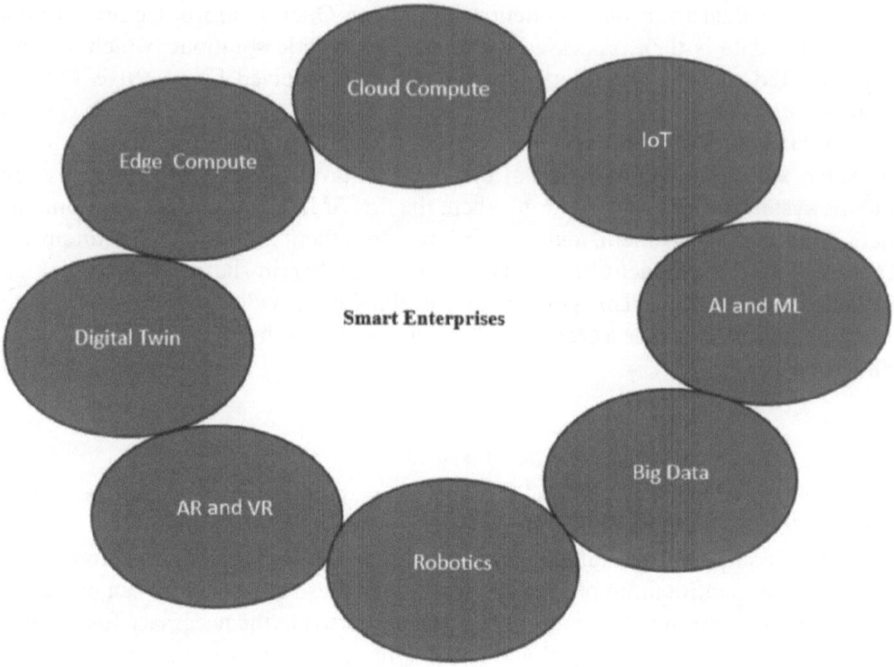

FIGURE 13.2 Technologies of smart enterprises.

13.4 IMPORTANCE OF DATA GATHERING IN SMART MANUFACTURING

The advent of Industry 4.0 will bring about an increase in automated interconnected assembly lines, specifically engineered for the fabrication of personalized and bespoke products as per immediate requirements. To facilitate this capability, there arises a demand for information-centric services, envisioning seamless communication among all physical, virtual, and human entities involved. Such an approach promises users in significantly sophisticated production systems, characterized by the seamless transfer of vast amounts of data [15]. However, making sure data from sensors is collected securely is very important for these systems to be useful. Researchers have suggested ways to make sure communication between IIoT devices is safe. But these ways are not always good enough at keeping things secure and private, and they essentially should have a lot of resources like figuring power and storage space. So, we need a simpler and more secure way for IIoT devices to talk to each other. That is why this paper [16] suggests a new way for IIoT devices to check each other's identity and communicate safely. Their way uses fewer resources and keeps the devices safe from different types of attacks. In places like the ocean, using technology to gather information from sensors and devices is not common yet, even though they had ways to do it for a long time [17]. This paper talks about a new idea called the Internet of Floating Things (IoFT), which is like the IoT but for the ocean.

We also discuss a system called DYNAMO, which helps collect data from sensors on boats and other marine equipment near the coast. To show why this is important, we talk about how we can use this data to study things like water pollution near mussel farms in Italy.

The intricate interplay of horizontal and vertical interactions must operate synergistically. A closer examination of current production processes reveals substantial opportunities for optimization taking place over the path to achieving a "Smart Factory" capable of accommodating high product changeability. Thus, one of the imminent production milestones involves establishing a virtual representation of tangible products within the manufacturing execution system (MES). This endeavor aims to ensure seamless traceability and comprehensive packing of process data across the entire charge chain.

Data mining, particularly, will play a pivotal role in pulling out valuable insights from the amassed information, uncovering meaningful correlations, and establishing guidelines and choices for the ensuing stages of the process [18]. In the realm of servo motor production, integrating magnetic field measurements of rotors provides a chance to collect information for data mining. This venture holds promise for enhancing failure detection, refining quality, and augmenting overall performance.

To evaluate the various facets mentioned, a pilot test series is slated. This series will not only facilitate the assessment of different aspects but also generate datasets for initial data mining investigations. The proposed concept for the initial rotor measuring test process has been delineated in existing literature [19].

13.5 ROLE OF AI IN DATA GATHERING

AI has made it possible for machines to learn from experience, react to new input data, and carry out jobs that are comparable to those performed by humans. AI has become more widely accepted in the education sector. Data security and privacy continue to be major concerns for consumers even though AI technologies and applications address many ethical issues [20]. Modern Industry 4.0 technologies such as big data, AI, and IoT are automating most industrial processes. Work cultures are changing from analog to digital because of these contemporary technological advancements. Of these, AI particularly has an important role in transforming company operations. Similar to this, AI is significantly influencing the development of human resources (HR), leading to the advent of HR 4.0 [21].

Recent advances in AI have made it possible to deliver services and apps in new ways with noteworthy improvements. Among the new uses that are being made possible by this technical progress is the metaverse. The metaverse is mostly dependent on processing enormous volumes of acquired data because of the cooperation of several cutting-edge technologies like IoT, DT, and big data. As security is a major concern in this situation, blockchain ensures transparent and safe data storage, which benefits the metaverse ecosystem greatly. However, typical blockchain elements [22] like consensus protocols and smart contracts will face major challenges in satisfying the increasing service requirements and scalability needs given the dynamic nature of such an ecosystem and the fast changes in its AI-driven services.

The COVID-19 pandemic serves as a concrete example of the challenges faced in this era, as its repercussions are experienced by virtually all nations. In addition to its health implications, the pandemic has also had a profound impact on the global economy. Small- and medium-sized businesses, as well as various sectors producing goods and services, have been heavily affected, alongside the tourism industry. AI and big data offer the capability to track viruses in real time, gather extensive public health data, and streamline vaccine distribution. Moreover, these technologies empower local decision-makers and policymakers with informed, evidence-based forecasts. It is essential to anticipate and tackle the threats posed by COVID-19 and other infectious diseases that disrupt public health and have wide-ranging impacts to safeguard the well-being and continuity of nations and states [23]. To eliminate confusion about military decision-making frameworks [24], this study argues that the phrase "AI-enabled" should be replaced with the more accurate and useful term "AI-automated." It highlights the present focus of researchers and government agencies—rather than creating systems that can make decisions on their own—on creating AI applications meant to help commanders and staff optimize the decision-making progression. The main goal line is to categorize the AI technologies that will be necessary to improve command and control (C2) systems in the future and then implement the necessary measures.

Ensuring privacy remains critical in healthcare applications, given the stringent privacy requirements within this sector. With the exponential growth in volume quality and granularity of health data collected via smart devices, there is a pressing need for novel mechanisms that can handle the challenges of large-volume and real-time data processing. While federated learning (FL) has traditionally been used to train AI models without accessing raw data, recent studies highlight its limitations in providing sufficient privacy guarantees. Meanwhile, differential privacy (DP) is a renowned method for protecting privacy because of noise addition, but it requires a careful balancing act between privacy and accuracy. In the point of health data streams, we present an end-to-end pipeline in this work that integrates DP and FL for the very first time. Furthermore, we suggest a bunching approach to make use of user commonalities with the goal of improving prediction accuracy while cutting down on the amount of time needed for model training [25].

AI enables a computer program to supplement or, in many cases, substitute human decision-making or actions. This often involves the acquisition and examination of data to make decisions more efficiently and/or accurately. While some AI is relatively straightforward, other AI applications rely on sophisticated ML procedures that empower machines to learn and improve from experience without requiring continual reprogramming by humans [26] (see Figure 13.3).

By employing a variety of data analysis methods, including statistical analysis and AI algorithms, startups can acquire valuable insights, navigate intricate business environments, and enhance their investment decisions. Research consistently underscores the potential of data science to optimize business processes, mitigate risks of failure, and increase the chances of long-term success. Furthermore, data science has a broad impact on many industries, as seen by its use in Industry 4.0 predictive maintenance and its revolutionary potential in the automobile sector. Commercialization demands the adoption of data science approaches as industry

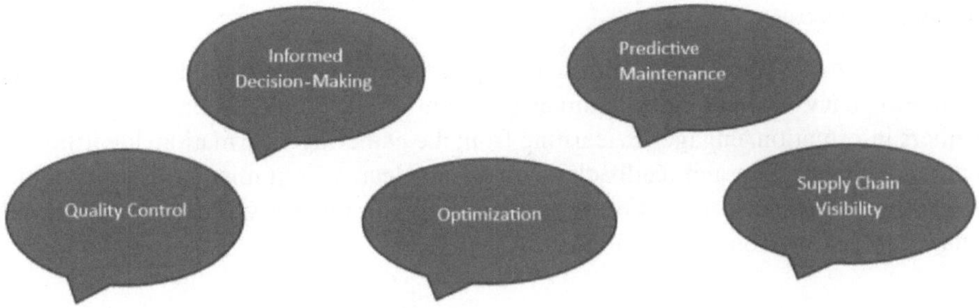

FIGURE 13.3 Elements of data gathering.

experts and analytical perspectives highlight the advancement of data sciences and even more [27].

AI in data gathering has impactful applications across various domains, which will enhance the precision of results from information provided by sensors. In the education sector, data gathering and processing will occur at every nook and corner, for example, we have to monitor students' performance and learning patterns. Several apps should be user-friendly to personalize the educational experience. Industries use this AI technology to collect and analyze large volumes of data from production sections, sensors, and machines to enhance productivity and overall performance. In health care, AI again gathers data from patients and enhances diagnoses, treatment suggestions, and disease monitoring. Startups with AI-powered tools will collect market data and user feedback and plot the performance metrics, which is very useful information for business strategies and decision-making. Most importantly, in military intelligence, this AI role is inevitable as it aids in assessing threats and optimizing defense strategies. In common, AI enhances data gathering by advanced sensors, analyzes complex problems with optimization tools, and provides insightful information across diverse sectors.

13.6 COMPONENTS OF AI-BASED DATA GATHERING ENVIRONMENT

AI is a software-based technology inspired by humans. It will mimic the intelligence of humans and perform the tasks that are generally required by human intelligence, like learning from mistakes and solving complex problems. Experience gained from mistakes will help to make better decisions and understand the natural languages. One important branch of AI is ML, which trains computer systems to recognize the forms in the data that are fed from sensors and will make decisions based on that pattern .Deep learning is another category of ML that processes complex data, such as speech or images. This is done by neural networks with several layers. Natural language plays a vital role in these ML and deep learning technologies, which is nothing but the capability of computers to understand, infer, and produce human languages such as speech or text. Like humans, computers can also evaluate and comprehend visual input such as pictures and films, known as computer vision [28].

13.6.1 Learning

In the education sector, AI systems help to pick up knowledge from data and enhance their efficiency without direct human intervention. As said earlier, AI technology gathers information, engages in learning from the gathering information, identifies the patterns of those data and feedback mechanisms, learns from mistakes, and provides rewards and penalties. Penalties will be nothing but adverse values associated with unaccountable outcomes or actions.

13.6.2 Reasoning and Decision-Making

The next important thing is reasoning and decision-making using AI. It utilizes the logical rules, probability theorems, and algorithms that we feed into it; based on that, it will derive conclusions and make an informed decision. There may be chances of dilemmas, even though AI models are expected to yield consistent results.

13.6.3 Problem-Solving

Strategies like planning, exploration, and enhancement help AI solve problems and beyond. Also, it gives customized solutions for specific problems, and at the same time, general methods help us to tackle a wide variety of challenges we face. It reduces the discrepancies between the current situation and the desired level of results called the problem-solving ability of AI [29].

13.6.4 Perception

AI also works based on perceptions. It was done through the observations of surroundings and getting data from sensors. They are loyal or genuine in their way of making perception-based decisions. Internally, the system analyzes the information from surroundings, gets decisions from previous experiences, and identifies entities with their characteristics and interactions. Perception is needed for tasks like visual comprehension and speech interpretations.

13.6.5 Language Processing

Language processing in AI encompasses the comprehension and generation of human language. Methodologies like natural language understanding, machine translation, and text synthesis empower AI to effectively interact with language. This enables the seamless operation of applications such as conversational agents, language translation platforms, and sentiment analysis tools.

13.7 BENEFITS OF AI-BASED DATA GATHERING IN SMART MANUFACTURING

The fourth industrial rebellion is one of the main forces behind advances in applied AI. As the summary of water and steam-powered technology to aid workers during the Industrial Revolution in the 1800s, manufacturing processes have advanced dramatically. The 1960s saw the introduction of Industry 3.0, the third technological revolution, with General Motors introducing the first industrial robot in 1961 [30]. Along with the rise in popularity of terms like AI and ML, Industry 4.0 has also led to a redefining of more traditional ideas like intelligent manufacturing, intelligent warehousing, intelligent production, intelligent supply chain, and intelligent factories to fit the new standard.

With the merging of deep learning, ML, and AI, many of these ideas in technology are being further refined. The term "intelligent" aptly describes the essence of these updated concepts, which aim to enhance intelligence across various domains. Technologies like CPS, which seamlessly integrate computer-based algorithms with the physical world, utilizing advanced sensors, play a crucial role in this revolution [31].

By automating manual operations within distinct physical units, automation has significantly changed a number of sectors. An increase in manufacturing production has been the goal. These days, enterprises use computers to monitor the activities of several manufacturing units concluding the use of AI and the IoT. By connecting physical components into an interconnected network and enabling information sharing between them, the IoT has completely changed several sectors. It has an impact on a number of industries, including industrial automation systems, business, healthcare, and transportation. IoT offers a multitude of industrial applications when paired with AI. Industry 4.0, which is characterized by machine-to-machine and internet-connected industrial units, has increased industry productivity and efficiency, particularly in supply chain management and production. Production units are observed every 10 milliseconds in the manufacturing industry to record product features. This technique generates a significant amount of characteristics. As humans are biased when classifying things as excellent or odd, assessing is a difficult job in labeled datasets [32].

In industrial production processes, ensuring safe, efficient, and eco-friendly operations and management are the main goals. Modern technologies have a strong reliance on human interaction, which has obvious practical limitations. The process industries are being significantly impacted by the burgeoning era of massive data, which presents previously unheard-of chances to realize intelligent production [33]. This kind of production requires machines that can perform cognitive activities and even innovate on their own, in addition to relieving humans of physically taxing jobs. ML and data analytics are essential to achieving this goal. This article looks at recent developments in ML and data analytics that are used to monitor, control, and improve industrial processes. It focuses on how easily understood and useful ML models are.

13.8 CHALLENGES AND CONSIDERATION

Deploying an integrated network infrastructure that enables smooth communication between engineering or machining units and end clients is a necessary step in the growth of a smart manufacturing system. Strong network connectivity is required for this, which is usually obtained over the internet. However, it is crucial to make sure that data and information sent over the internet are secure. Most of the data are shared wirelessly, so there may be a chance of data theft; to protect the data, this smart manufacturing system uses strategies like end-to-end encryption and unique global identification [34]. All the nodes that are sensors that are going to transfer the data need to be protected. When designing a system, this production should be given top priority, especially in smart manufacturing setups.

Another problem is integrating this intelligent manufacturing system into the current infrastructure [35]. Compatibility concerns between older and newer devices present various hurdles in adopting smart manufacturing technologies. Outdated machinery controlled by obsolete communication protocols may not align with contemporary devices operating on different protocols. Furthermore, establishing efficient machine-to-machine communication and ensuring system interconnectedness demand a robust communication infrastructure. Modern manufacturing systems often necessitate Internet Protocol version 6 (IPv6) connectivity to accommodate the growing number of concurrently connected devices.

The ability of various systems to understand and exploit one another's functionalities on their own is known as interoperability [36]. This feature makes it easier for systems to share information and data with each other, regardless of where the hardware or software came from. Interoperability appears on four levels in Industry 4.0 (I4.0): operational, systematic, technical, and semantic. CPS and I4.0 are aligned through operational interoperability, which pertains to I4.0's conceptual foundation. The foundation of systematic interoperability is made up of models, norms, rules, and procedures. The availability of tools and platforms for the technical and information and communication technology (ICT) environment, as well as related software, is included in technical interoperability. The goal of semantic interoperability is to facilitate knowledge sharing between individuals and across organizational levels [37].

Achieving efficient interoperability necessitates suitable communication protocols and standards. Variations in communication bandwidth, working frequency, communication mode, hardware competencies, and other factors determine the limitations of system interoperability.

A collaborative robot is a specialized robot intended to securely and physically engage with people within workspaces, promoting cooperation and bringing innovative paradigms in human–machine interaction (HMI) [38]. It is also known as a cobot or lightweight robotic system. The International Federation of Robotics states that HRC occurs when workers in an industrial context coordinate with robots to carry out certain tasks [39].

The work-related health and safety of those working on-site should be the first priority, and every attempt should be taken to maintain relevant standards of security and health at work and steer clear of potentially dangerous situations. To reduce a variety of workplace dangers, such as those connected to materials and substances, noise, vibration, radiation, mechanical, electrical, thermal, and vibration, special attention should be given while adopting CPS or industrial robotic systems .

Intelligent manufacturing systems should support operations in multiple languages, enabling the translation of human language instructions into machine language for directing the machinery accordingly [40]. To actualize the concept of intelligent manufacturing and mix AI and advanced technologies effectively into manufacturing schemes, they should have the capability to receive instructions directly from operators, whether conveyed verbally or in written form.

A careful analysis of the financial implications and return on investment is essential when switching from one state-of-the-art manufacturing system to another . The time required to recover the reoccurrence of investment and the production delays during the changeover are weighed against the additional capital needed to adopt the new technology.

13.9 CONCLUSION

In conclusion, the integration of software-based solutions like AI in applications such as smart manufacturing, healthcare, and educational institutions brought significant help in terms of data gathering, fast response, minimization of cost, etc. By using advanced sensors, IoT, and several AI-based algorithms, we will get a vast amount of data and evaluate and make better decisions. This enables complicated processes to have insightful information, optimize the operations, and lead to continuous improvement across their entire manufacturing process. The flawless integration of AI algorithms, advanced sensors, and other advancements like virtual simulations in the whole manufacturing process facilitates efficient data gathering, analysis and utilization, thereby increasing competitiveness in the smart manufacturing landscape. As technologies are evolving, we have to adapt to those situations and improve our domain to utilize them in a smart way.

REFERENCES

1. Andrea Bondin and Joseph Paul Zammit, "A New Age for Plant Maintenance: Making Use of Augmented Reality to Improve Maintenance of Production Assets," *2022 IEEE International Conference on Technology Management, Operations and Decisions (ICTMOD)*, 2022, pp. 1–6, doi: 10.1109/ICTMOD55139.2022.00045.
2. George Ball, Christopher Runge, Randy Ramsey, and Nicholas Barrett, "Systems Integration and Verification in an Advanced Smart Factory," *2017 Annual IEEE International Systems Conference*, 2017, pp. 1–8, doi: 10.1109/SysCon.2017.7934842
3. Wang Ruoxi, Halyna Beshley, Yan Lingyu, Oksana Urikova, Mykola Beshley, and Oleh Kuzmin, "Industrial 5G Private Network: Architectures, Resource Management, Challenges and Future Directions," *2022 IEEE 16th International Conference on Advanced Trends in Radioelectronics, Telecommunications and Computer Engineering (TCSET), Lviv-Slavske, Ukraine*. IEEE, 22–26 February 2022.
4. Zunk B.M., Soos J., Uitz I., Denger A., and Bader M. "The Influence of Human Motivation Factors on the Successful Implementation of Product Life Cycle Management Tools: Explorative Findings and Managerial Implications," *Manufacturing Technology*, 13(4), 580–586, 2013.
5. S. Arunarani, A. Prasanth, N. Pushpalatha, and M. Ouaissa, "Applications of Hyperautomation in Finance and Banking Industries," *Hyperautomation for Next-Generation Industries*, pp. 229–254, September 2024. doi:10.1002/9781394186518.ch9
6. Henning Petruck, Jochen Nelles, Faber Marco, et al., "Human–Robot Cooperation in Manual Assembly—Interaction Concepts for the Future Workplace," In: Chen J.

(ed.), *Advances in Human Factors in Robots and Unmanned Systems*. Cham: Springer International Publishing, pp. 60–71, 2020.

7. Naqliyah Zainuddin, Maslina Daud, Sabariah Ahmad, and Mayasarah Maslizan, "A Study on Privacy Issues in Internet of Things (IoT)," *2021 IEEE 5th International Conference on Cryptography, Security and Privacy (CSP)*, 2021, pp. 1–6, doi: 10.1109/CSP51677.2021.9357592.
8. Piyush Pant, Anand Singh Rajawat, S.B. Goyal, Deepmala Singh, Neagu Bogdan Constantin, et al., "Using Machine Learning for Industry 5.0 Efficiency Prediction Based on Security and Proposing Models to Enhance Efficiency," *2022 11th International Conference on System Modeling and Advancement in Research Trends (SMART)*, 2022, pp. 1–6, doi: 10.1109/SMART55688.2022.00098.
9. Dietmar Winkler, Alexander Korobeinykov, Petr Novák, Arndt Lüder, and Stefan Biffl, "Big Data Needs and Challenges in Smart Manufacturing: An Industry-Academia Survey," *2021 26th IEEE International Conference on Emerging Technologies and Factory Automation (ETFA)*, 2021, pp. 1–6, doi: 10.1109/ETFA45728.2021.9613600.
10. Qinglin Qi and Fei Tao, "A Smart Manufacturing Service System Based on Edge Computing, Fog Computing, and Cloud Computing," *IEEE Access*, 7, 86769–86777, 2019.
11. Fan-Tien Cheng, Hao Tieng, Haw-Ching Yang, Min-Hsiung Hung, Yu-Chuan Lin, and Chun-Fan Wei, "Industry 4.1 for Wheel Machining Automation," *IEEE Robotics and Automation Letters*, 1(1), 332–339, 12 January 2016.
12. Andrea Bondin and Joseph Paul Zammit, "A New Age for Plant Maintenance: Making Use of Augmented Reality to Improve Maintenance of Production Assets," *2022 IEEE International Conference on Technology Management, Operations and Decisions (ICTMOD)*. Institute of Electrical and Electronics Engineers (IEEE), 16 February 2023.
13. Gareth A. Gericke, Rangith B. Kuriakose, Hermanus J. Vermaak, and Ole Mardsen, "Design of Digital Twins for Optimization of a Water Bottling Plant," *IECON 2019—45th Annual Conference of the IEEE Industrial Electronics Society*. Institute of Electrical and Electronics Engineers (IEEE), 14–17 October 2019.
14. Chao-Chun Chen, Wei-Tsung Su, Min-Hsiung Hung, and Zhong-Hui Lin, "Map–Reduce–Style Job Offloading Using Historical Manufacturing Behavior for Edge Devices in Smart Factory," *IEEE Robotics and Automation Letters*, 3(4), 2918–2925, 15 June 2018.
15. Kay Burow, Marco Franke, Quan Deng, Karl Hribernik, and Klaus-Dieter Thoben, "Sustainable Data Management for Manufacturing," *2019 IEEE International Conference on Engineering, Technology and Innovation (ICE/ITMC)*. Institute of Electrical and Electronics Engineers (IEEE), 12 August 2019.
16. Barnana Baruah and Subhasish Dhal, "An Efficient Authentication Scheme for Secure Communication between Industrial IoT Devices," *2020 11th International Conference on Computing, Communication and Networking Technologies (ICCCNT)*. Institute of Electrical and Electronics Engineers (IEEE), 1–3 July 2020.
17. Diana Di Luccio, Angelo Riccio, Ardelio Galletti, Giuliano Laccetti, and Marco Lapegna, "Coastal Marine Data Crowdsourcing Using the Internet of Floating Things: Improving the Results of a Water Quality Model," *IEEE Access*, 8, 101209–101223, 22 May 2020.
18. DIMES-University of Calabria & DtoK Lab Srl, Rende, CS, Italy, "Making Knowledge Discovery Services Scalable on Clouds for Big Data Mining," *2015 2nd IEEE International Conference on Spatial Data Mining and Geographical Knowledge Services (ICSDM)*. Institute of Electrical and Electronics Engineers (IEEE), 15 October 2015.

19. Naman Mishra, Vandana Sharma, and Achyut Shankar, "Adoption of Sustainable Digital Technologies in Industry 4.0," *Sustainable Digital Technologies*. CRC Press, pp. 51–68, 2023.
20. Ghadeer Ismail Khalil, Hafsa Mohammad Sajjad, Manal Sohail, and Zahra Ishfaq, *Role of AI in the Education Sector in the Kingdom of Bahrains*. IEEE, 16 February 2023.
21. Pravin Vitthal Yadav, Umesh S. Kollimath, Sudarshan Arjun Giramkar, and Dnyaneshwar Tukaram Pisal, *HR 4.0: Role of AI in Transforming HRM*. IEEE, 10–11 October 2023.
22. Ouns Bouachir, Moayad Aloqaily, Fakhri Karray, and Abdulmotaleb Elsaddik, *AI-Based Blockchain for the Metaverse: Approaches and Challenges*. IEEE, 5–7 September 2022.
23. Ahmad Shahrul Mubarok, Armi Susandi, and Teddy Mantoro, *Applications of Artificial Intelligence and Big Data for Strengthening the Role of Intelligence in Human Security*. IEEE, 28–29 July 2022.
24. Salvatore Alessandro Sarcia' and Giordano Colo, *Organizing Structures and Information for Developing AI-Enabled Military Decision-Making Systems*. IEEE, 8 January 2024.
25. Sana Imtiaz, Sonia-Florina Horchidan, Zainab Abbas, and Muhammad Arsalan, *Privacy Preserving Time-Series Forecasting of User Health Data Streams*. IEEE, 19 March 2021.
26. Lucille Nalbach Tournas and Diana M. Bowman, *AI Insurance: Risk Management 2.0*. IEEE, 7 December 2021.
27. Lakshya Sandeep Khanna, Prathamesh Sachin Yadav, Sudhanshu Maurya, and Vrince Vimal, *Integral Role of Data Science in Startup Evolution*. IEEE, 30 January 2024.
28. Abdel Magid Hamouda, Khalifa N. Al-Khalifa, and Abdel Magid Hamouda, "Re-thinking Data Strategy and Integration for Artificial Intelligence: Concepts, Opportunities and Challenges," *Applied Sciences*, 13(12), Art. no. 7082, 13 June 2023. doi: 10.3390/app13127082
29. Marçal Mora-Cantallops, Salvador Sánchez-Alonso, Elena García-Barriocanal, and Miguel-Angel Sicilia, "Traceability for Trustworthy AI: A Review of Models and Tools," *BDCC*, 5(2), 4 May 2021.
30. Devol, George C. Jr., *Programmed Article Transfer*, U.S. Patent No. 2,988,237. U.S. Patent and Trademark Office, 1961. https://rb.gy/mut5o4.
31. Monostori L., Kádár B., Bauernhansl T., Kondoh S., Kumara S., Reinhart G., Sauer O., Schuh G., Sihn W., asnd Ueda K. "Cyber–Physical Systems in Manufacturing in CIRP Annals," *CIRP Annals*, 65(2), 621–641, 2016. https://doi.org/10.1016/j.cirp.2016.06.005
32. Sourabh Kumar, Saroj Kumar Chandra, and Ram Narayan Shukla, *Industry 4.0 Based Machine Learning Models for Anomalous Product Detection and Classification*. IEEE, 8 May 2023.
33. Chao Shang and Fengqi You, *Data Analytics and Machine Learning for Smart Process Manufacturing: Recent Advances and Perspectives in the Big Data Era*. Elsevier, vol. 5, no. 6, pp. 1010–1016, December 2019.
34. Klaus-Dieter Thoben, Stefan Wiesner, and Thorsten Wuest, "Industrie 4.0 and Smart Manufacturing—A Review of Research Issues and Application Examples," *International Journal of Automation Technology*, 11(1), 4–16, 2017.
35. SME, *"Smart Manufacturing: A Competitive Necessity,"* Society of Manufacturing Engineers, Dearborn, MI, 2018.

36. David Chen, Guy Doumeingts, and François Vernadat, "Architectures for Enterprise Integration and Interoperability: Past, Present and Future," *Computers in Industry*, 59(7), 647–659, 2008.
37. Yang Lu, "Industry 4.0: A Survey on Technologies, Applications and Open Research Issues," *Journal of Industrial Information Integration*, 6, 1–10, 2017.
38. Luca Gualtieri, Ilaria Palomba, Erich J. Wehrle, and Renato Vidoni, "The Opportunities and Challenges of SME Manufacturing Automation: Safety and Ergonomics in Human–Robot Collaboration," *Industry 4.0 for SMEs*, Springer, pp. 105–144, 2020.
39. Ritu Gautam et al., "Intelligent Manufacturing: Components, Challenges, and Opportunities," *Intelligent Manufacturing and Industry 4.0*, pp. 1–18, 2024.
40. Jens Uhlemann, Raquel Costa, and Jean-Claude Charpentier, "Product Design and Engineering—Past, Present, Future Trends in Teaching, Research and Practices: Academic and Industry Points of View," *Current Opinion in Chemical Engineering*, 27, 10–21, 2020.

14 Case Studies and Applications of AI in the Mechatronics Sector

Sivaranjani S., Uma J., Sathish Kumar D., and Bharani Prakash T.

14.1 INTRODUCTION

Everyone who is somehow connected to the profession has heard about the introduction of artificial intelligence (AI)—one of the greatest changes that are forward in all industries, including innovative mechatronics, which carries the elements of robotics and engineering control with mechanics, electric, and computer branches of engineering.

Introducing AI into mechatronic systems has enabled a new paradigm where machines can employ algorithms similar to the human processes of perception, learning, and making decisions. This text aims at exploring the many roles of AI in the field of mechatronics [1–5]. It evaluates the impact of AI in several industries such as manufacturing and healthcare, indicating the opportunities as well as challenges it presents [6–8]. The evolution of mechatronics was guided by the increasing demand for systems that could seamlessly blend intelligent control and mechanical and electrical systems in an efficient manner. All along, a rule-based control system and deterministic algorithms dominated the execution of a predictable set of tasks in the outlook of mechatronic systems [9–11]. Undoubtedly, there are challenges that affect traditional mechatronics systems, some of which arise from conventional modeling approaches to the integration of mechanical, electrical, and computer engineering systems [12–15]. The following are some of the critical challenges.

14.2 INTERDISCIPLINARY COMPLEXITY

Mechatronics involves the integration of multiple engineering disciplines, each with its own specialized knowledge and methodologies. Coordinating the efforts of mechanical engineers, electrical engineers, and computer scientists can be challenging, leading to communication barriers and integration issues.

- *Limited Adaptability:* The limitations of traditional mechatronic systems arise from their specifications toward particular processes/tasks, rendering them inflexible. This hinders their applicability and scalability, more so in cases involving dynamic and ambiguous environments.
- *Response Time and Efficiency:* There may also exist sub-optimal functional efficiency or speed of control algorithms in many traditional mechatronic

DOI: 10.1201/9781032615790-14

systems. Program slowness may translate into poor performance efficiency and some safety concerns, especially in sensitive operations.

- *Sensor and Actuator Limitations:* Sensors and actuators are paramount to the functioning of any mechatronic system. The daft's upper hand relies on the technological intricacy of the sensors in relation to accuracy, resolution, and operational range; also, the daft understudies the rate of activity operations, precision, and strength of actuators.
- *Reliability and Maintenance:* Mechatronic systems always incorporate mechanical parts, electrical circuit systems, and software-driven algorithms. The integration of all those systems puts these systems at the risk of reliability and maintenance problems which leads to system unavailability and considerable operational costs.
- *Cost and Time Constraints:* Mechatronic systems are developed over a significant period of time which can span from months to years while employing considerable financial resources and expertise. The cost of prototyping, testing, refining, and perfecting a design is significant.
- *Synchronicity with Older Systems:* There are some industries that are heavily dependent on older mechatronic systems which can be difficult to interconnect with more advanced systems or standards. Upgrades using AI should not be an afterthought as these may involve significant modification of the entire system.

Other times, tackling these problems calls for a creative mindset that takes advantage of AI, machine learning, and other emerging technologies. The ability of engineers to integrate AI concepts within mechatronic machines improves efficiency, responsiveness, and adaptability which sets the pace for the next generation of intelligent machines and devices. Today's mechatronic systems owe their diverse capabilities, including sensing, reasoning, and autonomy, to AI. Machine learning algorithms (an area of AI) process a lot of data to learn patterns, predict outcomes, and become more efficient. Machine learning's subset deep learning uses neural networks that employ deep hierarchies to model complex data and perform decision-making. The application of reinforcement learning algorithms can help mechatronic systems learn optimal strategies through exploration and feedback, enhancing the systems' flexibility and efficiency.

14.3 INCORPORATION OF AI IN DIFFERENT SECTORS OF MECHATRONICS

The domain of mechatronics includes a wide range of industries where the use of AI has its own challenges and opportunities. AI-powered robotic automation systems in the industrial sector use predictive maintenance and adaptive control to enhance production efficiency, boost throughput, and improve product quality. By combining AI-powered sensing, perception, and decision-making skills, medical equipment that are AI-enabled—such as surgical robots and prosthetics—improve precision, accuracy, and patient outcomes.

Autonomous cars are transforming the way people and products are moved by using AI algorithms to negotiate challenging situations, improve safety, and optimize traffic flow. The application of AI in various sectors of mechatronics is listed below:

- *Manufacturing Robotics:* Robotics applications of AI include pick-and-place, quality control, and assembly line optimization. Manufacturing processes may become more effective and adaptable by utilizing machine learning algorithms to enhance robot control and adaptability.
- *Autonomous Vehicles:* The development of drones, trucks, and self-driving automobiles is greatly aided by AI. AI systems analyze sensor data, make judgments in real time, and safely steer cars through challenging situations.
- *Predictive Maintenance:* The integration of AI in predictive maintenance systems enables the use of machine learning algorithms along with sensor data to resolve equipment-related issues in advance. This practice helps reduce the maintenance costs and downtime in mechatronic systems which include vehicles and industrial equipment among others.
- *Smart Grids:* AI optimizes the utilization of energy in smart grids by evaluating data from sensors and meters. Such artificial limitations include anticipating demand, detecting irregularities, and optimizing energy usage to improve power distribution efficiency and reduce energy consumption.
- *Healthcare Robotics:* In healthcare units, robots guided by AI are used to aid patients, perform surgeries, and assist in rehabilitation. Such models of robots can move with precision, gather pictures for medical analysis, and even use data in patient-specific treatment protocols.
- *Agricultural Automation:* Robotic harvesting, drones for crop monitoring, and autonomous tractors are examples of precision agriculture utilizing AI and mechatronics. These methods not only improve the efficiency of farming but also lower the consumption of resources while improving crop yield.
- *Supply Chain Management:* Logistics and supply chain activities are improved by AI applications such as demand forecasting, inventory management, and optimization of the transport route. Robotic and autonomous drone systems are mechatronics in package delivery and warehouse robotics.
- *Smart Buildings:* Building functions like lighting, security, and heating, ventilation, and air conditioning (HVAC) are managed by AI-controlled mechatronic systems. To maximize energy efficiency and occupant comfort, machine learning algorithms examine occupancy patterns and environmental data.
- *Consumer Electronics:* Voice assistance would not exist without natural language processing (NLP), and machine learning algorithms allow to enhance the performance of devices and tailor user experience, consumer electronics products, such as AI, are expanded in smartphones, wearables, and smart devices.
- *Space Exploration:* Robots and rovers operating on their own to explore far-off planets and celestial bodies require AI and mechatronics. These robots help human astronauts on space flights, gather scientific data, and negotiate difficult terrain.

14.4 CHALLENGES IN APPLYING AI IN MECHATRONIC SECTOR

To fully harness the potential of AI, the mechatronics industry must confront a number of obstacles. The following are a few of the main obstacles:

- *Interdisciplinary Complexity:* Integrating AI is a challenging endeavor as mechatronics integrates the fields of mechanical, electrical, and computer engineering. Effective design and implementation of AI-driven mechatronic systems need multidisciplinary collaboration and skills to bridge the gap between these disciplines.
- *Data Quality and Quantity:* Data can be said to be the input and output of AI algorithms, especially during training and inference. In the case of mechatronics, it is often a challenge to obtain sufficient amounts of usable, high-quality data—often within restricted contexts or specialized use cases. Successful implementations of AI-powered mechatronic systems hinge on several principles such as data relevance, correctness, and integrity.
- *Model Interpretability and Transparency:* There are instances when AI algorithms and especially deep learning models are known as black boxes due to the level of complexity and opacity they are associated with. It is extremely important to gain insight into how these models operate in decision-making, which is especially critical in safety-sensitive environments like medical devices or automated vehicles. Model interpretability and transparency are important not just for regulatory requirements but for user trust as well.

14.5 COMPUTATIONAL RESOURCES AND REAL-TIME CONSTRAINTS

- *Strength and Reliability:* A mechatronic system is likely to be used in diverse and rather unfriendly conditions which raise doubts, exposing the system to adversary conditions or volumetric variation. For AI to be able to perform efficiently in such conditions, the AI algorithms should be fault-tolerant. These AI-integrated mechatronic systems with embedded intelligence should be safe and reliable when deployed in unstructured domains.
- *Ethical and Regulatory Considerations:* Ethics arise in mechatronic systems that have integrated AI mainly based on accountability, fairness, and bias. Consider self-driven cars that need frequent operation in sensitive zones with human life involved. There is an ethical implication when using such autonomous vehicles. It is important to consider ethical aspects and legal requirements when using AI-integrated mechatronic systems.
- *Cost and Scalability:* The development and implementation of a mechatronic system with AI capabilities seem to be costly, making it imperative to invest in hardware, software, knowledge, and infrastructure. In addition, scalability bottlenecks may exist in AI technology deployment in large volumes or in optimizing other outputs. For AI to be ubiquitous, it is important to seek viable options and optimize the processes of the expansion of AI implementation.

14.6 CASE STUDY ON APPLICATION OF AI IN AUTOMATED VEHICLES

These case studies provide valuable insights into the application of AI in autonomous vehicles, showcasing diverse approaches and their respective impacts on safety, efficiency, and innovation in transportation.

14.6.1 Case Study 1: Waymo's Self-Driving Cars

Waymo, which has over a decade of experience developing self-driving systems, has its fleet of autonomous vehicles with millions of driving miles on regular roads sitting idle.

The rear seat allows passengers to feel the effects of an operational system, which many see as nothing but a matter of time. The main focus has always been on effortlessness; AI's applications are based on computers so that the road could be traveled independently as shown in what Waymo offers seamlessly.

14.6.1.1 Implementation of AI

- *Environment Perception:* The additional sensor uses a combination of LiDAR, radar, and cameras. AI algorithms also detect cutting-edge systems of processing native diagrams of a multitude of sensors above creating realistic renditions of 2D and 3D collision detection, complex obstacle avoidance motion planning, and path checking.
- *Decision Making:* Use of a greater deal of sensing data in AI algorithms makes decisions like merging into traffic or in and out of a junction or beset intersection and all other navigational obstacles based on learning algorithms.
- *Navigational Systems and Real-Time Control:* Waymo AI control system literally plans the scenario accurately based on geography as predicted by computer vision from its samples police jurisdictions road conditions and including reinforcement learning type algorithms.

14.6.1.2 Impact

- *Safety:* Waymo's fully autonomous vehicles have indeed shown a high level of safety. Hardly any accidents can be blamed on the technology, and because of AI in these vehicles potential hazards are promptly assessed, meaning the chances of a collision are drastically reduced.
- *Efficiency:* Waymo's autonomous vehicles use AI to maximize planning for specific routes and decrease traffic. This is an overall improvement in transportation efficiency and a decrease in the rate of fuel consumption.
- *Accessibility:* Potentially, autonomous vehicles will allow people who cannot drive due to different reasons such as age or physical impairment to have even better mobility. Waymo's autonomous vehicles use smart technology to provide self-driving vehicles for users of all ages.

14.6.2 Case Study 2: Tesla's Autopilot

Background: Tesla, renowned for its electric vehicles, has integrated AI-driven autonomous capabilities into its vehicles through the Autopilot system. This system utilizes AI algorithms to enable advanced driver-assistance features and autonomous driving capabilities.

14.6.2.1 Implementation of AI

- *Sensing and Perception:* All Tesla models come standard with a multicam system containing cameras, radar, ultrasonic sensors, SIMs, and GPS all integrated. AI detects objects, lane lines, and traffic signs utilizing sensor information.
- *Advanced Driver Assistance:* Driver-assistive systems such as adaptive cruise control, lane-keeping, and auto lane change are all AI-powered features that help the drivers in different driving scenarios. These features also operate on real-time sensor data, thereby improving safety and convenience.
- *Full Self-Driving (FSD):* Tesla's FSD package aims to achieve fully autonomous driving through AI. The system utilizes neural networks for object detection, path planning, and decision-making, with the goal of enabling hands-free driving in diverse environments.

14.6.2.2 Impact

- *Safety:* The Autopilot technology of the Tesla car incorporates technological advancements that assist drivers in staying in their lanes, are collision aware, and reduce the number of human mistakes that may occur.
- *Convenience:* With the aid of AI technologies like Autopilot, the driver's cognitive load gets decreased as well, making it easier to engage in long-distance driving and enjoyment.
- *Innovation:* The manner in which Tesla has sought to incorporate AI into its cars has instantiated a variety of new developments within the automotive sector, which has also notably accelerated the development of autonomous driving technologies.

14.6.3 Case Study 3: Uber's Self-Driving Technology

Background: As one of the prominent ride-hailing companies, Uber has made significant investments in self-driving vehicles to improve the efficiency and safety of its mobility services. Their self-driving cars use AI to move through urban areas and provide rides on request.

14.6.3.1 Implementation of AI

- *Sensor Fusion:* Using vision sensors like LiDAR and cameras, GPS and other data, Uber's autonomous vehicles self-combine all inputs to understand where they are. AI algorithms receive data from the sensors and identify objects, also detecting pedestrians and traffic signals.
- *Machine Learning:* Learning algorithms are integrated into Uber's self-driving car technology aimed at ensuring decision-making is enhanced and challenging driving situations are coped with. These algorithms improve by being exposed to human-driven data and actual situations.

- *Simulation:* Uber employs AI-powered simulation platforms to test and validate self-driving algorithms in virtual environments. Simulation enables rapid iteration and refinement of autonomous driving systems, ensuring safety and reliability.

14.6.3.2 Impact

- *Scalability:* The self-driving technology owned by Uber could help expand its ride-hailing services while eliminating laying off human drivers.
- *Safety:* While trying to deal with the same, Uber, with the help of self-driving technology, rests its case on addressing human error, which leads to accidents and improving road management.
- *Urban Mobility:* With time, autonomous taxi services will gradually transform urban mobility with easy access to quick, cheap, and efficient transportation means for city residents.

14.7 CASE STUDY ON APPLICATION OF AI IN MEDICAL MECHATRONICS

These case studies demonstrate the various and multifaceted roles of AI Second Nature within the field of medical mechatronics, which includes surgical robots and prosthetics, diagnostic imaging, and rehabilitation. Examples of AI technologies in healthcare ensure patients but also help develop the performance of doctors to the level of the art in medical assistance.

14.7.1 Case Study 1: Surgical Robots in Minimally Invasive Surgery

Background: Surgical robots have effectively changed the scope of minimally invasive surgical means comprising precision devices, dexterous actions, and relative in severity. Such robots incorporate the usage of AI for more sophisticated control, image processing, and decision-making during surgery.

14.7.1.1 Implementation of AI

- *Image Guidance:* AI allowance image processing algorithms are included in the surgical robots to improve the visualization and navigation processes. These methods make use of preoperative imaging data, for example, CT scans and MRI images, to assist during the course of such operatives.
- *Automation:* AI control systems integrated into surgical robots have enabled them to perform complex work fully or partially autonomously with a surgeon monitoring the process. Such machine learning techniques also enhance adaptive control and error detection and correction during the actual surgical procedure.
- *Safety and Precision:* AI algorithms responsible for the safety and precision of surgical robots are those controlling patient anatomy and instrument movement relative to the surroundings all in real time. These systems allow the correction and intervention of any anomalies or complications before they occur.

14.7.1.2 Impact

- *Improved Patient Outcomes:* Surgical robots provide a better surgical outcome as they improve surgical precision, accuracy, and consistency of actions, thus reducing complication and recovery time compared to conventional open surgeries.
- *Enhanced Surgeon Performance:* AI-based surgical robots make it easier for the surgeon to perform complex procedures by providing him with enhanced dexterity, stability, and precision; attributes improved AI-delivered surgical robots offer to the surgeons.
- *Reduced Trauma to Tissue:* Due to the use of surgical robots, trauma to the tissues is lessened, and there is less postoperative pain together with reduced recovery time of patients, which translates into enhancement in the quality of life and satisfaction of the patients.

14.7.2 Case Study 2: Assessment of a Health Issue Combining Diagnostic Imaging and Medical Robotics

Background: Medical diagnosis, treatment planning, and intervention became feasible because of the use of advanced imaging and robotics integrated with AI, which helps clinicians with visualization, analysis, and intervention. These technologies harness the power of AI for image analysis, data analytics as well as precision medicine.

14.7.2.1 Implementation of AI

- *Image Interpretation:* The data for X-rays, MRIs, CT scanning and ultrasounds are medical imaging AI's specialty, enabling clinicians in the investigation, diagnosis, and resolution of numerous medical conditions.
- *Data Analytics:* Employing AI, the data analytical applications combine, analyze, and interpret several patients' data, Electronic Health Records (EHRs), and scientific manuscripts allowing for the establishment of patterns and correlations which predict models for diagnosing, prognosing, and treating particular diseases.
- *Precision Medicine:* AI allows the designing of appropriate treatments and interventions for an individual patient by combining an individual's data, information about genes, and relevant clinical variables to ensure the treatment, medication, and surgical operations are suited to the patient's need.

14.7.2.2 Impact

- *Early Diagnosis:* AI-driven diagnostic imaging and medical robotics facilitate early detection and diagnosis of medical conditions, enabling timely interventions, improved prognosis, and better patient outcomes.
- *Treatment Optimization:* AI-powered data analytics platforms provide clinicians with evidence-based insights and recommendations for treatment planning, medication selection, and therapy optimization, leading to more effective and personalized patient care.
- *Surgical Precision:* The AI-powered surgical robots provide surgeons with real-time advice, support, and information while performing surgical

procedures. This is beneficial as it allows surgeons to perform surgeries in an accurate as well as safe way which results in less time being taken during surgery, minimal tissue damage, and quick recovery for patients.

14.8 CASE STUDY ON THE APPLICATION OF AI IN THE MANUFACTURING SECTOR

This case, in particular, shows how AI can be applied in the manufacturing sector, enabling predictive maintenance, automating quality control processes and optimizing supply chains. In this manner, the application of AI will help the manufacturers be more effective, reduce costs, and improve the quality of their products and overall customer satisfaction.

14.8.1 Case Study 1: Predictive Maintenance in Manufacturing

Background: The use of preventative maintenance which employs AI is now to be regarded as an important activity in the field of manufacturing as it helps to maintain the machines at the required efficacious level and avoid unnecessary interference with the manufacturing process.

14.8.1.1 Implementation of AI

- *Real-Time Data Acquisition:* AI-powered devices such as sensors constantly capture a range of parameters such as temperature, vibration counts, and energy utilization of the equipment.
- *Predictive Maintenance Machine Learning:* Depending on the recorded data, AI algorithms evaluate past maintenance records and sensor data to identify trends that could signal potential failures or decline in performance.
- *Predictive Maintenance Requirement or Spare-Part Analytics:* Maintenance prediction AI models predict when parts of the equipment will fail or require maintenance using a large number of analyzed data and allow the manufacturer to plan repair/maintenance activities ahead of time and thus reduce unnecessary loss due to downtime.

14.8.1.2 Impact

- *Minimize Production Downtimes:* Predictive maintenance has been demonstrated to cut past normal unplanned downtimes by up to 50% and as a result increase production uptime and optimization.
- *Maintenance Cost Cutting:* As the demand for maintenance is predicted ahead of time, the manufacturer can place a good maintenance strategy, minimize spare part stock levels, and eliminate exorbitant unscheduled breakdown maintenance costs.
- *Enhanced Predictive Maintenance Ability:* The AI predictive maintenance prevents pervasive failures in equipment reliability by addressing possible issues before they mature into full-blown failures, thereby, in turn, enhancing the time and life of the equipment.

14.8.2 Case Study 2: Quality Control with Computer Vision

Background: Computer vision, coupled with AI, is transforming quality control processes in manufacturing by automating defect detection and ensuring product quality consistency.

14.8.2.1 Implementation of AI

- *Image Capture:* Cameras capture images of manufactured products or components during various stages of production.
- *Multilayer Neural Networks:* Image data is processed by AI models trained to identify the presence of defects or any deviances from the set quality in the images.
- *Automated Imaging:* AI-enabled imaging systems would classify and reject defective items automatically, thereby ensuring that only reasonably good items are taken to the market.

14.8.2.2 Impact

- *Decrease in Defects:* The introduction of quality monitoring systems to control manufacturing with computer vision gives an impressive up to a 90% reduction in manufacturing defects which ultimately increases product quality effectiveness and customer satisfaction.
- *Costs:* The cost of production is always lowered with such improvements because of the increased rate of automation of the entire quality control process, enabling the reduction of manpower used for manual inspections as well as expenses for scrapping and reworking.
- *Improvement of Processes:* These AI systems control the quality of manufactured goods and enable the organization to gather good feedback on the manufacturing process, making it awesomely responsive to improvement.

14.8.3 Case Study 3: Supply Chain Optimization

Background: AI is changing how suppliers and manufacturers manage the supply chain functions of inventory management, demand forecasting, and logistics operations management in the manufacturing field.

14.8.3.1 Implementation of AI

- *Data Integration:* AI algorithms obtain information from various business functions—such as sales, inventory and suppliers, and market dynamics, all of which are needed for planning.
- *Machine Learning:* AI machine learning models utilize data to make predictions about future demand, assess the risks associated with the supply chain, and make recommendations on best stock levels.
- *Optimization Algorithms:* AI optimization algorithms align various supply chain activities, for example, such as order fulfillment, warehousing, and transport routing in an efficient and cost-effective manner.

14.8.3.2 Impact

- *Inventory Reduction:* Redundancies in inventories and stockouts are anticipated and minimized through AI's predictive demand technologies, maintaining desired levels at the most optimal costs.
- *Improved Customer Service:* AI practices ensure high consistent availability of products on time, thereby positively impacting customer satisfaction and leading to loyalty.
- *Cost Savings:* The integration of AI into the supply chains saves manufacture's costs associated with transport, holding inventory, and inefficiencies arising from operation.

14.9 CASE STUDY ON APPLICATION OF AI IN ROBOTIC AUTOMATION

These cases illustrate the multiple aspects related to the use of AI in robotic automation in industrial, commercial, and healthcare environments and how AI improves efficacy, safety, and originality in a profound manner.

14.9.1 Case Study 1: Boston Dynamics' Spot Robot

Background: Boston Dynamics has established itself as a pioneer in the field of robotics, and implemented technology that birthed the Spot robot. Spot is a mobile, dog-like robot that is capable of performing inspection, surveillance, and transportation tasks, all with AI for increased control and independence.

14.9.1.1 Implementation of AI

- *Dynamic Locomotion:* Spot's shift in gait and stance employed AI, making it more able to go over steps, rubble, and a host of other irregular terrains with ease
- *Perception and Navigation:* This robot is equipped with cameras, LiDAR, and depth sensors to form a better image of the space around the robot and the robot itself. Utilizing the data gleaned from its sensors, AI algorithms identify spatial limitations and obstructions, orient themselves, and traverse a three-dimensional post in a confined environment along the shortest possible path.
- *Manipulation and Interaction:* The AI-enabled control of Spot also allows for various manipulation tasks, such as turning doors, pushing or pulling objects, and even using specific tools. Additionally, more advanced grasping techniques, including specialized tactile sensors, have been incorporated to enhance its capabilities.

14.9.1.2 Impact

- *Industrial Automation:* The versatility and autonomy of Spot make it capable of several industrial automation applications including inspection, maintenance, and logistics in the construction, oil, gas, and manufacturing industries.

- *Safety:* Through AI, Spot can manage its operation and execution of tasks while avoiding obstacles which enhance safety in perilous working environments as tasks can be done remotely or autonomously.
- *Efficiency:* With Spot's ability to carry out tasks independently, manual input is kept to a minimum, and operational efficiency is increased which saves costs and improves productivity across different applications.

14.9.2 Case Study 2: Amazon Robotics' Fulfillment Centers

Background: Amazon Robotics actively leads in the automation of the warehouse by employing a controlled suite of AI-driven robotic systems to assist with the order fulfillment processes as we see in its distribution centers. These robots are capable of autonomously navigating warehouse environments, picking, packing, and transferring items.

14.9.2.1 Implementation of AI

- *Order Fulfillment:* The robotic system in Amazon's warehouse is equipped with sophisticated AI-powered technology, enabling automated order picking by optimizing routes that minimize time and maximize output in fulfillment centers. Additionally, predictive analytics and machine learning algorithms enhance the system's ability to anticipate demand and improve efficiency.
- *Collaborative Robotics:* Robots in the Amazon warehouses do not work independently but rather complement the processes being carried out by humans. AI-powered coordination algorithms help to ensure safe interactions between robots and human beings in workplaces that are shared.
- *Dynamic Adaptation:* The robotic systems at Amazon can adjust to changes caused by the layout of inventories, seasonal changes in demand, and operational restrictions instantaneously because of AI algorithms. Robotic skills and behaviors are trained, improved upon, and optimized in a continual fashion with the use of reinforcement learning techniques.

14.9.2.2 Impact

- *Scalability:* The AI-robotic systems at Amazon enable the company to scale its order fulfillment operations efficiently, allowing it to handle high volumes of orders during peak periods such as Black Friday and Cyber Monday. These systems can adapt to fluctuating demand, ensuring that operations remain smooth and timely even when order volumes surge.
- *Speed and Accuracy:* The AI-robotic systems at Amazon enhance both the speed and accuracy of order fulfillment by optimizing processes, reducing human error, and improving decision-making in real time. These systems ensure that orders are processed quickly, with precise picking and packing, which is crucial during high-demand periods.
- *Labor Optimization:* By automating repetitive and physically demanding tasks, Amazon's robotic systems free up human workers to focus on more value-added activities, such as quality control, problem-solving, and customer service.

14.10 FUTURE RESEARCH DIRECTION

Practical use of AI in mechatronics areas may be more oriented toward developing new solutions and meeting new problems. Here are some possible directions for research:

1. *Automated Learning and Control:* Automate the control of the mechatronic systems or implement even more advanced AI algorithms to autonomously design and optimize control strategies for mechatronic systems in real time. Consider, for example, the employment of reinforcement learning, adaptive control, or online learning algorithms that will enable the systems to adapt automatically to changed environments and system requirements.
2. *Human–Machine Interaction:* Investigate the design and development of mechatronic systems with integrated AI that adhere to human-centered design characteristics, including ease of use, safety, and efficient human–machine interaction. Focusing on options including NLP, gesture recognition, and enhancement of affective computing may contribute to more desirable and easier-to-use interfaces.
3. *Trustworthy AI Systems:* Explainable AI (XAI) for the design of AI mechatronic systems can also improve our understanding of the use of AI by providing the necessary level of transparency. Going into research activities should focus on developing appropriate XAI methods that will guarantee AI systems including autonomous systems to explain the reason behind the decision made.
4. *Edge AI and Distributed Intelligence:* Examine the possibilities and advantages of edge computing and architecture for distributed intelligence that allows mechatronic systems driven by AI to be able to perform entirely and effectively in different spatial contexts. This embraces studies on low-power AI algorithms, federated learning approaches, and edge-aware control strategies that can make use of local data processing with decision-making capabilities.
5. *Multi-Agent Systems and Swarm Robotics:* Explore the role and deployment of AI techniques in the management and control of mechatronic swarms in performing autonomous coordinated collaborative tasks. Research activities should seek to create decentralized decision-making processes, swarm optimization methods, and self-organization processes for effective and large-scale application of swarm robotics in different areas.
6. *Ethical and Societal Implications:* Deliberate on the ethical, legal, and social implications of the use of AI in the implementation of mechatronic systems. Research activities should seek to investigate ethical design principles, fairness, bias mitigation approaches, and policy strategies to ensure AI-enabled mechatronic systems are ethical in their operation and do not undermine human value.
7. *Energy-Efficient and Sustainable Solutions:* Explore the AI-based methods aimed at improving energy consumption, environmental factors, and other aspects of sustainability in mechatronic systems. This comprises an investigation into the development of energy-efficient control techniques, renewable

energy sources integration, and lifecycle evaluation technologies to achieve lower emissions from the mechatronics applications.

With these research directions in mind, the use of AI in the mechatronics sector can also be further progressed and new issues can be solved paving the way for innovations that can benefit society in a more effective way.

14.11 CONCLUSION

This study highlights the significant influence of AI on the mechatronics industry via case studies and applications. AI-driven solutions have completely changed sectors, fostering productivity, efficiency, and creativity in everything from factory robots to autonomous cars, predictive maintenance, healthcare robotics, and more. But to fully utilize AI in mechatronics, it will be necessary to solve a number of ethical, legal, and technological issues while encouraging transdisciplinary creativity and cooperation. To shape the future of automation and human–machine s, this introduction lays the groundwork for future research into the revolutionary effects of AI on mechatronics.

REFERENCES

1. Krafcik, J., and Waymo Team. (2020). "Building the World's Most Experienced Driver: The Waymo Self-Driving Car." *IEEE Intelligent Systems*, 35(4), 87–95. doi: 10.1109/MIS.2020.2998071
2. Kaplan, A. (2020). "Tesla's Autopilot: A Step towards Autonomous Driving." *International Journal of Automotive Engineering*, 10(3), 427–438. doi: 10.20467/IJAE.10.3.427-438
3. Alphabet, A. (2021). "Advancing Autonomous Technology: Uber's Approach to Self-Driving Cars." *Journal of Transportation Engineering*, 147(2), 102–115. doi: 10.1061/(ASCE)TE.1943-5436.0001358
4. Pushpalatha, N., Batcha, S., Yesvanthkrishna, V., Rathinamala, S., Devi, B. P., and Gomarhi, V. (2023). "Gesture-Based Control of a Bionic Arm—A Prototype Design," *2023 7th International Conference on Computing Methodologies and Communication (ICCMC), Erode, India*, pp. 856–859. Institute of Electrical and Electronics Engineers (IEEE). doi: 10.1109/ICCMC56507.2023.10084175
5. Manikandan, N., Anjaneyulu, D. S. S. R., Anwar, S., Kumar, V., Gupta, R., and Sood, A. K. (2022). "Performance Analysis of DGA-Driven Botnets Using Artificial Neural Networks," *2022 10th International Conference on Reliability, Infocom Technologies and Optimization (Trends and Future Directions) (ICRITO)*. IEEE.
6. Achouch, M., Dimitrova, M., Ziane, K., Sattarpanah Karganroudi, S., Dhouib, R., Ibrahim, H., and Adda, M. (2022). "On Predictive Maintenance in Industry 4.0: Overview, Models, and Challenges." *Applied Sciences*, 12(16), 8081. https://doi.org/10.3390/app12168081
7. Hütten, N., Alves Gomes, M., Hölken, F., Andricevic, K., Meyes, R., and Meisen, T. (2024). "Deep Learning for Automated Visual Inspection in Manufacturing and Maintenance: A Survey of Open-Access Papers." *Applied System Innovation*, 7(1), 11. https://doi.org/10.3390/asi7010011.
8. Richey, R.G., Jr., Chowdhury, S., Davis-Sramek, B., Giannakis, M., and Dwivedi, Y.K. (2023), "Artificial Intelligence in Logistics and Supply Chain Management: A Primer

and Roadmap for Research." *Journal of Business Logistics*, 44, 532–549. https://doi.org/10.1111/jbl.12364

9. Chaudhary, A., Sharma, V., and Alkhayyat, A. (Eds.). (2024). *Intelligent Manufacturing and Industry 4.0: Impact, Trends, and Opportunities* (1st ed.). CRC Press. https://doi.org/10.1201/9781032630748
10. Hariharan, R., Pushpalatha, N., Rubikaa, M., Jeeva, M., Kunguma Divya. M., and Jayakumar, S. (2022). "An Intelligence Gathering Cyborg Aided by the Internet of Things (IoT)," *2022 8th International Conference on Smart Structures and Systems (ICSSS), Chennai, India*, pp. 1–6. doi: 10.1109/ICSSS54381.2022.9782266
11. Prasath, S. V., Pushpalatha, N., Gunapriya, D., Kumar, P. M., Santhosh, R. T., and Srinivasan, S. (2022). "Automated Agronomic Bot for Green Ailment Scanner," *2022 5th International Conference on Contemporary Computing and Informatics (IC3I), Uttar Pradesh, India*, pp. 935–940. doi: 10.1109/IC3I56241.2022.10073042
12. Kowsalya, M., Pradeep, R., Elakya, A., and Gunapriya, D. (2022). "PLC Based Critical Composition Control for Industries," *2022 8th International Conference on Advanced Computing and Communication Systems (ICACCS), Coimbatore, India*, pp. 1636–1639. doi: 10.1109/ICACCS54159.2022.9785226
13. Pusphalatha, N., Gunapriya, D., Bharath, A., Chandru, S., Sanjeevkumar, S., Vigneshwaran, S. (2024). "A Sustainable Agricultural Field Surveillance System Involving Intelligent Irrigation System for Ecological Balance," *2024 International Conference on Science Technology Engineering and Management (ICSTEM), Coimbatore, India*, pp. 1–7. doi: 10.1109/ICSTEM61137.2024.10560948
14. Maheshwaran, M., Kiruthiga Devi, V., Gunapriya, D., Pushpalatha, N., Karthik, S. S., and Selvi, A. (2024). "Machine Learning-Based Pre-Stroke Detection System," *2024 International Conference on Science Technology Engineering and Management (ICSTEM), Coimbatore, India*, pp. 1–5. doi: 10.1109/ICSTEM61137.2024.10560875
15. Vinoth Kumar, P., Priya, S., Gunapriya, D., and Batumalay, M. (2024). "Novel Battery Management with Fuzzy Tuned Low Voltage Chopper and Machine Learning Controlled Drive for Electric Vehicle Battery Management: A Pathway towards SDG." *Journal of Applied Data Sciences* [Online], vol. 5.3, 936–947, Web. 7 August 2024.

15 Future Research Directives of AI in Mechatronics Engineering

Thirunavukkarasu S., Taruun Prasath S., Santhiya G., Anand R., and Nithya Shree S.

15.1 INTRODUCTION

The future of mechatronics engineering will be shaped by the partnership of humans and robotics. As industries begin to change, human–machine collaboration should be essential to producing effective and flexible solutions for upcoming demands. In industries, AI-driven communication system enables quick task execution within the specified timeframes and it contributes to boosting productivity and accuracy [1]. In the sector of automotive industries, the ML model improves the chance of fault tolerance and improves the safety of humans by providing an intelligent system [2]. In healthcare, AI transforms the mechatronics application's design and functionality of prosthetics limbs, thereby remarkably improving the quality of life of individuals with physical disabilities. These enable real-time tracking and adaptive control (AC) based on the needs [3]. Moreover, the combination of reinforcement learning (RL) and AC has emerged to provide a satisfying approach to address the problem and struggles of robot path planning and navigation. It forms the spatial awareness and ability to adjust their behavior in real time [4]. The AI models in the development of textile industry offer sustainability and improve the manufacturing of textile products with quality assurance and customization [5]. Infrastructure projects heavily depend on the construction robotics and technology. Mechatronics with deep learning (DL) technologies will address the problems of skilled labor scarcity and guarantee safety in high-risk areas [6]. Mechatronics-based renewable energy systems with AI algorithms can forecast the weather and modify operations to maximize energy output. The system leads to a shift in energy management toward future advancement, more adaptability, and efficient energy systems [7]. The fusion of AI with mechatronics in autonomous vehicles provides transformative synergy, assuring these systems to give the unmatched level of safety, efficiency, and cognitive intelligence to control [8]. In building designs, it offers the approach to managing energy and enables sustainability in the built environment. These intelligent systems enable dynamic adaptability for the changing of weather conditions and ensure long-term environmental and economic benefits in achieving urban and rural developments [9]. It enables farmers to monitor, analyze, and optimize their productivity efficiently; for example, System for Agriculture Autonomous

DOI: 10.1201/9781032615790-15

Monitoring and Decisional Assistance (SAAMDA) gives predictive and analytical power and supports small-scale farmers too [10]. Recent advancements in mechatronics have enabled the development of intelligent systems for agriculture, environmental monitoring, and automation, integrating sensors, control units, and decision-making algorithms to enhance sustainability and operational efficiency [11-15].

15.2 HUMAN–ROBOT COLLABORATION IN MECHATRONICS

The human–robot collaboration is the strength for humans and robots to achieve higher efficiency, precision, and adaptability to focus on the complex task. It acts as a bridge to connect human creativity and decision-making seamlessly. The main objective is to maintain the safety and trust between the robots and humans to act based on their actions and personalize the behaviors of robots to the intuitive interaction through gesture and voice control. Core enabling technologies include artificial intelligence (AI) for sensors and actuators for processing the data in real time to provide a precise data and control algorithms for smooth interaction, and human–machine interfaces (HMIs) like touchscreens and voice controls.

This can be applied in the field of manufacturing, healthcare, logistics, construction, and service-based industries for enhancing productivity, quality, and user experience. Future trends include cognitive robots with well-advanced reasoning, personalized collaboration, and social awareness with emotional intelligence. It ensures the IoT-integrated environment for seamless data sharing, an advanced safety mechanism to prevent the collisions and provides significant reliability. These advancement promises to redefine human–machine synergy in industries.

Figure 15.1 shows the human–robot cooperation in a futuristic mechatronics environment, stressing the cooperation between human engineer and humanoid robot in high-tech sectors.

15.3 AI-BASED COMMUNICATION SYSTEM

The integration of AI in the mechatronics systems has upgraded them to communicate with their kind and human operators. This way of communicating promises

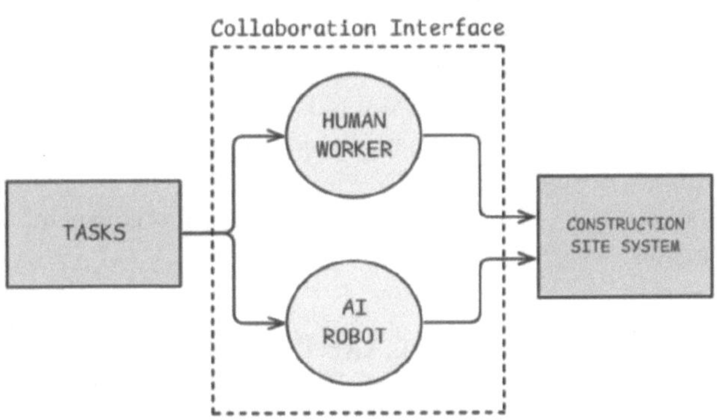

FIGURE 15.1 Human robot.

efficiency, responsiveness, and adaptability for the industries. In mechatronics, communication relies on interconnected systems where AI models fed with different algorithms process and transfer data to ensure accurate and timely decision-making. Important elements such as data acquisition and processing through sensors and IoT devices will retrieve the data from the environmental areas and respond immediately using the techniques of machine learning (ML) and DL.

These systems can utilize wireless protocols like Bluetooth, Wi-Fi, and 5G networks, along with wired networks, to facilitate reliable data transfer with minimal latency. Sensors and actuators play a vital role in gathering the data in real time, such as pressure, proximity, and temperature, while actuators execute tasks based on the given data. AI includes natural language processing (NLP) for the sake of human–machine communication and predictive analysis for optimizing the quality of the machine. The future of AI-based communication in mechatronics looks very promising with the rollout of 5G networks expecting a fast communication and bandwidth for high-demand applications. A collaborative AI system will enable better synergy among machines in vast areas. Quantum communication holds the potential to create a new revolution in data transmission with unpredictable data sharing and security.

Figure 15.2 shows an AI-powered communication system with robotic arms and humanoid robots collaborating seamlessly on an automated assembly line with a large display interface displaying real-time analytics and data.

15.4 ML APPROACH IN AUTOMOTIVE INDUSTRIES

ML has emerged as a transformative technology in the automotive industry, which created a revolution in how vehicles are operated, interact, and evolve. The advanced algorithms are capable of learning from data; ML supports the automobiles to perform tasks for human intervention, ensures safety measures during travels, and provides a comfortable user experience. Autonomous driving is among the best futuristic changes where ML enables the vehicle to analyze the real-time data with the help of sensors, recognize objects, and detect the road's condition.

Supervised models are trained to provide lane detection and object classification. RL optimizes decision-making in dynamic traffic environments. Beyond autonomous vehicles, the ML provides power predictions by analyzing sensor data to anticipate mechanical failures, allowing timely interventions and reducing downtime. In ML-driven vehicles, it enhances the features like adaptive cruise control, automated parking, and collision avoidance by continuous data analysis from the environmental data. Additionally, ML enables a personalized in-car experience by analyzing the driver's behavior and acting based on it.

Figure 15.3 shows how AI is being used in the automotive sector to help mechatronics robots manufacture automobile.

15.5 AI ENABLED IN HEALTHCARE PROSTHETICS

Mechatronics combined with AI in the field of healthcare prosthesis had changed by offering expanded functionality, better patient comfort, and flawless human–machine interaction. Using cutting-edge mechatronics systems, prostheses with AI-driven

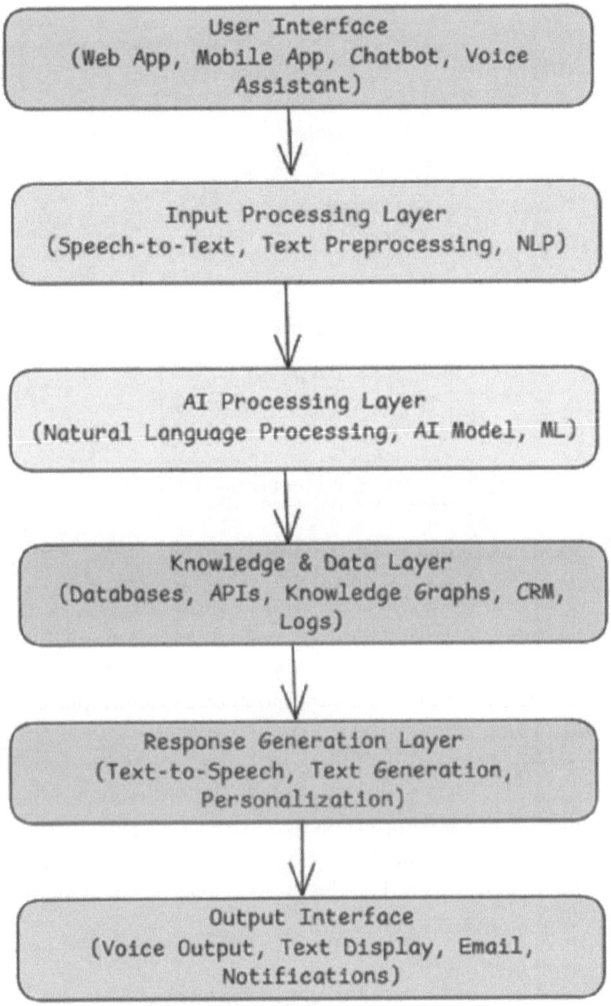

FIGURE 15.2 AI-powered communication.

technologies follow the action of natural limb movement, adapt to user demand, and totally satisfy the quality of life of the individuals with disabilities. The smart system integrates sensors, actuators, AI algorithms, and communication technologies to provide the people with prosthesis responsive and tailored treatment.

The main ability of prosthesis assisted by AI is adaptable control. Using inertial measuring units (IMUs), pressure sensors, and electromyography (EMG) outputs, models of ML study the data from the sensors contained in the prostheses. It guarantees natural and appropriate control and helps to enable the real-time modification of user movement patterns. They can discriminate between walking, jogging, climbing stairs, etc., and act in line with it. Predictive capabilities on healthcare prosthetics can function based on the user intent or environmental conditions to reduce delays and

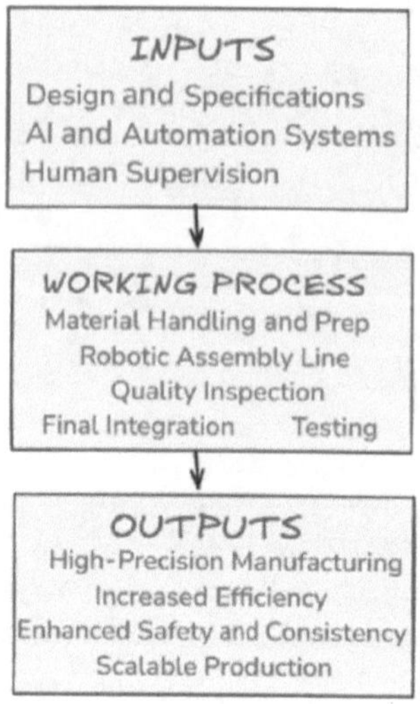

FIGURE 15.3 AI-powered automobile manufacturing robot.

improve stability. To guarantee safety, the AI system detects anomalies, that is, gadget malfunctions or misalignments. Another benefit of AI-driven prostheses is customizing and personalizing based on the user's biomechanics and preferences, therefore matching their way of life.

These customizations help to boost the user's confidence and satisfaction. Storing that healthcare data with cloud computing and IoT facilitates remote monitoring to track the performance data for future aspects.

Figure 15.4 shows futuristic healthcare for a prosthetic limb fitted with sensors for a patient. As it is an AI-powered device, it showcases prosthetic performance and analytics and tracks patient health.

15.6 RL AND AC IN ROBOTICS NAVIGATION

RL and AC will be acting as part of the transformation of robotics navigation by enabling robots to operate autonomously in different terrains and uncertain environments. These techniques empower robots to learn optimally to create their own navigation strategies through a trial-and-error approach, adapt to changing conditions, and execute the given task within the given timeframes. This reduces human interference in the task. The combination of RL and AC into mechatronics plays a major role for several applications, such as autonomous vehicles, drones, and robotic assistants.

RL enables robots to form different navigation patterns to reach the assigned destination by interacting with the environments. Many algorithms are initialized to retrieve information to create possibilities of optimal paths by avoiding obstacles, navigating in tight spaces, and managing energy consumption. These are adaptable in unstructured environments like crowded cities and cluttered indoor spaces. AC improves the robust and precise navigation. It automatically adjusts the robot's control parameter

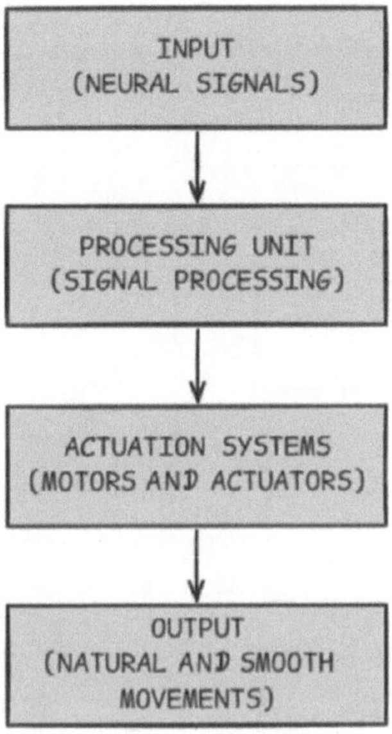

FIGURE 15.4 AI-powered prosthetic limb fitted with sensors.

based on the environmental data gathered from the sensors to maintain the stability from external factors.

The fusion of RL and AC improves long-term learning and focuses on the quick reflexes of the robots and different circumstances to get the maximum effort to achieve the high-level tasks.

Figure 15.5 depicts the illustration of a futuristic robotics lab where RL and AC enable precise robotic navigation. It is an advanced AI-driven solution for intelligent navigation.

15.7 AI ENABLES SUSTAINABILITY IN TEXTILE INDUSTRIES

AI in the sector of textile industries drives sustainable development by integrating intelligent systems in manufacturing, quality control, and supply chain management. The fusion of AI with mechatronic systems ensures efficient usage of resources, waste reduction, and enhanced environmental stewardship, which promotes sustainable development practices. In the production of textiles, AI-enabled mechatronics systems optimize processes based on the real-time analyses from the sensors in machines. A smart monitoring system helps to monitor the energy consumption to reduce it.

Water usage during dyeing, weaving, and washing operations is also monitored. ML algorithms predict the machine's performance and accordingly allocate scheduled maintenance to prevent resource wastage and machine failures. The AC system in the textile machinery ensures precise control of speed, temperature, and tension, leading to consistent product quality with reduced material waste. In the phase of

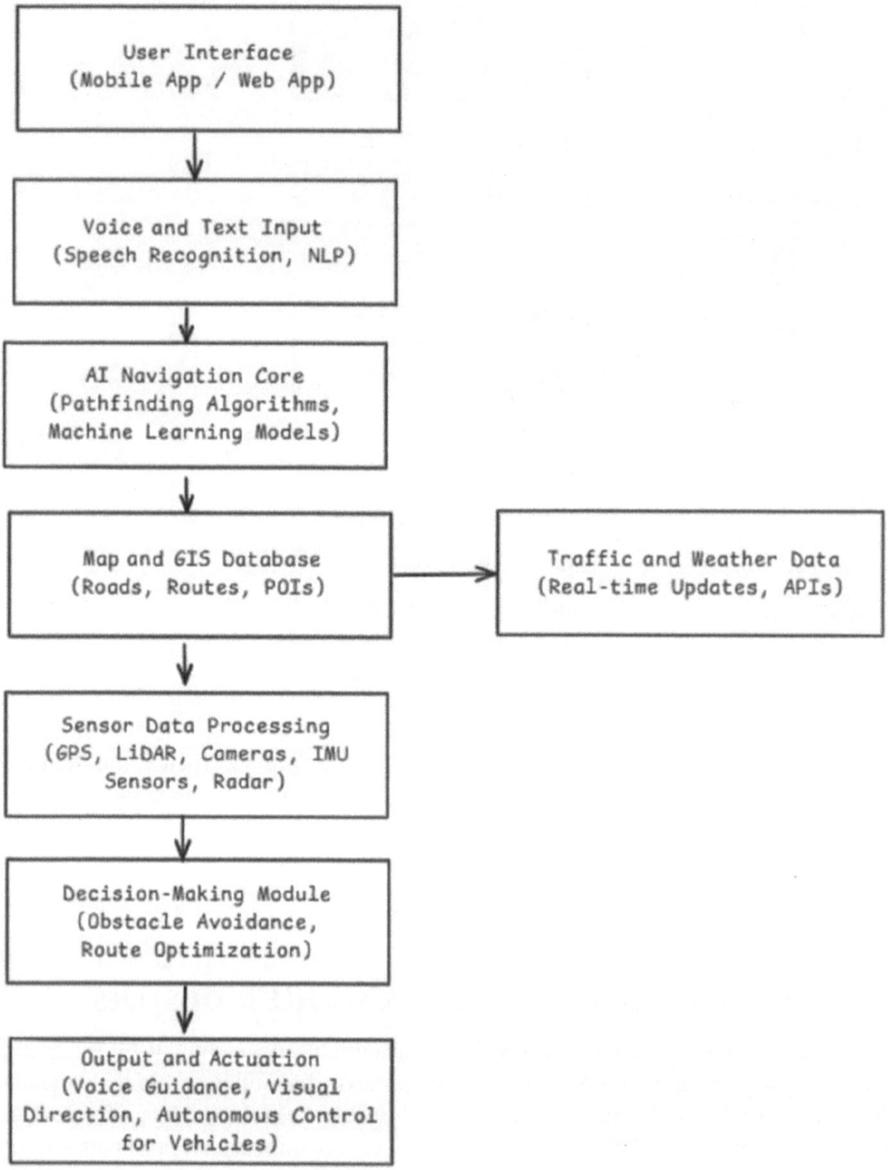

FIGURE 15.5 Advanced AI-driven navigation system.

sustainability, it ensures the identification of eco-friendly materials based on properties like durability and recyclability. An advanced AI-based sorting system will be useful to separate different fabrics based on the quality and categories.

In the department of quality control, an AI-powered vision system detects the defective fabrics at the very early stages, which ensures high-quality products. This

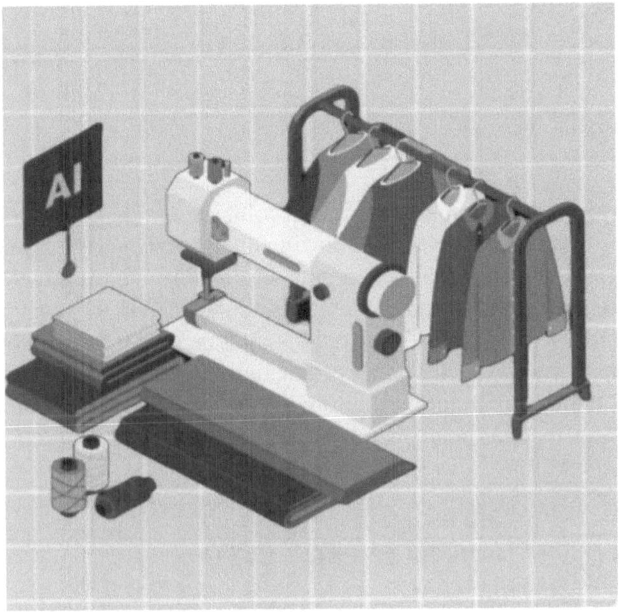

FIGURE 15.6 AI for textile industry.

behavior helps to reduce waste from rejected resources and raises running efficiency. Monitoring textile manufacture from raw materials to finished goods enhances supply chain management as well. It will enable future autonomous manufacturing to be sustainable, improve carbon footprint tracking, and support green textile technologies.

Figure 15.6 shows automated textile industry machinery driven by AI-powered mechatronics tools that effectively weaves environmentally friendly materials

15.8 MECHATRONICS IN CONSTRUCTION SITES

AI and mechatronics combined in building sites increase the accuracy, efficiency, and safety of the work. Combining mechanical systems, electrical control, and intelligent systems allows contiguous operations to address the challenging problems in the building settings. AI-powered mechatronic systems enhance the performance of the construction machinery based on the real-time data from the sensors. For instance, automated earth-moving equipment and robotic excavators can optimize digging and ensure minimal resource consumption to avoid waste.

ML algorithms give predictive maintenance by analyzing equipment data and increasing the lifespan of expensive machinery. During precision tasks such as bricklaying and welding, robots can be equipped with advanced vision systems and AC to do the task with a consistent result of more accuracy. These robots can track and act on the building plans and execute intricate operations while adjusting to real-time environmental conditions. Drones with AI features can further support aerial surveys, mapping sites, and monitoring progress.

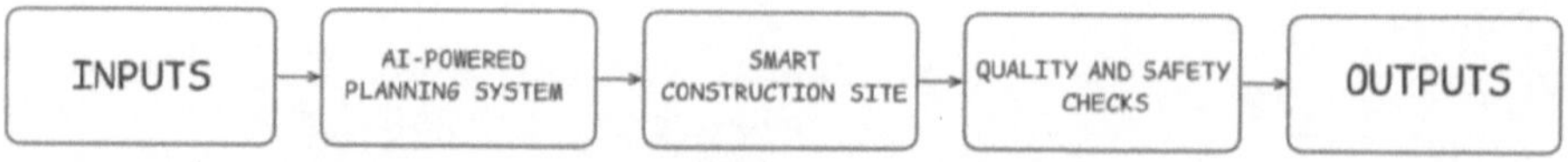

FIGURE 15.7 AI in construction sector.

A vehicle with an autonomous system with collision avoidance lowers the risk factor for the mishaps in crowded areas. By allowing energy-efficient equipment operations and the best material use, AI also supports sustainable building methods. Through advanced waste management and recycling strategies, smart systems guarantee little environmental impact. The future innovations may include a fully autonomous environment for construction sites for better productivity, sustainability, and safety.

Figure 15.7 depicts the future of building sites furnished with completely autonomous drones and machines to operate effectively alongside humans to finish the building operations.

15.9 AI INTEGRATION IN ENERGY SYSTEM

The convergence of mechatronics with AI in energy systems infuses the energy system with more intelligence by means of real-time decision-making, predictive maintenance, and adaptive optimization for the expanding needs of sustainability in an efficient manner. AI-enhanced mechatronics systems maximize conventional operations and renewable energy sources. For example, blade angles and rotational speeds in wind power plants maximize energy generation. In a solar power plant, these systems align photovoltaic panels with respect to the sun's trajectory to increase the energy yield through solar.

Energy storage systems, such as advanced battery systems, can be monitored to determine the charge-discharge cycles, predict performance degradation, and optimize energy allocation based on the demands. For example, in grid-level storage, AI-driven controllers integrate with mechatronic actuators to balance the energy supply and act as a solution to reduce energy waste and improve the grid's reliability. During distribution, AI-powered mechatronic systems self-regulate the networks. It also detects faults, reconfigures distribution paths, and maintains an uninterrupted power supply. This improves adaptability even in fluctuating load conditions.

Figure 15.8 shows how mechatronics and AI energy management might help to maximize resources and sustainability for next generations.

15.10 AUTONOMOUS VEHICLES: AI AND MECHATRONICS SYNERGY

The development of autonomous vehicles is driven by their seamless integration of AI and mechatronics. An autonomous car allows a synergy between car and driver to move, detect, and interact with their environment to open the path for safer trips. Mechatronic parts like light detection and ranging (LiDAR), radar, ultrasonic sensors, and cameras record real-time surrounding data. An AI system handles these facts by

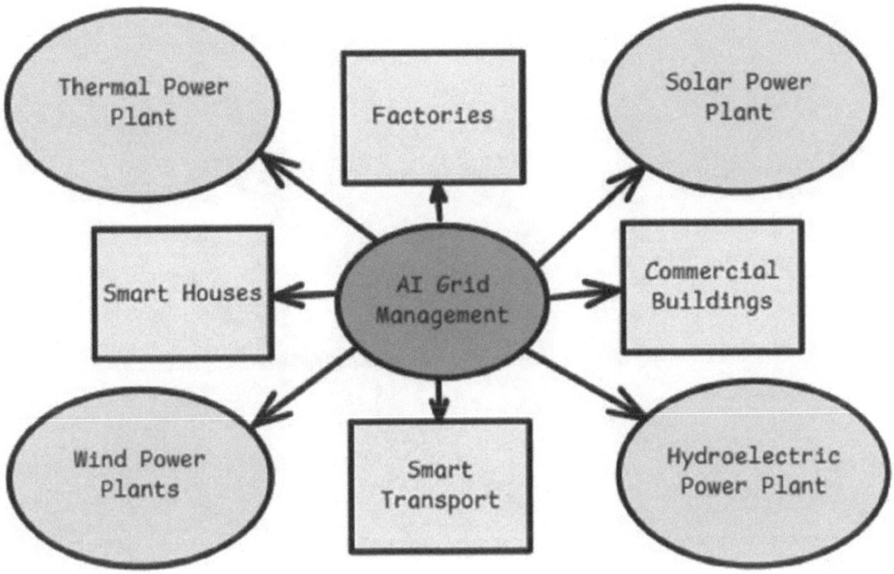

FIGURE 15.8 AI for energy sector.

visualizing the journey path by spotting impediments and traffic pattern prediction. By use of constant learning from real-world situations and environment simulation, RL models maximize the driving strategies of the cars. Path-planning systems help to prevent collisions and traffic congestion.

Mechatronics guarantees in autonomous automobiles perfect control and electrical subsystems. Maximum-level performance of actuators managing acceleration, braking, and steering translates commands into useful actions. Furthermore, sophisticated communication systems are supporting efficient traffic management and vehicle-to-infrastructure (V2I) interactions as well as vehicle-to-vehicle (V2V) interactions.

In autonomous cars, dependability and safety rules should be given top priority. These fusion approaches help to produce strong, redundant systems that reduce failures. Mechatronics of diagnostics guarantees the best performance. They not only transform transportation but also set the groundwork for mobility and environmentally friendly cars driven on sustainable energy sources.

Figure 15.9 showcases the autonomous vehicles powered by the AI and mechatronics synergy to navigate in the urban environment. They are featured with sensors, cameras, and LiDAR systems to emphasize precision, safety, and autonomous transportation.

5.11 MECHATRONICS FOR PRECISION FARMING

Mechatronics is rather important for modern farming tools and equipment. Mechanical systems, sensors, actuators, and electronics form essential combinations. These systems are supposed to manage tasks including sowing, soil analysis, irrigation, and

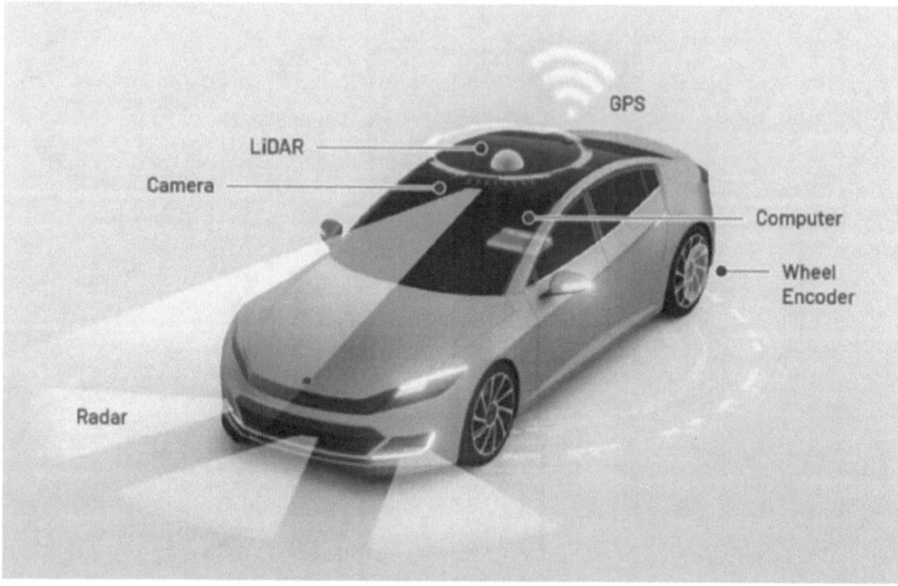

FIGURE 15.9 Autonomous vehicle powering system.

harvesting. AI enhances the running performance of mechatronic systems by means of data processing from the nearby farms. Using ML approaches, AI investigates inputs including soil moisture, nutritional levels, weather variables, and plant health. This approach allows predictive modeling that tells farmers about plant scheduling, fertilizer treatment, and insect management.

One of the prominent applications is autonomous agricultural vehicles. Tractors and drones using GPS and LiDAR may perform minimum human-involved tasks including plowing, spraying, and crop monitoring. Drones could help to monitor large regions, track waste, and spot plant diseases. Moreover, supporting environmentally friendly farming practices are AI and mechatronics. Using AI, for example, will enable maximum water distribution dependent on real-time sensor data, therefore helping to considerably lower water consumption. In the same line, automated nutrient delivery systems guarantee suitable fertilizer ratio application to reduce environmental impact.

Figure 15.10 shows AI-enabled drones swooping over a large swath of land. Using sensors and cameras to gather data on soil condition and plant development, and also tracking plant illnesses, the drones precisely monitor crops.

5.12 CONCLUSION

Combining AI with mechatronics helps to shape resilience, sustainability, and future breakthroughs by bridging the gap between human machines to produce better, more efficient answers for world problems. This chapter discusses about few domains where mechatronics and AI cooperatively improved each other. Future innovation in this area should help address many global issues by means of more creativity.

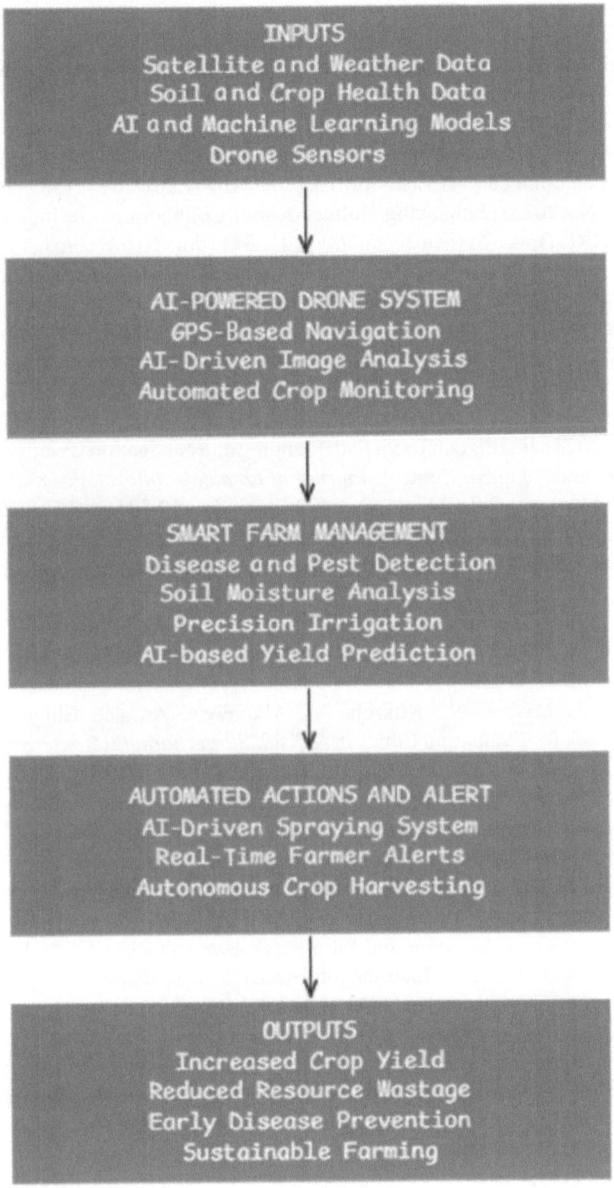

FIGURE 15.10 AI-enabled drones in agriculture sector.

REFERENCES

1. Paolanti, M., Romeo, L., Felicetti, A., Mancini, A., Frontoni, E., and Loncarski, J. (2018). "Machine Learning Approach for Predictive Maintenance in Industry 4.0," *2018 14th IEEE/ASME International Conference on Mechatronic and Embedded Systems and Applications (MESA), Oulu, Finland*, pp. 1–6. Institute of Electrical and Electronics Engineers (IEEE). doi: 10.1109/MESA.2018.8449150
2. Vemuri, N. (2023). Enhancing Human–Robot Collaboration in Industry 4.0 with AI-Driven HRI. *Power System Technology*, 47, 341. doi: 10.52783/pst.196
3. Prince Antony Joel, J., Joshua Samuel Raj, R., and Muthukumaran, N. (2021). "Cognitive and Cybernetics Based Human Adaptive Mechatronics System in Gait Rehabilitation Therapy," *2021 Third International Conference on Intelligent Communication Technologies and Virtual Mobile Networks (ICICV)*. Institute of Electrical and Electronics Engineers (IEEE). https://doi.org/10.1109/ICICV50876.2021.9388423
4. Zhu, C. (2023). Intelligent Robot Path Planning and Navigation Based on Reinforcement Learning and Adaptive Control. *Journal of Logistics, Informatics and Service Science* Vol. 10, No. 3, pp. 235–248, ISSN 2409-2665. doi: 10.33168/JLISS.2023.0318
5. Elnashar, E. and Elnashar, A. (2023). Artificial Intelligence Applications in the Textile Industries Using Mechatronics. *Latest Trends in Textile and Fashion Designing*, Vol. 5, 914–927. 10.32474/LTTFD.2023.05.000215
6. You, K., Zhou, C., and Ding, L. (2023). Deep Learning Technology for Construction Machinery and Robotics. *Automation in Construction*, Vol. 150, 104852. doi: 10.1016/j.autcon.2023.104852
7. Sharma, V., Mishra, N., Kukreja, V., Alkhayyat, A., and Elngar, A. A. (2023). "Framework for Evaluating Ethics in AI," *2023 International Conference on Innovative Data Communication Technologies and Application (ICIDCA)*, IEEE.
8. Ignatious, H. A., Sayed, H.-E., and Khan, M. (2022). An Overview of Sensors in Autonomous Vehicles. *Procedia Computer Science*, Vol. 198, pp. 736–741, ISSN 1877-0509. https://doi.org/10.1016/j.procs.2021.12.315
9. Das, U., Sharma, V., Das, M., Mishra, S., Iwendi, C. and Osamor, J. (2023). "Vehicular Propagation Velocity Forecasting Using Open CV," *2023 4th International Conference on Computation, Automation and Knowledge Management (ICCAKM), Dubai, United Arab Emirates*, pp. 2–7. Institute of Electrical and Electronics Engineers (IEEE). doi: 10.1109/ICCAKM58659.2023.10449587
10. Gooroochurn, M.. (2024). "Mechatronics Implementation of Passive Building Elements to Improve Thermal Comfort and Promote Energy Efficiency in Buildings." *Artificial Intelligence, Engineering Systems and Sustainable Development: Driving the UN SDGs*. Emerald Publishing Limited, pp. 171–182. https://doi.org/10.1108/978-1-83753-540-820241013
11. Andritoiu, D., Bazavan, L. -C., Besnea, F. -L., Roibu, H., and Bizdoaca, N.-G. (2018). "Agriculture Autonomous Monitoring and Decisional Mechatronic System," *2018 19th International Carpathian Control Conference (ICCC), Szilvasvarad, Hungary*, pp. 241–246. Institute of Electrical and Electronics Engineers (IEEE). doi: 10.1109/CarpathianCC.2018.8399635
12. Pushpalatha, N., Gunapriya, D., Bharath, A., Chandru, S., Sanjeevkumar, S., and Vigneshwaran, S. (2024). "A Sustainable Agricultural Field Surveillance System Involving Intelligent Irrigation System for Ecological Balance," *2024 International Conference on Science Technology Engineering and Management (ICSTEM), Coimbatore, India*, pp. 1–7, doi: 10.1109/ICSTEM61137.2024.10560948

13. Loganathan, N., Selvanayakam, A., Anbarasu, P., and Logeswaran, T. (2024). IoT-Based Energy Meter Monitoring System for Sustainable Development. *Journal of Environmental Protection and Ecology*, Vol. 25, No. 1, pp. 182–191.
14. Krishna, V. Y., Roshan, B., Kracose Nishanth, J., Anbarasu, P., and Mohan Raj, C. (2024). "Smart Reservoir Navigating Waters with Integrated Automation," *2024 10th International Conference on Advanced Computing and Communication Systems (ICACCS), Coimbatore, India*, pp. 86–91. Institute of Electrical and Electronics Engineers (IEEE). doi: 10.1109/ICACCS60874.2024.10717175
15. Babu, W. R., et al. (2024). "Space-Based GPS Solar Tracking System," *Proceedings of the International Conference on Signal Processing and Computer Vision (SIPCOV-2023)*, Advances in Engineering Research, pp. 371–382. Institute of Electrical and Electronics Engineers (IEEE). doi: 10.2991/978-94-6463-529-4_33

Index